Fifth Edition

Geology
of
Oregon

Elizabeth L. Orr

William N. Orr

KENDALL/HUNT PUBLISHING COMPANY
4050 Westmark Drive   Dubuque, Iowa 52002

Cover: Ripple-marked sand dunes on the Oregon Coast resemble a gigantic fingerprint [photo by Gary Tepfer].

Table of Contents

Dedicated to the graduates

Acknowledgments

Between the fourth and fifth editions of *Geology of Oregon,* the importance of global tectonics to the state has been ingrained even deeper. Riding on the leading edge of the moving North American plate, Oregon reflects the underlying mechanism of the plate collision boundary in virtually all aspects of its geology. In the seven years since the fourth edition was written, an irregular but continuing drumbeat of earthquakes reminds us of the forces beneath our feet and the need to prepare for catastrophic changes be they quakes, volcanoes, tsunamis, floods, or massive earth movements. Most profound is the discovery of irrefutable evidence that these catastrophes have been visited on the state quite regularly over the past few thousands of years as well as the absolute surety that similar disasters will occur in the near and distant future.

A book of this nature, covering a state of such geologic diversity, could not have been written without the research efforts of everyone who has contributed to our understanding of the past. We would like to express our thanks to all of these unnamed and named individuals: John Armentrout, Ewart Baldwin, Marvin Beeson, Howard Brooks, Bob Carson, Mike Cummings, Greg Harper, Paul Hammond, Klaus Neuendorf, George Priest, Tracy Vallier, Beverly Vogt, and Bob Yeats. Others provided literature, comments, and espresso. In this regard we thank Richard Heinzkill, Gene Humphries, Allan Kays, Carol McKillip, and Cindy Shroba, as well as the helpful staff of the Map and Science libraries at the University of Oregon.

Eugene, Oregon, 1999

Introduction

It is difficult to do justice to Oregon's unique and varied geologic past with the broad brush strokes of a summary. As the state experiences a variety of processes from erosion to volcanism and even glaciation, each leaves its unmistakable signature recorded in the rocks and strata.

Plate Tectonics

Because of Oregon's position on an active continental margin, tectonic processes have played a profound role in the state. With global or plate tectonics, land masses are riding atop great plates of the earth's crust that are constantly in motion. As they move, the plates can pull apart, collide, or slide past each other. When a plate rifts or breaks up, molten rock pours up as lava from cracks that open in the stretched crust. As the lava cools in a rift zone under the sea, a submarine mountain range or ridge is constructed along the rupture. Currently an extensive and growing system of undersea ridges exists along the floor of the world's oceans. When plates merge and collide, the heavier one dives beneath the overriding slab. Once this occurs, the veneer of sediments atop the descending slab is scraped off and accreted to the upper plate. At the same time, rocks carried down or subducted beneath the overriding plate are melted to be recycled. Some of the melted rock makes its way through the overriding plate to emerge as volcanics. The remaining magma crystalizes beneath the surface forming a foundation to the volcanic chain.

Developed on moving crustal plates, volcanic island archipelagos are typically involved in accretionary tectonics. Carried on an oceanic plate, the island chain will eventually collide with a landmass on an oncoming plate where it is attached or accreted to the edge to become part of the mainland. As part of the accretion process, plate fragments making contact with the mainland rotate to obtain a "fit". The contact point between two colliding plates is often marked by a long trough or trench. Between this trench and the island chain a shallow forearc basin may develop. Behind the archipelago, a backarc basin may subside between the volcanic chain and the mainland. Parallel to the

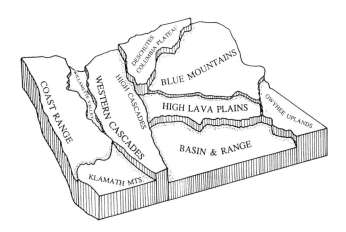

archipelago, both of these troughs derive volcanics from the island chain, although the backarc basin receives sediments from erosion of the adjacent mainland as well.

Each moving tectonic plate has a leading and trailing edge. Along the leading edge, where collision and subduction are taking place, there are large-scale earthquakes and widespread volcanic activity. The leading edge, with its active continental margin, contrasts sharply with the passive margin of the trailing edge. Along passive margins of continents earthquakes and volcanic activity are subdued to nonexistant. Here the main geologic processes are erosion and deposition.

Much of Oregon's early geologic history concerns plate movements and the accretion of exotic terranes. Situated on the leading edge of the North American plate, Oregon has repeatedly been the site of multiple collisions with smaller continental plates that were swept up and added to North America. It is estimated that two-thirds to three-fourths of the state is composed of rocks and sediments that originated elsewhere in the Pacific basin. Fragments of these terranes make up the geologic collage of Oregon. Throughout these collisions, volcanic events and earthquakes mark the mileposts of changing landscapes, climates, and flora and fauna.

Paleozoic

Although the assembly of Oregon's component fragments began in the lower Mesozoic over 200 million years ago, pieces of the state were being fabricated in far-flung ancient oceans as far back as 400 million years before the present. The oldest known rocks in the state are Devonian, and during this period Oregon was a broad embayment of an ocean with a shoreline to the east across Idaho and Nevada. On the Farallon oceanic plate sliding beneath the North American continental plate, a line of volcanic islands accumulated limestones and even a few terrestrial deposits containing plant fossils in an offshore tropical latitude. Throughout the Paleozoic sediments continued to develop in this distant oceanic environment. At some point in time the landmass collided with and was attached to the western margin of the North American plate.

Remnants of the terranes are found today in the center of the state in eastern Crook County and in parts of the Klamath Mountains to the southwest. In eastern Oregon a few scrappy exposures of Devonian limestone and a flint-like rock called chert contain Oregon's oldest fossils. Within these rocks remains of microscopic creatures known as conodonts as well as specimens of brachiopods and tropical corals are

preserved. A few Mississippian brachiopods and corals as well as a diverse Pennsylvanian flora of fern and fern-like plants that grew in an ancient upland are also found in rocks of central Oregon. Permian fossil protozoa, corals, crinoids, molluscs, and brachiopods from the same region reflect both marine continental shelf as well as deep water conditions. Devonian rocks may occur in the terranes of the eastern Klamath Mountains, although the strata here has been so thoroughly altered by heat and pressure that most of the evidence has been destroyed.

Mesozoic

At the beginning of the Triassic, events in Oregon reflect a period of collision and accretion. It was about this time that North America began breaking away from Europe and Africa to create the early Atlantic Ocean. As the Atlantic grew progressively larger, the ancient Pacific became smaller. During this process, the continental shelf off the Atlantic Coast expanded as sediments were dumped there from the intensive erosion of the mainland. Simultaneously the West Coast grew by accreting large and small exotic blocks to its leading edge. In the Mesozoic of Oregon, the major areas of continental accretion were the Blue Mountains and Klamaths. Although today these regions

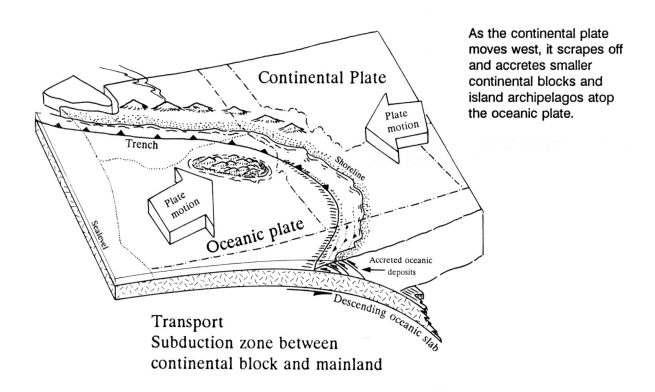

As the continental plate moves west, it scrapes off and accretes smaller continental blocks and island archipelagos atop the oceanic plate.

Transport
Subduction zone between continental block and mainland

are across the state from each other, there are indications that during the Mesozoic they may have been side by side near present day Idaho. Here they projected as low islands above waters of the deep ocean that covered Oregon.

Packets of imported rocks or terranes, bounded by major faults, make up the bulk of the Blue Mountains and Klamaths. Evidence that the rocks in these areas are not native to Oregon or North America is confirmed by fossils found here. Triassic coral reefs in the Wallowas of eastern Oregon are particularly significant in that they represent a type only previously known from the same period in the European Alps. Such a similarity suggests a strong affinity between the two areas. Oregon's oldest fossil vertebrate remains are those of ichthyosaurs from the Wallowa Mountains. Like tuna, these huge streamlined animals, dwelling in a open ocean, were sustained speed swimmers that fed primarily on fish and squid or molluscs. The nearly complete skeletons of these fish-like reptiles resemble specimens known today from the western Pacific and southwest China. Nearby Triassic ichthyosaur fossils from Nevada, moreover, are altogether different from those in Oregon.

Fossil plants from the Jurassic period 180 years ago are known from Douglas County in the Klamath Mountains and Hells Canyon of the Snake River in the Wallowas. Like fossils from all of Oregon's terranes up

to this time, these are almost certainly exotic. The variety of cycads, conifers, and ginkgos in these floras indicate an uplands environment dotted with shallow lakes and marshes. Horsetails, quillworts, and the abundance of large ferns point to a moist climate. Such a diverse cosmopolitan flora, found across Oregon as well as Alaska, Asia, Europe, and Antarctica, suggests widespread tropical and subtropical conditions. Alternately it may mean that microcontinental blocks with a similar flora were dispersed to all corners of the world.

The skeletons of tiny glassy microfossils of protozoa known as radiolaria have been used very effectively in the past few years to correlate Permian, Triassic, and Jurassic rocks in Oregon as well as to attempt to trace some of these terranes back to their point of origin. Radiolaria are significant for their potential in dating ophiolitic rocks. Ophiolites represent a sequence of oceanic crust beginning with igneous mantle rocks and ascending to undersea lava flows. Capped by radiolarian-bearing deep sea silts, clays, and cherts, ophiolites make up much of the Klamath and Blue mountains.

By Cretaceous time, 80 million years ago, two-thirds of Oregon lay beneath a broad seaway with a shoreline extending diagonally northwest by southeast across eastern Washington and Idaho. Depressions or basins between the newly attached exotic blocks and

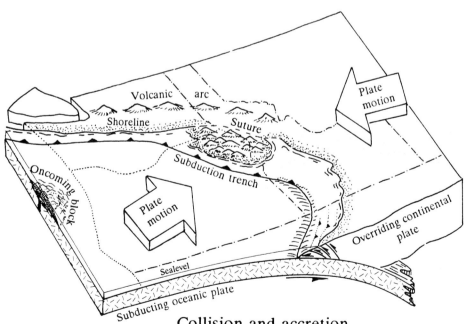

Collision and accretion
New subduction zone on outboard
side of accreted block

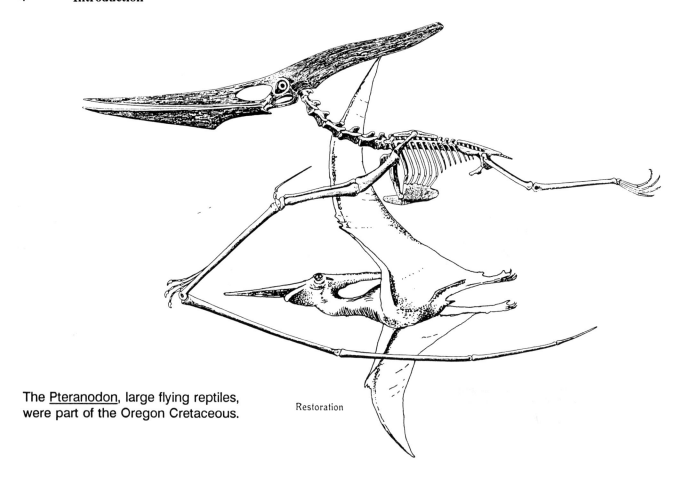

The <u>Pteranodon</u>, large flying reptiles, were part of the Oregon Cretaceous.

Restoration

the mainland were filled by Cretaceous sands, muds, silts, and volcanic material that created, for the first time, "home-grown", local rocks and fossils in the state. The rapid deposition of these sediments in the Mitchell basin of central Oregon and in several smaller basins in the Klamaths was optimum to preserve a record of the rich plant and animal life that existed here. The majority of Cretaceous rocks containing fossils in Oregon are marine in origin. Terrestrially derived sediments yield cycads, ginkgos, and palm. The fern-like tree, <u>Tempskya,</u> known from Baker County, is represented by silicified chunks that identify the characteristic matted trunk structure in cross-section. Even though scattered marine Cretaceous fossils have been described from Crook, Grant, and Wheeler counties, the best preserved material is found in southwest Oregon in the Klamath Mountains where bottom dwelling clams, and snails, as well as cephalopods reflect shallow water continental shelf conditions. Sandstones and mudstones in Wheeler County represent shallow Cretaceous seas that covered eastern Oregon. Here bones of swimming reptiles, the fish-like ichthyosaur, as well as those of the flying reptile, <u>Pteranodon</u>, with a 10 foot wing span, have been found in the Mitchell basin.

Cenozoic

Throughout the Cretaceous, the shoreline steadily retreated westward until the Cenozoic Era, 60 million years ago, when it curved diagonally across the state taking in the Klamath Mountains and central Oregon before sweeping north into Washington. The Eocene epoch marked the beginning of a complicated series of tectonic events that would completely alter the face of the state. Moving slowly westward, the North American plate was approaching a volcanic hot spot that separated the Juan de Fuca and Kula plates. Volcanic activity and ocean spreading combined to create a north-south chain of volcanoes across the hot spot off the western coast. Shortly after formation, this island archipelago collided with North America to become the foundation for the Oregon and Washington Coast Range. After collision the subduction zone that had delivered the block to North America was abandoned, and a new trench appeared on the western, oceanic side of the island chain.

Subsidence of the new Coast Range block formed two troughs, a forearc basin over the volcanic block and a backarc terrestrial basin east of the island archipelago. Initially sediments eroded from the adjacent Klamath Mountains were carried into the open ocean basin atop the volcanics of the island arc.

ranged as high as 60 to 80 degrees Fahrenheit. Fossil fig, laurel, and palm from the coastal plain also reflect a warm climate with as much as 60 inches of rainfall annually. Adjacent to the coastal margin, an open embayment with swampy depressions was responsible for coal deposits near Coos Bay. Foraminifera in middle Eocene sediments of the basin include many of the larger disc-shaped forms only inhabiting tropical latitudes today, whereas late Eocene formations yield a more temperate, continental shelf assemblage.

Beginning in the early Eocene, a backarc terrestrial basin developed in central Oregon. As the Coast Range block rotated and expanded westward, the thinning crust behind the block erupted with extensive volcanics. Great outpourings of lava from 44 to 32 million years ago covered the region with volcanic debris that blocked streams, created ponds, and caused mudflows when loose ash mixed with water. Rapid burial in this tropical climate was favorable to preserving fossils of crocodiles, rhinoceroses, horses, sheep-like oreodons, and large hoofed mammals such as uintatheres and titanotheres that roamed the water-ways, lush forests, and grasslands of the early Tertiary.

A rich flora of tropical leaves, nuts, seeds, and fruits, preserved from this region, shows the lack of any real climatic change east to west across Oregon. At this time the state had a low, wide coastal plain extending out to the shoreline, and moisture laden air masses from the Pacific were free to move in an easterly direction without the impediment of either a Coast Range or Cascade Mountain barrier. As climatic and volcanic conditions altered in the late Oligocene and Miocene, semi-tropical vegetation supported a population of animals preserved in local basins that would look somewhat familiar, although none of the species exactly correspond to those living today. Oligocene floras, dominated by the dawn redwood, Metasequoia, became distinctly more temperate in aspect.

Tempskya, a unique fern-like tree, is composed of many branching stems encased in a mat of roots that form a mass resembling a trunk.

Eventually, however, the easterly Idaho batholith provided material that was transported by a well-developed riverine system to be deposited in the forearc trough. Sandy bars formed along the shallow waters of the shelf and slope, while fine-grained sediments were carried into the deepest region of the trench. Shallow marine deltas are common in the southern end of the basin in the early and middle Eocene, while intermediate and deep water settings in the northern part are better developed in the late Eocene.

A wide variety of ocean environments from the surf zone to deep bathyal existed, each with its own distinctive suite of marine invertebrates. Fossils from these sediments record warmer Eocene climates that cooled toward the Miocene. The presence of tropical molluscs as the snail Turritella show temperatures

Eocene warm water foraminifera
(protozoa) are the size and shape of coins.

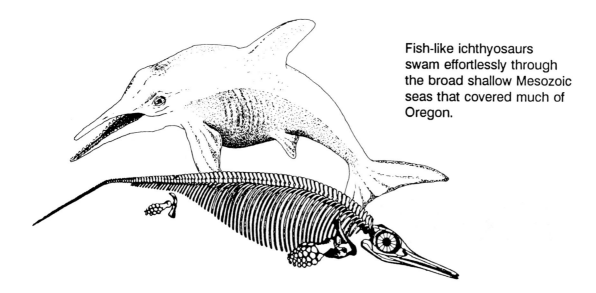

Fish-like ichthyosaurs swam effortlessly through the broad shallow Mesozoic seas that covered much of Oregon.

Around 40 million years ago, construction of the ancestral Western Cascades began with volcanism triggered by the renewed subduction of the Farallon plate beneath the North American plate. Lying just to the east of the shoreline, a broad volcanic arc produced thick accumulations of lava and ash that would eventually develop the foundations of the older Cascade range. Even as they were growing, the mountains were being actively eroded by streams carrying the volcanic material westward to the nearby ocean basin.

Near the end of the Oligocene and into the early Miocene, 20 to 25 million years ago, the forces of crustal stretching and the movement of tectonic plates triggered faulting, extensive volcanism, and the development of the Basin and Range block mountain topography characteristic of southcentral Oregon. An east-west stretching action expanded this area of Oregon severely, and the resultant thinning of the semi-brittle crust produced faults that provided a route for the massive extrusions of the Columbia River, Steens, and Strawberry Mountain volcanics. Vast amounts of lava flooded much of eastern Oregon creating an extensive level basalt plateau. With only crests of of the Blue Mountains projecting, outpourings of the Grande Ronde basalts here reached their maximum 15 million years ago. Accompanied by the northward tilting of the Columbia plateau, lava poured into eastern Washington and through the gap in the ancient Cascade Range cut by the Columbia River to reach the sea. Once plateau volcanism ceased, the late Miocene saw renewed eruption in the Western Cascades where andesitic lavas formed the broad platform which would later accommodate the Pliocene High Cascades on the eastern flank.

Concurrent with the development of the Western Cascades, the Coast Range began to form as a broad north-south fold in the late Miocene. As the older Western Cascades were tilted and uplifted, a west-facing ramp formed effectively cutting off moisture to eastern Oregon, thereby profoundly changing the earlier wet, tropical climate to the drier one of today. Gradually plants such as Metasequoia that required a large amount of rainfall disappeared to be replaced by oaks, box elder, willow, and cottonwood. As the inland forests gave way to grasslands, the hoofed mammal population changed from large browsing animals to fleet-footed grazing species as well as small shrews, moles, mice, and squirrels.

The appearance of the Pliocene High Cascade peaks relatively late in geologic time began with extensive outpourings of basalt that in turn triggered the creation of discontinuous north-south grabens or troughs. As the lavas were extruded, the region beneath the High Cascades drained as it continued to sink into a series of depressions bounded by faults that extended most of the length of the range.

The Ice Ages of the Pleistocene brought glaciers then widespread floods to Oregon beginning almost 2 million years ago when continental ice masses advanced into Washington and Montana. Ice dams in Montana, that created enormous lakes by repeatedly blocking the the upper drainage of the Columbia River, failed periodically releasing huge floods that scoured eastern Washington and the Columbia channel. On their way to the ocean, the rushing waters ponded up to produce temporary lakes at Wallula Gap, The Dalles, and in the Willamette Valley. Floodwaters that washed into the valley carried suspended sediments along with large erratics or boulders on chunks of ice that came to rest at heights up to 400 feet above sea

Semitropical and temperate leaves and seeds of walnuts, grape, <u>Palaeophytocrene</u>, and <u>Chandlera</u> from the Clarno period of eastern Oregon 34 million years ago.

level. The colder temperatures also brought glacial ice and erosion to the higher peaks of the Blue Mountains, Steens, and Cascades. Extensive glacier complexes were responsible for carving out the distinctive U-shaped valleys, daming small lakes, and leaving behind long moraines. Where glacial activity was accompanied by active volcanism, ash, lava, and mud flowed down the slopes of the mountains to mix with glacial deposits on the aprons of the peaks.

Continuous volcanic activity resulted in further changes in the Cascades. Explosive episodes of volcanism produced Newberry Crater and Crater Lake. As late as 200 years ago, volcanism built small domes and cones on the sides of peaks here. Mt. Hood has been particularly active. Crater Rock, a dome near the summit of the mountain erupted in the late 1700s and early 1800s, while gas, smoke, and pumice have been observed historically.

The slow uplift and tilting of the Coast Range block, beginning in Eocene time, continued into the Pleistocene as the descending Juan de Fuca plate accumulated a wedge of accreted sediments along the edge of the North American continent. Eastward tilting of the coastal mountains also produced a basin that would become the Willamette Valley, the southern end of a much larger structure extending into Washington called the Puget Sound-Willamette lowland. Along the Oregon coastline, tilting from pressure of the growing accretionary wedge combined with the rise and fall of

sea level produced a series of step-like Pleistocene erosional terraces. During glacial stages, volumetric fluctuations of the ocean occurred when ice, formed on land, lowered sea level. As the ice melted to re-enter the ocean, sea level rose substantially. By Pliocene time the western shoreline lay only slightly to the east of where it is today.

Pleistocene vertebrate fossils are thoroughly scattered across the state. Famous for its vertebrate remains, Fossil Lake, just northeast of Christmas Lake valley, is now a dry lakebed that was part of a much larger system of ice age lakes scattered over this region. The fossil bones of large number of species of water fowl show that the lake was a major stop-over for migratory birds as well as being the habitat for many freshwater fish. In addition, loose sands of the old lake are littered with bones of elephants, horses, camels, cats, wolves, dogs, peccary, and bears, as well as a diversity of rodents.

Ice Age mammals roamed throughout western Oregon where thick forests and plentiful rainfall resulted in the development of swamps and sloughs in almost every natural low spot or sedimentary trap. Animals would occasionally stumble or be washed into the ancient bogs where they were buried. Very often skeletons were preserved with most or all of the bones intact. Among the large Pleistocene mammals, the most awesome were the large, ox-sized ground sloths which weighed up to a ton. Standing upright, these bipedal animals reached 10 feet high. The great claws of these massive creatures suggest they were harmless root diggers as well as browsers.

A major enigma with respect to these fossil mammals of the Pleistocene is their disappearance over a very short period. This event took place about 11,000 years ago in Oregon as well as throughout the rest of North America. This interval correlates with the opening of an ice free corridor through Canada to Alaska and Asia across the Bering Strait. Much of the Strait was still emergent from the sea at this time, so a clear avenue for animal migration was open. Among the migrants was prehistoric man. With weapons and group hunting techniques, humans could have made short work of even the largest mammals which were probably driven to the waterways by increasing drying and warming conditions that marked the end of the Pleistocene.

Holocene

Recent geologic evidence points to catastrophic prehistoric earthquake activity along the Oregon offshore trench separating the Juan de Fuca and North American plates. Unlike other active margins in the Pacific, the subduction zone here binds up for centuries before releasing suddenly. As rocks of the Oregon Coast Range suddenly snap from their stressed position during an earthquake, coastal regions drop down several feet. Periodic catastrophic earthquake activity in the past is marked by distinctive evidence of rapid land subsidence the length of the coast. Large-scale earthquakes are also recorded offshore where multiple turbidity currents, shaken loose by the event, deposit their distinctive layers across the continental shelf and slope.

A History of Geologic Study in Oregon

The currency of science is the published literature. In this regard, the record of publications reflects the history of the development of geological sciences in Oregon.

Exploring Expeditions

The study of geology in the Pacific Northwest began obliquely with the Lewis and Clark Expedition commissioned by President Thomas Jefferson in 1803 to explore the west to the Pacific Ocean. President Jefferson, in a lengthy letter of instructions, exhorted the party to look for the *"remains and accounts of any [animals] which may be deemed rare or extinct. The mineral productions of every kind ... [and] volcanic appearances."* Jefferson was convinced that animals did

not become extinct, and, since he had seen bones of the giant sloth, <u>Megalonyx</u>, in Virginia, he felt that these huge beasts had migrated westward instead of disappearing from the earth. However, no one with training in geology accompanied the expedition, and the resulting volume, *History of Expedition under the Command of Lewis and Clark to the Sources of the Missouri, thence across the Rocky Mountains, and down the Columbia,* published in 1814, is of minimal geologic relevance.

Thirty years went by, including nearly 10 years of debate in Congress, before a bill was passed on May 14, 1836, to outfit an exploring expedition to the southern polar regions as well as along the Pacific islands and northwest coast of this continent. One

VIEW LOOKING TOWARD SADDLE MOUNTAIN FROM YOUNGS RIVER BELOW OLNEY

View looking toward Saddle Mountain from Youngs River below Olney, Clatsop County. Illustration by Joseph S. Diller, 1896.

Silver Lake looking southeast by E.D. Cope, 1889.

purpose of this American exploring expedition was to test the theory, presented in Paris in 1721 and debated in America years later, that holes in the polar regions led to the hollow interior of the earth. Theories such as this, combined with the desire for political and military expansion, economic gain, and scientific curiosity about western North American, led Congress to allocate funds for an expedition of six ships commanded by Charles Wilkes. The *Vincennes, Peacock, Relief, Porpoise, Sea Gull,* and the *Flying Fish* left Norfolk, Virginia, on August 17, 1838, with a team of scientists which included geologist James Dwight Dana. Dana, then 25 years of age, taught at Yale University.

Rounding Cape Horn, the ships criss-crossed the Pacific, landing near Puget Sound on May 2, 1841. Arriving later than the others, the *Peacock,* with Dana aboard, was destroyed on a bar at the entrance to the Columbia River, although everyone survived the mishap. The Wilkes party explored between Puget Sound and Vancouver before crossing the Washington Cascades, where they conducted surveys of the Columbia River gorge. In August, 1841, Commander Wilkes directed a party of men, led by Lieutenant George Emmons, to journey overland across the Klamaths to San Francisco where they would meet up with the *Vincennes* on September 30. The entire expedition returned to New York in June, 1842.

Dana accompanied the San Francisco group, and his report provided a concise, geologic account of the region through which he travelled. Leaving the "Willammet" district, the party traversed the Umpqua Valley and climbed what Dana called the Umpqua Mountains, "*a most disorderly collection of high precipitous ridges and deep secluded valleys enveloped in forests*". He was obviously referring to the Klamaths and Siskiyous immediately before the group encountered the Shasta River.

Dana observed and wrote extensively on the geologic features of Oregon, commenting on the mountain ranges bordering the oceans, the evidence of volcanic activity, describing the, "*abundance of basaltic rocks over its [Oregon] surface*", the granites of the Klamath area, and other rocks as he passed through. Coastal units, assigned to the Tertiary, are listed, and molluscs collected from the Astoria Formation were subsequently identified by Timothy Conrad, renowned invertebrate paleontologist with the Philadelphia Academy of Sciences. Descriptions of fossil cetacean bones, fish, crustacea, foraminifera, echinoids, and plants were incorporated into the report.

With his concluding remarks Dana assessed Oregon's potential. "*Although Oregon may rank as the best portion of Western America, still it appears that the land available for the support of man is small ... only the coast section within one hundred miles of the sea ... is all fitted for agriculture. And in this coast section there is a large part which is mountainous, or buried beneath heavy forests. The forests may be felled more easily than the mountains, and not withstanding their size, they will not long bid defiance to the hardy axeman of America. The middle section is in some parts a good grazing tract; the interior is good for little or nothing*".

The magnificent volume X of the *United States Exploring Expedition During the Years 1838, 1839, 1840, 1841, 1842, under the Command of Charles Wilkes* contains Dana's "Geological Observations on Oregon and Northern California." These 145 pages constitute the first formal geologic observations in this area. Unfortunately Congress was in a parsimonious mood when it came to funding publication of the final 18 volumes and 11 atlases of the Expedition so that they took 30 years to complete. Ultimately Congress only allowed 100 copies of the set to be printed. One volume was intended for each state of the nation, and the remaining copies were to go to foreign governments. Three private copies were given out, one of which was to Dana who received $16,000 in pay for writing volumes 7, 10, 13, and 14. At the outset, 30 of the volumes were destroyed by fire and never replaced. An angry Dana had 100 copies of Volume X printed from his own funds, distributing them to private

U. S. GEOLOGICAL SURVEY
ANNUAL REPORT 1883 PL. LXXXV.

A sketch of Lake Abert, Lake County, by I.C. Russell, 1884.

libraries. In Oregon there is an "unofficial" copy at the University of Oregon Library in Eugene, although the accompanying atlas of illustrations is missing.

Western Journeys

Charles Fremont, while not trained as a geologist, cannot be ignored in the early examination of the West. Certainly a person of curious and quixotic nature, Fremont could be called a "scientific explorer." Fremont had married Jessie Benton, daughter of the influential Senator Thomas Hart Benton from Missouri. After that Fremont had no difficulty obtaining appointments and funding from Congress for his trips across the continent including one in 1843 to survey Oregon Territory. The purpose of this, his second of three western journeys, was political as well as scientific. The expedition was charged with reporting on the topography and collecting geologic and botanical specimens in order to gain any knowledge of the interior which might prove useful in substantiating the claim of the United States government to the "*valley of the Columbia*".

Departing in the spring of 1843, the Fremont party of 39 men included the German map maker, Charles Preuss. Once the group had crossed the Rockies, it journeyed down the south bank of the Columbia to The Dalles. After a brief side trip to Ft. Vancouver for supplies, the party explored the region east of the Cascades to Klamath Lake and down into California. While Fremont's journal contains some notes of geologic interest, the main contributions from his trip were some fossil plant specimens from the Cascades sent to James Hall, New York State paleontologist, along with a map of Oregon and California.

In 1848 when Congress restricted Dana's Volume X to 100 copies, it generously funded the publication of 20,000 copies of Fremont's "Geographical Memoir upon Upper California" and Preuss' "Map of Oregon and Upper California." In actuality, most of the map was a synthesis of data on Oregon and California from previous trips and other sources. An earlier "Map of the Exploring Expedition to the Rocky Mountains in the year 1842 and to Oregon and California in the years 1843-4" by Preuss in 1845 was well-drawn and accurate, but the Willamette Valley and Coast Range were left blank. These maps, as well as others of this period, were topographic only and lacked geology.

Pacific Railroad Surveys, 1853-1855

In 1853 an act of Congress was once again instrumental in furthering geologic knowledge of the Pacific Northwest. The act was designated to employ "*such portion of the Corps of Topographical Engineers ... to make such explorations and surveys ... to ascertain the most practical and economical route for a railroad from the Mississippi River to the Pacific Ocean*". Congress budgeted $150,000 for this project and sent out "*a mineralogist and geologist, physician, naturalist...*" under the leadership of the Topographical Engineers. These

U. S. GEOLOGICAL SURVEY   SEVENTEENTH ANNUAL REPORT   PART I   PL. XII

A view looking north from Forest Grove, Washington County, by J.S. Diller, 1896.

explorations were known as the Pacific Railroad Surveys from 1853 to 1855, and the geologists employed to survey Oregon and northern California were John Evans and John S. Newberry.

The first railroad survey of the northwest funded by the U.S. government set out from St. Paul in 1853 with John Evans hired as the geologist of the expedition that was overseen by Isaac Ingalls Stevens, newly appointed as Governor of Washington territory. Evans, whose training was in medicine, "*was intrusted with the geological reconnaissance of Oregon...*". Once in Oregon Territory, Evans began his work on the Columbia River, exploring much of Washington before finding himself 150 miles south on the Umpqua River. He then traversed to Empire, near Coos Bay, before returning north by way of the Willamette Valley and Vancouver. Unfortunately the "Geological Survey of Oregon and Washington Territories" by Evans was never received by Stevens and was thought to have been lost. A partial log of Evans' trip was found 50 years later and deposited at the Smithsonian. The log was never officially published, although typewritten copies entitled the "Route from Fort Vancouver, 1854-56" are available. With so many gaps, however, and only references to the lithology in the notes, the report is only of limited geologic value.

A professor of geology and chemistry at the School of Mines at Columbia College, New York,

Newberry was titled the geologist and botanist of the survey which left San Francisco in 1855. This party was under the leadership of Lieutenant R.S. Williamson and Lieutenant Henry L. Abbot of the Army Corps of Topographical Engineers.

Once over the Sierra and Klamath Mountains, the men spent a month crossing and recrossing the Cascades near the Three Sisters permitting Newberry to study the geology of the Cascade and John Day regions in great detail. The variety and color of the John Day beds impressed Newberry. "*Some are pure white, others pink, orange, blue, brown, or green. The sections made by their exposure have a picturesque and peculiar appearance...*". Following the "Mpto-ly-as" (Metolius) and Deschutes rivers, Newberry reported on the stratigraphy and structure of the canyons. The warm springs of the Deschutes were examined and analyzed separately by E. N. Horsford, who accompanied the party. Newberry was the first to differentiate between the coastal chain in California and the Oregon Cascades. He correctly identified glacial features in the Cascades and erosional evidence in the Columbia River canyon. His most significant economic contribution was an account of the coal mining possibilities of "Coose Bay", Vancouver Island, and Cape Flattery.

Volume VI of the Pacific Railroad Survey included the "Reports on the Geology, Botany, and Zoology of Northern California and Oregon", made to the War Department, in which John Newberry's "Geological Report on Routes in California and Oregon" with its beautifully tinted plates appeared in 1855.

U.S. Geological Survey

With the end of the Civil War in 1865 the government began funding numerous excursions to the West in order to carry out observations in the natural sciences. Many publications resulted from the four great western surveys conducted between 1867 and 1879 under Clarence King, Ferdinand V. Hayden, John Wesley Powell, and George M. Wheeler. Subsequently the tasks and many of the scientific workers of those surveys were incorporated in the newly created U.S. Geological Survey within the Department of the Interior.

Representative of the earliest published studies dealing with Oregon are Leo Lesquereux's 1883, "Description of Miocene Species of California and Oregon" published as part of the Hayden Survey and Russell's slender volume, "A Geological Reconnaissance in Southern Oregon", 1884, from the Wheeler survey. Joseph S. Diller's report on the "Coos Bay Coal Field, Oregon," was written as part of the 19th annual report of the Survey. Diller also produced the first three Oregon folios between 1898 and 1903 for the U.S. Geological Survey's "Geologic Atlas of the United States."

Turn of the century

About this time Thomas Condon had begun an informal study of Oregon geology and paleontology while stationed at The Dalles as a missionary. Condon made frequent trips to the John Day fossil beds, collecting mammal bones and leaves. Condon's geologic studies eventually forced him to choose between the ministry and a career in geology, a decision that came when he accepted an academic appointment at Forest Grove College in 1873.

The increasing attention given to Oregon geology precipitated the creation of the office of State Geologist in the fall of 1872. The appointment fell to Thomas Condon, and it was his duty "to make geological examination of different parts of the State from time to time," a job which was to pay $1000 in gold annually. Several years after his appointment, the University of Oregon opened at Eugene, and Condon accepted a position there as a professor of geology in 1876.

Condon, in his capacity as geology professor and State Geologist, living and working for years in Oregon, was in the position to synthesize a comprehensive theory on the geologic development of the state.

Condon's ideas were published as *The Two Islands and What Came of Them* in 1902. This book appeared about 100 years after Lewis and Clark had reached the Northwest and 53 years after Dana's "Geological Observations...". The significance of the book is that it attempts to summarize knowledge and scientific expertise on Oregon geology in order to draw conclusions about Oregon's geologic past, a task much more difficult than Dana's because of the increased information current in the 1900's. In his introduction Condon states "*But this large body of information is so scattered that few have the time to collect enough of it to form a continuous unity of its history*". One of Condon's greatest contributions to expanding knowledge of Oregon's geology was an outgrowth of his own intense interest in science. Condon carried on a voluminous correspondence with several prominent eastern scientists, as John Newberry, who had seen Condon's Miocene fossil leaves and requested more material. Condon had sent specimens and corresponded with Spencer Baird of the Smithsonian, Edward Cope of the Academy of Natural Sciences, Joseph Leidy of Philadelphia, O.C. Marsh of Yale University, and John C. Merriam of the University of California. As a result of these activities the scientific community was not only aware of Oregon's geologic resources, but many took the time to come to Oregon to see the geology first hand. Articles on Oregon paleontology and geology proliferated as a consequence of these activities.

Condon was not alone in pushing forward the boundaries of geologic knowledge in Oregon. Among those who worked in the northwest, mineralogist and stratigrapher Joseph S. Diller stands out as having contributed significantly to the geologic record. Diller produced 56 papers and maps on Oregon alone while employed by the U.S. Geological Survey. Among his important works are "The Bohemia Mining Region of Western Oregon" (1900) as well as a number of outstanding map folios of southwestern Oregon. The "Port Orford Folio" was issued in 1903 and the "Riddle Folio" in 1924.

Diller's efforts produced the first detailed geologic maps of regional sections of Oregon. Up to this point a number of maps on the geology of the United States had been composed, as William Mclure's "A Map of the United States, Colored Geologically" (1809), but, in spite of the title, these covered only the eastern part of the country. "Grays Geological Map of the United States" by Charles H. Hitchcock, professor of geology at Dartmouth, issued in 1876, is typical of the maps being published which depicted the geology of the country in broad strokes. The hand-colored Hitchcock map shows the Cascades as volcanic, the

Willamette Valley and coastline as Tertiary, some Cretaceous rocks in the Klamaths and southeastern Oregon, and the remainder of Oregon (Coast Range, eastern Oregon, Klamaths) as "Eozoic", that is preSilurian. From the 1850's to after 1870, the first regional geologic maps of the Pacific states included only areas of California and a number of profiles of Vancouver Island.

Most of the early maps that even featured Oregon were concerned with the Coast Range only. Ellen Condon McCornack's *A Student's Geological Map of Oregon with Notes* in 1906 is the first complete geologic map where Oregon is not part of a larger geographic unit. On this map the geology, from preCretaceous through Pleistocene, is not colored but is depicted by symbols. In drawing up the map, McCornack *"tried to gather knowledge from all available sources,"* at that time relying on publications by Condon, Diller, Lindgren, and Russell. The accompanying

25 pages of text provide a summary of Oregon's geologic history.

Diller's record was nearly matched by that of Edward Drinker Cope, vertebrate paleontologist, whose work focused on the John Day and Great Basin regions. "The Silver Lake of Oregon" (1889) and "The Vertebrata of the Tertiary Formations of the West" (1883) were produced by Cope who often collected fossils under difficult conditions. At one point Cope was conducting his field work with a team of four mules and a wagon and was reduced to eating bacon, stewed tomatoes, and crumbled biscuits, even feasting on cold canned tomatoes for breakfast.

Once again, an overview of Oregon geology was synthesized this time by two geologists, Warren Dupre Smith, a field geologist, and Earl L. Packard, a paleontologist, at the University of Oregon. "The Salient Features of the Geology of Oregon" in 1919 was issued as an article in the Journal of Geology and

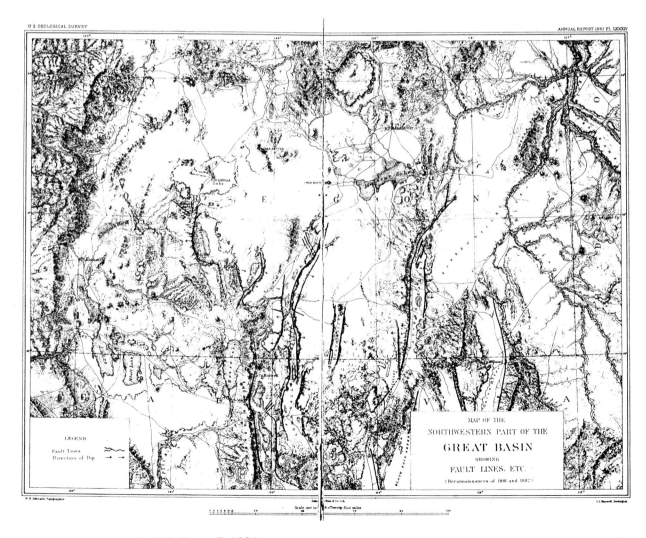

A map of the Great Basin by I.C. Russell, 1884.

as a University of Oregon Bulletin. This short account of 42 pages takes *"an inventory of our knowledge"* of Oregon geology presented in a fairly technical manner. Most of the text deals with stratigraphy, although there is a short section on the geologic history and economics of the state. Smith, serving as chairman of the Geology Department some time after Thomas Condon, received numerous requests for copies of the publication, which was not distributed commercially but which he sent out without charge.

Classic papers by Ira Williams, "The Columbia River Gorge; its Geologic History", with its outstanding photographs, appeared in 1916 as did J. Harlan Bretz's controversial papers, "The Channeled Scablands of the Columbia Plateau" (1923) and "The Spokane Flood beyond the Channeled Scablands" (1925). Geologists dismissed Bretz's ideas and were slow to accept the notion of large-scale floods.

Following Decades

Several geologic steps forward in the state took place in the 1930's. The first of these occurred with the beginning of the school year in 1932, when a new department of science at State College in Corvallis began to train students in geology. A shy, well-known paleontologist, Earl L. Packard, was the reluctant dean of the new science school. Prior to this time, geology classes had been taught at Corvallis by the School of Mines. In 1932 the school of geology along with other science departments was transferred from the University of Oregon at Eugene to Corvallis where all science programs were combined. The purpose of the reorganization was to reduce duplication in class offerings during times of financial stress. The decision was based on recommendations made after a survey of Oregon higher education had been conducted. In the sciences the major argument soon emerged over the issue of which school should offer "pure" science as opposed to "applied" science.

As a result of this arrangement, 86 sets of science journals were moved from the University of Oregon to the library at Corvallis. These books were never fully processed at State College, and the transfer of books ceased in 1933 when the University of Oregon librarian, O.F. Stafford, Dean Packard's official representative at the University of Oregon, refused to allow Beilstein's *Handbook of Organic Chemistry* to be moved from the Eugene campus. After a decade of unhappiness, protest, and conflict, the State Board of Higher Education restored the natural sciences to Eugene in 1942.

Yet another geologic milestone was marked in 1937 with the initiation of a state geology department.

Ficus, Berberis, Platanus, and Quercus drawn by John Newberry, 1898.

Some years earlier, in 1911, the State Legislature had created a state Department of Geology, nearly 40 years after Thomas Condon's appointment as the State Geologist. This Oregon Bureau of Mines and Geology was first incorporated into the Oregon Agricultural College at Corvallis, but, when beset with financial and bureaucratic problems, the Bureau disbanded in 1923. In spite of these troubles the Bureau had contributed significantly to the geologic literature by issuing bulletins, short papers, maps, and miscellaneous publications.

In an attempt to rectify the situation, the Legislature created the present Department of Geology and Mineral Industries in 1937 with a biennial operating budget of $60,000 and an additional $40,000 to

Batcheler.   *Cope*   *Condon*   *Pass. 5/50*   *Bald.Mtns*   *Jefferson.*

A view of the Cascade Range by E.D. Cope, 1888.

grubstake loans of $50.00 a year to any prospector who applied and met the requirements. The Department's first bulletin in 1937 contained federal placer mining regulations. Clyde P. Ross' "The Geology of Part of the Wallowa Mountains," in 1938, sold for $.50, and Henry C. Dake's "The Gem Minerals of Oregon", for $.10 in the same year. During 1938 the Department issued the "Press Bulletin" monthly to be replaced a year later by the "Ore.-Bin", sent free to libraries, universities, colleges, and legislators. Earl K. Nixon, first Director of the Department, stated the purpose of the publication was to "*advise the public of the work of the Department and of new and interesting developments in mining, metallurgy, and geology*". In 1979 the "Ore-Bin" became "Oregon Geology", the title reflecting a change in emphasis in the geology of the state in more recent years.

During this period, Howel Williams' "The Ancient Volcanoes of Oregon" (1948) and Ralph Chaney's "The Ancient Forests of Oregon" (1956) were the culmination of many years of work in Oregon. A significant force in Oregon paleobotany for decades, Chaney wrote and published extensively in the state until his death in 1971 at the age of 81 years.

Ewart Baldwin, a professor of geology at the University of Oregon, was another geologist who worked and published extensively in the state. His first edition of the *Geology of Oregon* appeared in 1959, just 57 years after Condon's synthesis of Oregon geology. As with Condon, Baldwin had lived in Oregon for a number of years and was faced with the task of gathering and compiling information from an overwhelming

number of sources. His first edition written in a non technical manner, "*is concerned mainly with the historical geology of Oregon...*". The initial edition of Baldwin's book was issued and distributed by the University of Oregon bookstore selling for $2.25 each.

Beginning in the late 1960 s and continuing to the present, the theories of continental drift and plate tectonics had the greatest impact on geologic research and literature in Oregon. Gaining acceptance only after years of controversy, continental drift theory was synthesized in 1960 by Harry Hess of Princeton University who compiled facts about the creation and movement of the sea floor in a hypothesis that he called "geopoetry".

From that point foreward, the concept of plate tectonics took almost 10 years to cross the continent to the West Coast where it was ushered in with Robert Dott's, 1969, "Implications for Sea Floor Spreading". Dott briefly reported on the "*widely acclaimed hypothesis of sea floor spreading, and especially its latest refinement, the lithosphere plate hypothesis...*" touching briefly on its impact in Oregon. He felt, rightly enough, that these "*rare simplifying generalizations of knowledge (will) provide powerful new bases for formulating and rapidly testing questions about the earth*". In 1970 Tanya Atwater's paper "Implications of Plate Tectonics for the Cenozoic Evolution of Western North America", an extensive treatment, appeared. As knowledge of tectonics expanded, the theory that much of the western margin of North America was made up of accreted terranes was supported by David L. Jones, N. J. Silberling, and John Hillhouse, 1977, in "Wrangellia - a Displaced Terrane in Northwestern North America" and William P. Irwin's "Ophiolitic Terranes of California, Oregon, and Nevada" (1978).

As early as 1965, before plate tectonics or accretion theories had been fully accepted, William Dickinson and L.W. Vigrass, in their "Geology of the Suplee Izee Areas, Crook, Grant, and Harney Counties, Oregon," puzzled over the geology of central Oregon. These problems were readily explained in 1978 by "Paleogeographic and Paleotectonic Implications of Mesozoic Stratigraphy and Structure in the John Day Inlier of Central Oregon" by William Dickinson and Thomas P. Thayer. The paper by Howard C. Brooks and Tracy L. Vallier in 1978, "Mesozoic Rocks and Tectonic Evolution of Eastern Oregon and Western Idaho", summarized tectonic knowledge of the Blue Mountains at this point.

Cascadia, the Geologic Evolution of the Pacific Northwest, by Bates McKee appeared in 1972. The versatile Bates McKee, pilot, hockey player, yachtsman as well as geologist, began teaching at the University of Washington in 1958. Energetic and enthusiastic, McKee gathered together material for his book, taking sabbatical leave from 1970 to 1971 in order to complete the manuscript. *Cascadia* was written for the layman, student, and geologist to provide *"an introduction to the evolution of the Northwest landscape"*. Oregon geology is included along with that of Washington, British Columbia, and parts of Idaho although some of the geologic details are cursory because such a wide region is covered. McKee died in 1982 in a plane crash while doing field work in the Cascades.

Notable regional compilations at this time focused on the Cascades and western Oregon. The classic paper by Dallas Peck and others "*Geology of the Central and Northern Parts of the Western Cascade Range in Oregon* (1964) was supplemented by *Geology and Geothermal Resources of the Central Oregon Cascade Range* in 1983, edited by George R. Priest and Beverly F. Vogt. Parke D. Snavely and Holly Wagner summarized the "Tertiary Geologic History of Western Oregon and Washington" in 1963. John E. Allen's *The Magnificent Gateway*, a geologic guide to the Columbia River, was issued in 1979. Other significant contributions that assembled geologic material were the *Mineral and Water Resources of Oregon* in 1969, published by the U.S. Geological Survey, as well as the *Handbook of Oregon Plant and Animal Fossils* by William and Elizabeth Orr in 1981, which collected all of the paleontological material for the state under one title. Two major additions were the COSUNA charts and the state geologic map. The *Correlation of Stratigraphic Units of North America* appeared in 1983, a compilation of regional stratigraphy, along with the "Geologic Map of Oregon West of the 121st Meridian", 1961, by Francis G. Wells and Dallas L. Peck, and its companion "Geologic Map of Oregon East of the 121st Meridian, 1977, by George Walker.

Geologic Literature on Oregon

The first bibliography of Oregon geology was C. Henderson and J. Winstanley's *Bibliography of the Geology, Paleontology, Mineralogy, Petrology, and Mineral Resources of Oregon,* published in 1912. The number of articles listed in this bibliography reflects the interest generated by early explorations which had opened Oregon to a multitude of geologic workers. Reviewing the citations in the bibliography, it is clear that the early publications give sweeping overviews of the region, while the subject matter of papers after the 1870s had narrowed to more localized topics. The majority of publications during this period focused on paleontology, followed by those on physiographic features.

In 1936, Ray C. Treasher and Edwin T. Hodge produced the *Bibliography of the Geology and Mineral Resources of Oregon* which appeared just after Dorothy E. Dixon's *Bibliography of the Geology of Oregon* in 1926. Examination of these volumes shows a striking increase in the number of articles on Oregon geology. Treasher and Hodge list 2,155 publications as compared to Dixon's 1,065 articles, and Henderson and Winstanley's 493 items in 1912.

With these bibliographies, publications on mines and minerals, followed by those on paleontology and regional topics, constituted the bulk of geologic literature. During this period there was a definite emphasis on economic geology. Gold was treated in 302 articles, coal in 110, copper in 98, and iron in 91. This interest is reflected in the publications dealing with counties that have a high amount of mining activity as Baker County (118), Klamath (92), Josephine (81), and Jackson (79), whereas Lane County appeared in only 17 publications and Multnomah in 9. Individual paleontologists were the most prolific researchers, led by John C. Merriam's 44 papers on vertebrate paleontology and Ralph Chaney's 28 on paleobotany.

From the 1940s to 1960s, the subject matter of geologic literature showed gradual changes and trends. The preponderance of literature was on minerals, followed by those on formations and regional areas, while the number of papers on paleontology had declined. There were a total of 2,069 papers listed in the *Bibliography of the Geology and Mineral Resources*

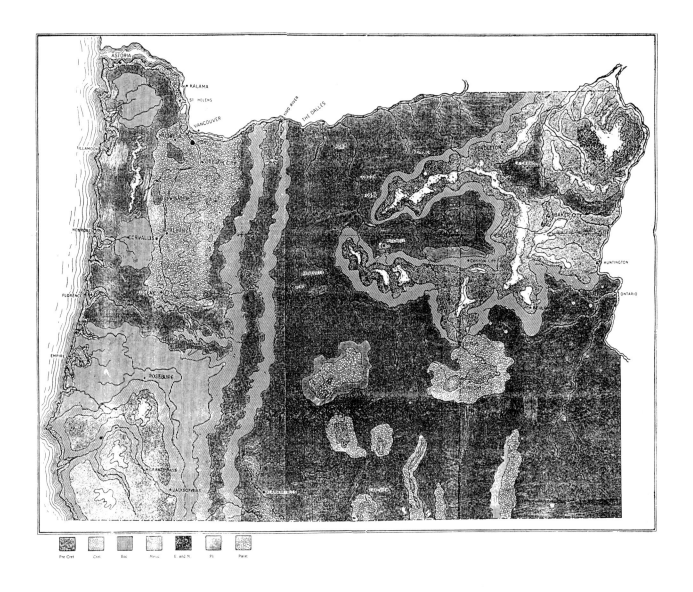

The first geologic map exclusively of Oregon by Ellen Condon McCornack, 1906.

of Oregon for these years. The emphasis was still on the mining industry in both geologic literature and research with 77 articles on gold, 78 on chromite, and a surge of papers on oil and gas (113) led by the State Department of Geology's reports on Oregon's mineral industry along with other department publications. Those counties most often cited in the literature were Clackamas, Lake, Lane, Lincoln, Douglas, Baker, and Grant, which were being closely examined for their economic resource potential as well as being surveyed for regional geologic data. Characteristically Lincoln County was reviewed for fossil localities, landslide hazards, coastal geology, agates, and a variety of other subjects. By the end of the 1950s and 1960s, research in fields of Oregon geology were becoming more diversified. Groundwater, volcanology, structural geology, and petrology were expanding fields of study.

Topics in geologic literature were spread fairly evenly over broad areas from the 1960s to 1980s. Approximately 975 citations were on formations, 850 on areas of regional interest, 735 on minerals, 563 on paleontology, 334 on ground water, 286 on geomorphology, 260 on structure, 250 on volcanology, 235 on tectonics, and 199 on petrology. The total number of papers for Oregon geology listed in the *Bibliography of the Geology and Mineral Resources for Oregon* was 4,549 for the 20 year period.

Areas generating the most interest were in Malheur County (160), Grant (117) and Harney (121) counties in central Oregon, Klamath (118) and Lake (146) counties in the south, and Coos (111) and Curry (130) counties on the coast. Much of the activity was in regional mapping, structure, stratigraphy, mineral resources, and paleontology. Volcanism ranked high during the period. As evidence of this interest, the Columbia River basalts led with 227 citations, followed by the John Day Formation with 101, and the Clarno Formation with 78. Coastal regions were examined for mineral resouces, coastal processes, paleontology, and stratigraphy.

After 1980 research was further refined into narrow categories as tectonophysics, strontium isotopes, clay mineralogy, diagenesis, and organic materials. However, the lines between the individual fields became blurred with the development of interdisciplinary fields of research. Delineation of formations (1145) led in citations in the literature, followed by economic geology (994), regional topics (779), volcanology (622), paleontology (540), and groundwater (471). The number of papers on aspects of tectonics had doubled after 1980. The total number of publications for the period from 1980 to 1990 was 2,244.

It is interesting to note that research focused on the same counties as earlier (1960-1980) with the exception of Deschutes County where intensive explorations at Newberry Crater contributed significantly to the compiled data. In Malheur County the preponderance of articles were written on the McDermitt Caldera, in Klamath County hydrogeothermal exploration and Crater Lake dominated, whereas volcanism, mineral resources, and some geothermal research accounted for the emphasis in Lake County. Several geologic maps as well as maps of mineral resources were produced on Curry County.

Finding solutions to geologic problems today is really no more difficult than in the past. In most recent years focus has been on seismicity and earthquake hazards, geothermal research, economic resources, and environmental concerns. The geologic literature will reflect these trends as they continue to mature.

Suggested Readings:

Allen, John E., comp., 1947. Bibliography of the geology and mineral resources of Oregon (supplement), July 1, 1936 to December 31, 1945. Oregon Dept. Geol. and Mineral Industries, Bull.33, 108p.

Anderson, Frank M., 1932. Pioneers in the geology of California. Calif. Div. Mines, Bull.104, 24p.

Ashwill, Melvin S., 1987. Paleontology in Oregon: workers of the past. Oregon Geology, v.49, no.12, pp.147-153.

Barksdale, Julian D., 1984. Memorial to Elliott Bates McKee, Jr., 1934-1982. Geol. Soc. Amer., Memorials, v.14, p.1-3.

Clark, Robert D., 1989. The odyssey of Thomas Condon. Portland, Oregon Historical Soc. Press, 569p.

Cope, Edward D., 1888. Sketches of the Cascade Mountains of Oregon, v. 22, p.996-1003.

----- 1889. The Silver Lake of Oregon and its region. American Naturalist, v.23, no.275, p.970-982.

Davidson, George, 1869. Coast pilot of California, Oregon, and Washington territory. U.S. Coast Survey. Pacific Coast, 262p.

Diller, Joseph S., 1896. A geological reconnaissance in northwestern Oregon. U.S. Geol. Survey, 17th Ann. Rept., pp.441-520.

Faul, Henry, and Faul, Carol, 1983. It began with a stone: a history of geology from the stone age to the age of plate tectonics. New York, Wiley, 270p.

Haskell, Daniel C.,1968. The United States Exploring Expedition, 1833-1842 and its publications 1844-1874. New York, Greenwood, 188p.

Holmes, Kenneth L., 1968. New horizons in historical research: the geologists' frontier. Ore Bin, v.30, no.8, pp.151-159.

McCornack, Ellen Condon, 1928. Thomas, Condon, pioneer geologist of Oregon. Eugene, Univ. of Oregon Press, 355p.

McBride, James Noland, 1976. Science at a land grant college: the science controversy in Oregon, 1931-1942, and the early development of the college of science at Oregon State University. Ms, Oregon State Univ., 208p.

Merrill, George P., 1924. First hundred years of American geology. New Haven, Yale Univ. Press, 773p.

Neuendorf, Klaus K.E., ed., 1978. (GeoRef, comp.) Bibliography of the geology and mineral resources of Oregon (sixth supplement), January 1, 1971 to December 31, 1975. Oregon Dept. Geol. and Mineral Indus., Bull.97, 74p.

-----1981. (GeoRef, comp.) Bibliography of the geology and mineral resources of Oregon. seventh supplement. January 1, 1976 to December 31, 1979. Oregon Dept. Geol. and Mineral Indus., Bull.102, 68p.

-----1987. (GeoRef, comp.) Bibliography of the geology and mineral resources of Oregon

eighth supplement), January 1, 1980 to
December 31, 1984. Oregon Dept. Geol.
and Mineral Indus., Bull.103, 176p.

Nevins, Allan, 1955. Fremont; pathmarker of the
west. New York, Longmans, 689p.

Roberts, Miriam S., comp., 1970. Bibliography of the
geology and mineral resources of Oregon
(fourth supplement). January 1, 1956 to
December 31, 1960. Oregon Dept. Geol.
and Mineral Indus., Bull.67, 88p.

Roberts, Miriam S., Steere, Margaret L., and Brookhyser, Caroline S., comp., 1973. Bibliography of the geology and mineral resources of
Oregon (fifth supplement). January 1, 1961
to December 31, 1970. Oregon Dept. Geol.
and Mineral Indus., Bull.78, 198p.

Russell, Israel C., 1883. A geological reconnaissance
in southern Oregon. U.S. Geol. Survey, 4th
Ann. Rept., pp.431-464.

Schneer, Cecil J., ed., 1979. Two hundred years of
geology in America. Hanover, New Hampshire, University of New Hampshire, 385p.

Socolow, Arthur A., ed., 1988. The state geological
surveys; a history. [n.p.] Amer. Assoc. of
State Geol., 499p.

Steere, Margaret L., comp., 1953. Bibliography of
the Geology and Mineral Resources of Oregon (Second Supplement), January 1, 1946
to December 31, 1950. Oregon Dept. Geol.
and Mineral Indus., Bull. 44, 61p.

-----and Owen, Lillian F., comp., 1962. Bibliography
of the geology and mineral resources of Oregon (third supplement), January 1, 1951
to December 31, 1955. Oregon Dept. of
Geol. and Mineral Indus., Bull.53, 96p.

Stevens, Isaac I., 1853. Report of explorations for a
route for the Pacific Railroad near the forty-
seventh and forty-ninth parallels of north
latitude from St. Paul to Puget Sound. In:
Reports of Explorations and surveys to
ascertain the most practicable and economical route from the Mississippi River to the
Pacific Ocean...1853-4. Vol.l, 1855, 651p.

Thwaites, Reuben G., ed., 1955. Original journals of
the Lewis and Clark Expedition, 1804-1806.
8 vols. New York, Antiquarian Press.

Tyler, David B., 1968. The first United States Exploring Expedition (1838-1842). Philadelphia, The American Philos. Soc., 435p.

Blue Mountains

Physiography

The Blue Mountains physiographic province in northeast Oregon is defined on the east by the Snake River Canyon, on the south at Ontario in Malheur County, on the north by the Washington State line, and to the west by an irregular line running near Pendleton, Prineville, Burns, and back to Ontario. Topography of the Blue Mountains intensifies eastward beginning with the low hills of the Ochoco Mountains in Wheeler County and rising to glaciated summits of the Wallowa Mountains in Wallowa County. The western portion of the province is part of a wide uplifted plateau, while the eastern section contains a striking array of ice sculpted mountain peaks, deep canyons, and broad valleys.

The multiple origins of the Blue Mountains are evident in the topography. It is not a cohesive mountain range but a cluster of smaller ranges of various orientations and relief. To the northeast are the Wallowa Mountains; the Elkhorn and Greenhorn mountains are centrally located; and the Ochoco, Aldrich, and Strawberry mountains are in the southern

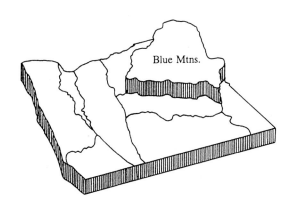

part of the province. The highest peaks of the province are the Wallowas, an immense oval-shaped range 60 miles long and 30 miles wide. Within the Eagle Cap Wilderness, the most rugged portion of the Wallowas, nine peaks rise over 9,000 feet and seven others over 8,000 feet. A number of broad valleys as Grande Ronde Valley, Baker Valley, Virtue Flat, Sumpter Valley, Lost Valley, and Bonita Valley lie between the mountain ranges.

Several extensive watersheds drain the Blue Mountains. The Grande Ronde, Imnaha, Wallowa, and John Day are the longest rivers. Of these, the John Day River, one of the most lengthy in Oregon, cuts across the province in a northwesterly direction before making a sharp turn toward the north near Clarno, entering the Columbia River close to Rufus in Sherman County. The John Day runs 280 miles from its headwaters in the Blue Mountains.

Geologic Overview

The Blue Mountains are geologically one of the most fascinating areas in Oregon. The unique aspect of the province is its patchwork origin of separate massive prefabricated pieces of the earth's crust. Permian, Triassic, and Jurassic rocks were swept up and accreted to the late Mesozoic shoreline, which at that time lay across eastern Washington and Idaho.

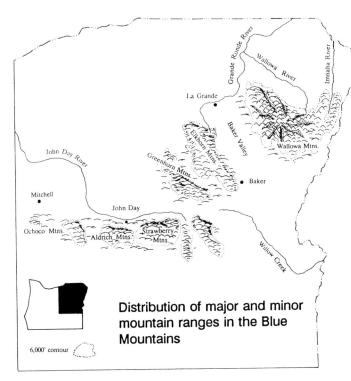

Distribution of major and minor mountain ranges in the Blue Mountains

Within these terranes the oldest rocks in the state are to be found.

Although major events have been sketched out, details of the geologic history are still only poorly known. Geologists are now aware of the exotic nature of most of the older rocks here, which have been mapped and described as well as dated with fossils and radioactive decay techniques. What remains is to determine where each of the blocks or terranes originated as well as how and when they were transported and accreted to North America.

Metamorphism, intrusion, and volcanic activity annealed these exotic blocks to North America where they became the foundation of northeast Oregon. These remnants of the earth's crust in the Blue Mountains offer an opportunity to study in detail a "collision boundary", the point where multiple large-scale blocks or tectonic plates have collided and merged with each other. Here a natural outdoor laboratory is available to view the wreckage of ancient islands that were cast up on Oregon's prehistoric shores.

Following the accretion of the major exotic terranes here, a vast shallow seaway of late Mesozoic or Cretaceous age covered most of the state depositing thousands of feet of mud, silt, and sand. This oceanic environment of warm quiet embayments and submarine fans was populated by ammonites and other molluscs as well as by marine reptiles, while tropical forests of giant tree ferns grew along the shore.

With uplift and gentle folding of the crust, the shoreline, which ran northward through central Oregon, retreated westward beyond the present day Cascades, and a long interval of volcanism ensued. Thick

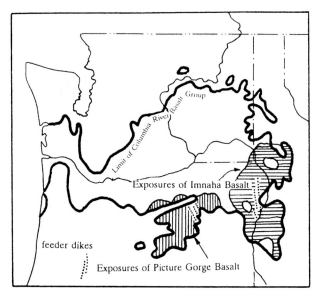

Distribution of Miocene Columbia River Group lava flows in the Blue Mountains (after Anderson, et al., 1987)

lavas intermittently oozed from fissues and vents throughout the region during the Eocene and Oligocene. Volcanic activity was interspersed with periods of sedimentation and erosion when local basins or depressions in the lava were filled with sediments deposited there by streams. Reptiles, mammals, fish and plants living in the tropical climate of this time died and were buried in the volcanic sediments. Originating from fissures in the Blue Mountains flow after flow of Miocene lavas eventually covered almost a quarter of the state. Areas of central and northeast Oregon were a wasteland of dark basalt up to a mile thick. As the lavas cooled and shrank, long vertical columns formed within the layers. Today the harder basalts cap and preserve many of the colorful rock displays.

Widespread glaciation in the Wallowas during the Pleistocene gave the region its final unique appearance. Beginning about 2 million years ago, the climate rapidly grew colder with the advance of a continental ice sheet southward as far as northern Washington, Montana, and Idaho. Although the Wallowas weren't entirely covered by ice, glaciers built up in valleys formerly occupied by streams. At the height of the Ice Ages, nine major glaciers spread through the mountains, only to melt and retreat with warmer temperatures.

Geology

Only in the past two decades has it become apparent that the Blue Mountains province is actually a number of smaller pieces of terranes which originated in an ocean environment to the west. The word, terrane, implies a suite of transported similar rocks that are separated from adjacent rocks by faults. Seas covered all of Oregon as far back as Triassic time, 200 million years ago, when these exotic blocks or terranes, traversing the Pacific Ocean basin, began to collide with the North American West Coast. As each of these island blocks arrived, they successively docked against the ancient North American craton, producing layered bands of exotic terranes. The craton itself represents the North American core or foundation of rocks native to this continent.

Scattered along the North American West Coast, these terranes are much more complex than previously thought. New evidence suggests that after collision movement of terranes by faulting was predominantly southward up to late Triassic time, then northward during the early and middle Jurassic, southward in the late Jurassic and early Cretaceous, and then finally northward from the late Cretaceous onward. With such a complexity of movements, it is little wonder that West Coast rocks look like a well-shuffled deck of cards.

Each of the terranes which make up the Blue Mountains contains its own distinctive suite of rocks and fossils. The combination of these characteristics in the province represents a wide variety of ancient environments from marine to nonmarine and from tropical to temperate. Fossils from the terranes as well as the orientation of the magnetic minerals in the rocks suggest that many of these blocks may have formed well out in the western Pacific Ocean as far south as Mexico City at 18 degrees north latitude before moving north and eastward to attach to North America. Today in the Blue Mountains these exotic terranes are exposed in linear strips that extend in a southwest to northeast trend.

Individual Terranes

Five major terranes have been recognized in the Blue Mountain province. These are the Olds Ferry island archipelago of volcanics and sedimentary rocks, the Izee terrane, an assemblage of layered rocks from the forearc basin between the island archipelago and the trench of the subduction zone, the Baker terrane, a deep ocean floor crust environment, the Grindstone terrane from a shallow ocean backarc basin, and the Wallowa volcanic archipelago terrane.

The Olds Ferry terrane extends northeast by southwest in a curved alignment of isolated exposures from Ironside Mountain in Malheur County through Huntington in Baker County, up to the Cuddy Mountains of western Idaho. Older rocks of this terrane, designated the Huntington Formation, include a thick sequence of volcanics and sediments of a type today associated with volcanic island chains or archipelagos like those in the north and western Pacific. A variety of volcanics, pyroclastics, and ash, as well as silica-rich rhyolites, and lesser amounts of basalt and andesite, make up the formation. Between the volcanic layers, fossils from thin marine beds of sandstones, siltstones,

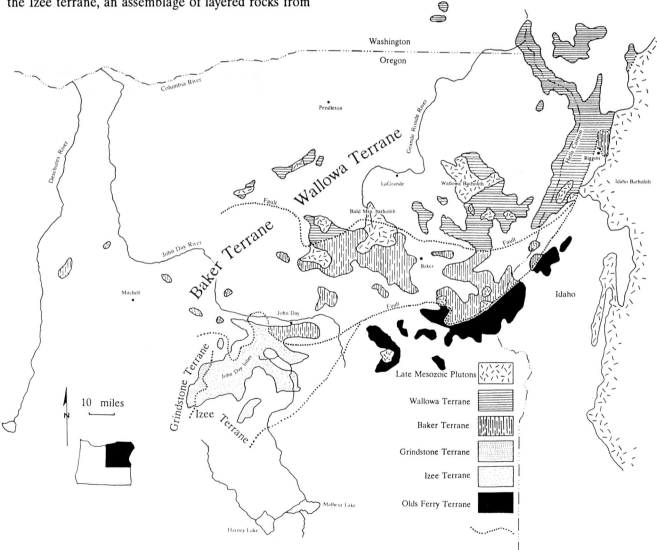

Distribution of major exotic terranes in the Blue Mountains (after Silberling and Jones, 1984)

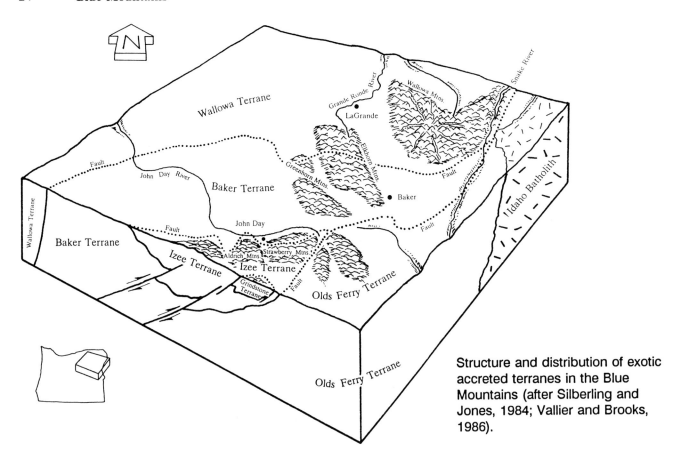

Structure and distribution of exotic accreted terranes in the Blue Mountains (after Silberling and Jones, 1984; Vallier and Brooks, 1986).

and limestones have been dated at late Triassic to middle Jurassic. Significantly, the Olds Ferry terrane includes volcanics that invaded the formations almost as rapidly as they formed, suggesting the marginal ocean island archipelago was in very close proximity to the volcanic source vents. In contrast to other Blue Mountain terranes, Olds Ferry terrane rocks show only minimal folding, which probably took place near the end of the Jurassic.

The Izee terrane, directly to the west of the Olds Ferry terrane, is separated by a fault which runs between the two. This terrane is exposed in what is called the John Day inlier, an erosional window cut through younger rocks revealing the older strata below. Composed of a number of formations, the Izee terrane represents the environment of a shallow marine forearc basin between a volcanic island archipelago and the oceanic subduction trench. Fossils are abundant in this twelve mile thick forearc basin sequence in which the limestones, mudstones, silts and sandstones, have been subdivided into distinct marine environments.

Folding of the Izee terrane began during the late Jurassic and gradually intensified due to compressional, thrusting events. With the intrusion of lower Cretaceous granites, rocks of the Izee terrane have experienced several episodes of folding and refolding. The resultant picture of folds within folds has greatly complicated the interpretation of the John Day inlier.

Imbedded in the western edge of the Izee block, but quite distinct from it, is the Grindstone terrane. The smallest of all northeastern Oregon terranes, the Grindstone is limited to a narrow area of eastern Crook County called the Berger Ranch where a mere 200 foot thickness of Devonian limestones, cherts, and argillites represent Oregon's oldest rocks. Corals and fishtooth-like microfossil remains called conodonts from these beds have been dated as lower Devonian, 380 million years old. In the same vicinity, the Mississippian Coffee Creek and Pennsylvanian Spotted Ridge formations include over 2,500 feet of limestones, mudstones, sandstones, and cherts. A shallow-water tropical invertebrate fauna of corals and brachiopods are found in the Coffee Creek Limestone, but the overlying, lower Pennsylvanian Spotted Ridge nonmarine sandstones once supported coastal plain ferns and fern-like foilage similar to those found in the late Paleozoic coals of the central and eastern United States. Nine hundred feet of fossiliferous Permian age Coyote Butte limestones, cherts, and sandstones are the uppermost formation of the Grindstone terrane, although isolated blocks of radiolarian-bearing cherts in the vicinity are as young as lower Triassic. Because of poor exposures and the difficulty of tracing the rock

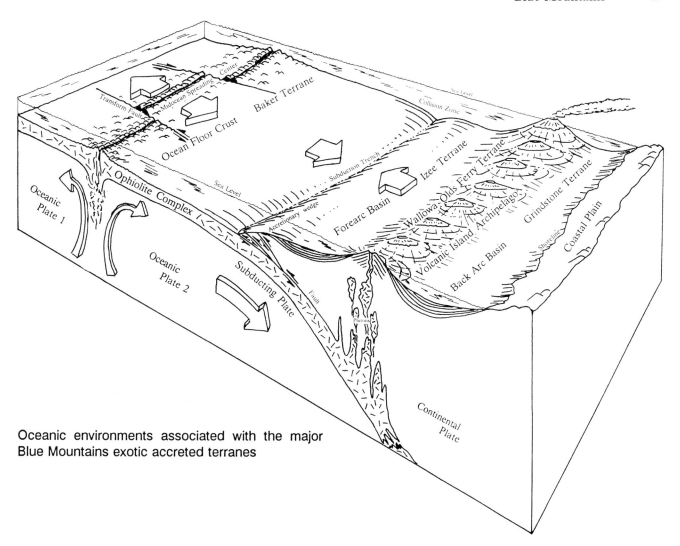

Oceanic environments associated with the major
Blue Mountains exotic accreted terranes

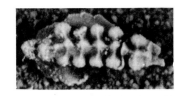

Fishtooth-like microfossils, conodonts from the
Devonian limestone in the central Oregon Grindstone
terrane, are less than 1/32 inch in length (courtesy of
N.M. Savage)

units, it is impossible at present to associate details of
this suite of ancient rocks to the rest of the rocks in
the Blue Mountains. Clues to the origin of the Grindstone terrane are evident in the similarity of rocks and
fossils of the Coyote Butte Formation to the Bilk
Creek exotic terrane north of Winnemucca, Nevada.

The diversity of formations in the Grindstone
terrane make up what is called a mixture or melange of
rocks. Interpreted as once forming the boundary
between two converging crustal blocks, melanges are
areas where different geologic environments have been
telescoped or jammed together by the colliding plates.
As with the Olds Ferry and Wallowa terranes, the
Baker and Grindstone terranes have sometimes been
combined because of similarities between the melanges
in them. The position of the Grindstone on the western side of the Izee terrane seems to support this but,
contrasting rocks between the two terrances would
argue against it.

The Baker terrane lies just northwest of the Izee terrane extending in a curved arc from isolated exposures in Wasco County, Oregon, to the Cuddy Mountains in Washington County, Idaho. Separated from the Izee and Olds Ferry terranes by a fault, the Baker terrane is substantially older than the Izee forearc basin terrane, and protozoan, mollusc, and coral fossils found in its various formations range from Permian through upper Triassic. Microfossil fusulinids and radiolaria from thin limestone layers of the Baker melange reflect not only the age but the possible source of these rocks. These varieties of single-celled microfossils reflect a "Tethyan" or tropical western Pacific environment. The discovery of these exotic fossils over 30 years ago was one of the first hints that not all of Oregon was homegrown.

The Baker terrane is made up of several formations beginning with the oldest Burnt River schist which is succeeded by the Elkhorn Ridge argillite. Both of these units represent metamorphosed, deep ocean floor mud and chert originally deposited in an environment some distance from the land. Another component of the Baker terrane exposed due south of John Day is the Canyon Mountain complex, interpreted as both an island arc and deep ocean floor environment associated with a midocean spreading ridge. With an abundance of easily sheared rocks, the Canyon Mountain complex forms a melange that has been faulted, broken, and shattered during the compression that is characteristic

of a plate collision at a subduction zone. This complex mixture of rocks has been recognized as an "ophiolite", a term which refers to the abundance of the metamorphic rock serpentine. A derivative of gabbro and perioditite, serpentine is easily broken and sheared by faults when subjected to pressure. "Ophiolite" refers to

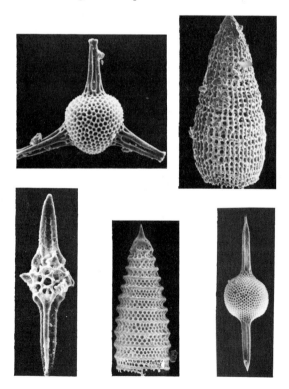

Mesozoic radiolaria from eastern Oregon. Specimens are less than 1/64 inch in length (courtesy of Emile Pessagno)

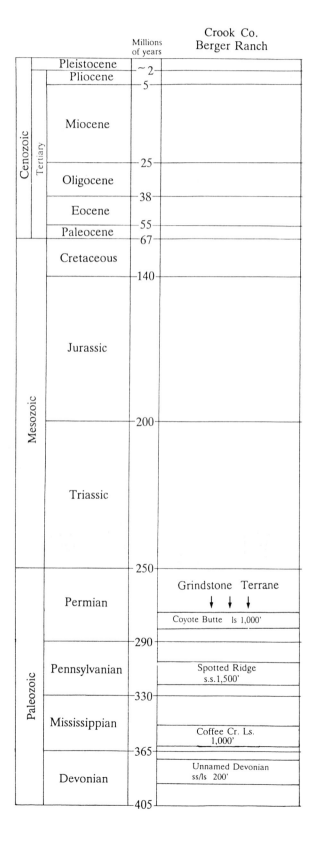

the similarity of these polished faulted rock surfaces to the texture of snakeskin. The term ophiolite is comprehensive, used to describe thick sections of raised deep sea floor. An ophiolite sequence is a predictable arrangement of deep ocean crust with olivine rich peridotities at the base, followed by gabbro and basalts, and capped by pillow lavas and deep ocean shale and cherts. Along with serpentine, the Canyon Mountain ophiolite complex contains argillites, cherts, gabbro, diorite, and volcanic tuffs.

As the weak rocks of the Baker terrane were shoved against the mainland, they were squeezed into what is called an accretionary wedge. In today's oceans these wedges are the sites of severe folding and fault-

E. Blue Mtns. ~Baker Co.	NE. Blue Mtns. ~Wallowa Co.	SW. Blue Mtns. ~Grant Co.	SE. Blue Mtns. ~Malheur Co.
Glacial debris	Glacial debris	Glacial debris	
Picture Gorge Basalt / Imnaha Basalt 1,000'	Columbia River Group	Rattlesnake ash flow tuff / Strawberry Volcanics / Mascall Fm. 1,500' / Picture Gorge Basalt	
John Day Fm. 3,500'	John Day Fm.	John Day Fm.	
Clarno Fm. 3,000'	Clarno Fm.	Clarno Fm.	
Hudspeth Sh. 10,000" — Gable Creek Congl. 9,000'		Bernard Fm. 1,500'	
Baker Terrane	Coon Hollow Fm. Mudstone 1,800'	Izee Terrane — Lonesome Fm. 10,000' / Trowbridge Fm. 3,000' / Snowshoe Fm. 3,000'	Olds Ferry Terrane
	Wallowa Terrane	Mowich Group: Hyde Fm. 1,500' / Nicely Sh. 30' / Suplee Fm. 30' / Robertson Fm. 300'	Jurassic Flysh 10,000' / Weatherby Fm.
Elkhorn Ridge Argillite 5,000'	Lucile Gp. 2,500': Hurwal Fm. 1,000' / Martin Bridge Ls. 1,500'	Caps Ck. / Graycock Fm. 450' — keller Cr. Shale 4,500' / Aldrich Mt. Gp.: Rail Cabin Argillite 1,800' — Laycock Fm. 10,000' / Fields Cr. Fm. 15,00'	Huntington Fm. 10,000'
	Clover Cr. Greenstone (upper) — Seven Devils Volcanic Group (Upper) 10,000': Doyle Cr. Fm. 1,500' / Wildsheep Cr. Fm. 1,500'	Brisbois Fm. 5,000' / Begg Fm. 9,000' — Vester Fm. 10,000'	
Canyon Mt. Ophiolite Complex	Clover Cr. Greenstone (lower) — Seven Devils Volcanic Group (lower) 10,000': Hunsaker Cr. Fm. 9,000' / Windy Ridge Fm. 1,500'		
Burnt River Schist 5,000'+			

Stratigraphy of Blue Mountains terranes overlain by "native" formations that were deposited after accretion took place (Correlation of Stratigraphic Units of North America, 1983)

ing, and Baker terrane rocks are no exception. The terrane is a series of enormous coherent blocks separated from each other by large-scale shear fault zones. The fine-grained rocks of this terrane were tightly compressed into folds during the late Triassic and early Jurassic.

The northern-most terrane of the Blue Mountains province is the Wallowa volcanic arc represented by a five mile thick sequence of volcanic archipelago rocks ranging from Permian to Jurassic in age which were accreted behind the Baker terrane. Isolated exposures of the Wallowa terrane appear northeast of Walla Walla, Washington, as well as south of Lewiston, Idaho, but continuous sections of this exotic block are located in the Snake River Canyon where they can be traced for over one hundred miles.

The older, volcanic rocks of the Wallowa terrane accumulated along the axis of an ancient island archipelago. Fossiliferous sedimentary rocks of the Martin Bridge Limestone and shales and siltstones of the Hurwal Formation were later deposited on top of volcanic rocks of the arc. The Martin Bridge and Hurwal formations span the Triassic-Jurassic boundary and reflect a shallow, warm shelf environment. Fossils of molluscs, corals, and swimming reptiles within the Martin Bridge and Hurwal are similar to those found in rocks of the eastern Pacific as well as from central Europe to the Himalayan Mountains.

Over the Hurwal Formation, fossil wood, leaves, plant fragments, and invertebrates of the Coon Hollow conglomerates and sandstones record the nonmarine and nearshore environment that existed on the Blue Mountains island arc during the middle Jurassic. Growth rings in petrified tree wood point to a temperate, seasonal climate with abundant rainfall and a short dry season before winter begins. The commonly occurring ferns and quillworts grew along stream banks and in marshes. Molluscs and corals are found mixed with plant fossils in shallow coastal sediments. As granites of the Wallowa batholith intruded these formations, heat and pressure locally altered them to marble and slate in late Jurassic and Cretaceous time.

The Wallowa terrane is heavily sheared by faults developed during the accretion against North America during the late Mesozoic. The rocks are so heavily deformed that whole folds are locally overturned. Another major structural event affecting the Wallowa terrane is shearing or faulting along three separate zones visible in the Snake River canyon. The 25 mile long north-south Oxbow shear zone shows displacement where the west side moved south and the east side move north. During the early Jurassic the total distance all three faults moved was several hundred miles.

Part of the Wallowa terrane, the Seven Devils Group of rocks has been compared to widespread terranes collectively called "Wrangellia", for extensive exposures in the Wrangell Mountains of Alaska. This large terrane is a disjunct group of similar volcanic and sedimentary rocks of Triassic age and older stretching southward from the Wrangell Mountains in southern Alaska to include Chichagof Island, Queen Charlotte Islands, and Vancouver Island in Canada. The southern most piece of this string might be the Wallowa Mountains. The scenario for the development of Wrangellia implies two separate and distinct tectonic processes. During Triassic time, the various pieces of Wrangellia were a single island block situated somewhere in the tropical Pacific. The pieces were slowly being transported toward North America on a conveyor belt-like tectonic plate where they collided with the West Coast. With the docking event, the block was sheared up by large-scale faults similar to the San Andreas. Movement along these faults "translated" or conveyed slices northward one by one. The Seven Devils volcanic rocks are typical of those from ocean margin island archipelagos, whereas the remainder of Wrangellia to the north has volcanic sequences frequently found in midocean, deep water environments. However, the overall similarity between the rocks and fossils of all the regions of the terrane suggests there is a strong affinity between northeastern Oregon and Wrangellia.

Cretaceous

In the later Jurassic and early Cretaceous between 160 and 120 million years ago, newly joined terranes of the Blue Mountains region were invaded by granitic batholiths of granodiorite and gabbro to cement them together with a granitic glue. These intrusions include the Bald Mountain and Wallowa batholiths as well as several smaller bodies. A batholith is a mass of coarse-grained crystalline rock which melted upward from deep within the earth's mantle into the surrounding layers above. As the liquid magma slowly cooled, large crystals formed. Batholith roofs exposed by erosion have a surface area greater than 40 square miles, and these two batholiths in northeastern Oregon certainly exceed that. The Bald Mountain batholith, northwest of Baker in the Elkhorn Mountains, extends over 144 square miles, while the Wallowa batholith, forming the core of the Wallowa Mountains, covers 324 square miles and is the largest in Oregon.

It was during the early Cretaceous that the last exotic crustal blocks were attached or accreted to

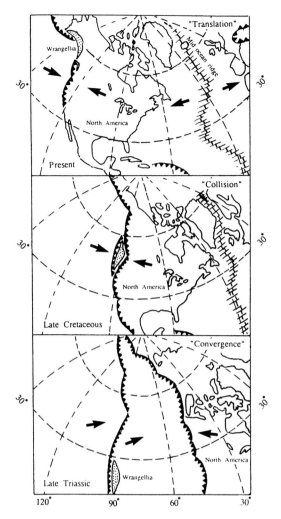

Convergence, collision, and translation model for aquisition of the Wrangellia terrane by North America. The North American plate, moving in a northwesterly direction, captures and sweeps up the Wrangellia terrane prior to translating it along faults to Alaska (modified after Moore, 1991)

flying reptiles or Pterosaurs. No dinosaur remains have been recovered in the state even though there are abundant dinosaur fossils in coastal plain or shallow marine sediments from Montana, Utah, Colorado, and Wyoming. Because similar coastal marine environments have been identified at several locations across Oregon, it seems to be only a matter of time before diligent fossil hunting will unearth dinosaur remains. Once the large terrane blocks had been annexed to the mainland by the end of the Cretaceous period, the sea had withdrawn to the west, and erosion began to attack the newly arrived rocks.

Cenozoic

With the beginning of the Cenozoic era, 65 million years ago, the Blue Mountains area of Oregon was above sea level undergoing uplift and subsequent erosion as well as ongoing, but intermittent, volcanic activity. The Cascades and Coast Range had yet to form, and the ocean shoreline lay east of where the Cascades are today. Because Paleocene rocks are not known from the Blue Mountains, it is assumed that this epoch was mostly one of erosion. Tertiary volcanoes of the Clarno and John Day periods as well as flows of the later Columbia River group covered the region with widespread, thick deposits of lava and ash interspersed with fossiliferous sediments containing remains of plants and animals.

North America. After this final stage of convergence, deposits of the later Tertiary, which covered the region, are more or less homogeneous from east to west. The same formations are found in the same sequence over the area. Following the tectonic activity of the middle Mesozoic, shallow Cretaceous seas spread over wide expanses of Oregon. The ancient shoreline during this time was a meandering diagonal line from the Blue Mountains southeast into Idaho. Divers molluscs living in the Cretaceous seas included many species of coiled, shelled cephalopods or ammonites which would eventually prove useful in age dating these rocks. Other inhabitants preserved in rocks of the Cretaceous ocean in central Oregon were the fish-like ichthyosaurs and

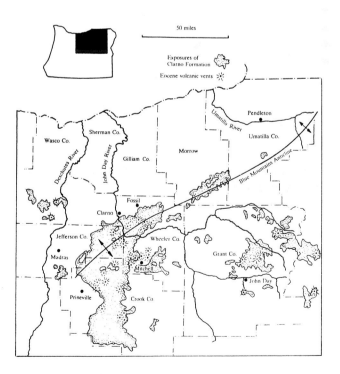

Distribution of Clarno Formation volcanics, mudflows, and vents in eastern Oregon (after Walker and Robinson, 1990)

the North American plate. After the oceanic slab was carried down and melted, liquid rock on the surface appeared as andesitic lavas of the Clarno Formation.

As the loose ash from these volcanoes mixed with water, mud flows or "lahars" were formed. Lahars flowed like viscous syrup to entomb both plant and animal fossils in a broad band across eastern Oregon. Individual, uninterrupted deposits of Clarno mudflows up to 1,000 feet thick occur near Mitchell in Wheeler County. Preserved in the sediments of the mudflows, broad leaf plants record the wet, tropical climate of eastern Oregon during the Eocene epoch. Avocado, cinnamon, palm, and fig created a lush woodlands interspersed with open, grassy plains and freshwater lakes. A diversity of large land mammals, tapirs, rhinoceros, titanothere, and oreodons shared the stage with freshwater fish and snails. A peculiar aquatic rhinoceros, not unlike a modern hippopotamus, wallowed in shallow waterways to be preyed upon by crocodiles and other large carnivores. The Clarno is perhaps most famous for its "nut beds" where many species of tropical and hardwood nuts, fruits, and seeds are preserved in soft volcanic muds.

By early Oligocene time the Clarno volcanic episode had subsided to be replaced by eruptions of a new complexion in the John Day period about 36 million years ago. Only a brief interval of erosion separates the two events. Most of the John Day volcanoes were just to the west of the Blue Mountains province near the present-day Cascade Range. Great volumes of ash and dust erupted into the sky to settle and be carried by streams and deposited in the broad John Day basin. Volcanic vents at this time often generated incandescent pyroclastic ash flows or "ignim-

Eocene palm leaf Sabalites from the Clarno Formation is over 2 feet long (photo courtesy Oregon Department of Geology and Mineral Industries)

Tertiary volcanic events in eastern Oregon opened in Eocene time around 44 million years ago with andesitic and rhyolitic lavas erupting from a chain of volcanic vents in the western Blue Mountains. Lavas, accompanied by mud flows and successions of multicolored soils as well as tuffaceous sediments deposited in lakes or by streams, make up the colorful blend of rocks comprising the Clarno Formation. The source of the Clarno andesitic volcanic activity is not well understood but apparently relates to what is called the Challis arc of Eocene volcanic vents extending diagonally northwest by southeast across Washington and Idaho. Adjustments in the tectonic plates to the west at this time produced a considerable bulk of material being subducted from the west toward the east beneath

Many John Day volcanoes issued "ignimbrites" or incandescent clouds of ash that flowed across the landscape.

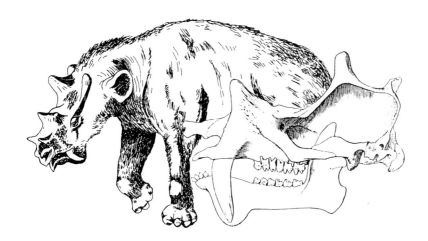

About the size of a small elephant, the uintathere was a browsing ungulate with long canines as well as three symmetrical pairs of bony projections on the skull.

Similar to a large rhinoceros in size and structure, brontotheres had frontal horns of solid bone.

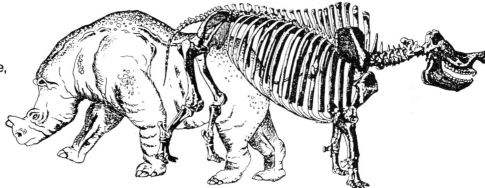

Cypress swamps of the John Day period were inhabited by the cow-sized rhinoceroses, while Metasequoia grew along stream banks.

Stratigraphy and structure in the vicinity of Picture Gorge and Sheep Rock in Grant County

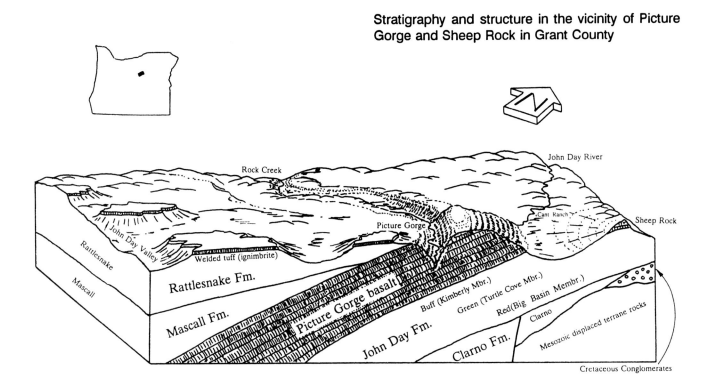

brites". These hot, gas-charged clouds of ash poured down the flanks of volcanoes then across the landscape for miles, destroying everything on the way, before settling to the ground and annealing into a solid crystal-studded, glassy mass. Within the western part of the basin, vents pumped out silica-rich rhyolites, ash flow tuffs, and lavas, to build up domes, whereas higher temperature alkali-olivine basalt and andesite lavas erupted from volcanoes farther east.

Rapid deposition of volcanic ash and mud in every depression on the landscape was optimum to preserve remains of the semi-tropical vegetation and mammals which lived in the John Day basin. Because this period of fossil accumulation dates back only to the Oligocene and Miocene, from 18 to 36 million years ago, all of the plants and most of the mammals look somewhat familiar, although none of the species exactly correspond to those living today.

As climatic and volcanic conditions altered, the nature and color of volcanic sediments of the John Day changed as well. Highly oxidized, deep red ash layers of the lower John Day Big Basin member are only poorly fossiliferous. These are succeeded by the pea green Turtle Cove layer dominated by clay and rich in fossils. The Turtle Cove is, in turn, followed by the fossiliferous, cream or buff-colored layers of the Kimberly and Haystack rock members. Throughout the deposition of these layers in streams, lakes, and floodplains, volcanoes dispersed incandescent clouds of glowing ash that melted and flowed together. John Day fossils in the

green and buff intervals reflect a succession from lakes and rivers giving way to more open grasslands as the episode progressed.

Following the John Day deposition, the soft rocks of the formation were lightly dissected by erosion before the vast basalt flows of the Columbia River lavas covered much of the Blue Mountains region in the middle Miocene between 17 and 12 million years ago. Very fluid basaltic lavas began to erupt and pour from cracks and fissures in northeastern Oregon, southeastern Washington, and western Idaho. As much as 400 cubic miles of lava were extruded in single flows, and eruptions lasted for weeks at a time. Even though some later flows can be traced all the way to the Pacific Ocean, most of the Columbia River basalts in central Oregon are from dikes near Monument and Kimberly, where they covered a distance of less than 20 miles.

Throughout the Columbia basin over 40 separate flows of the Miocene basalt are recorded. However, it is primarily the Picture Gorge lavas that overlie the John Day Formation. Lavas of this group were very liquid, and the eruptions rapidly filled in valleys and low spots within the Blue Mountains province. Prior to the eruption of the Columbia River basalts, the Miocene topography had a relief of as much as 1,000 feet from the valley floors to the crests of the hills. As volcanism continued, flows lapped over each other until only the highest exposures of earlier rocks projected above the youngest layers of basalt.

WEST

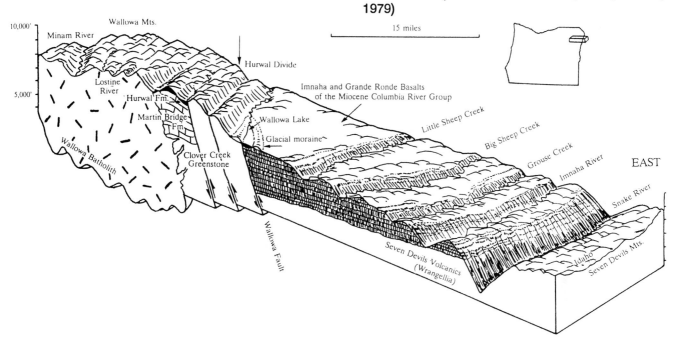

Structure and stratigraphy of the eastern flank of the Wallowa Mountains from the Minam River to the Snake River (after Smith and Allen, 1941; Walker, 1979)

The extent and volume of the Columbia River lavas tend to overshadow important Miocene eruptive provinces that were developing at the same time in the southern Blue Mountains where Sawtooth Crater, Strawberry Volcano, and Dry Mountain evolved as three separate volcanic centers. These centers in turn line up with a fourth volcano of the same age at Gearhart Mountain to the southwest in the Basin and

Range province. Lavas from these four volcanic regions were predominantly a stiff, slowly flowing andesite variety that accompanies very explosive eruptions. Vents in the vicinity of Strawberry and Lookout mountains were the most numerous, extruding the thickest and most extensive lavas. Covering a total of 1,500 square miles, the Strawberry volcanics are over 1 mile thick at Ironside Mountain in Malheur County.

By late middle Miocene time, the Oregon oceanic shoreline had retreated far to the west near the present day coast. Ash from late Miocene and Pliocene volcanoes in the Cascades fell over an increasingly cooler temperate landscape where streams and rivers intensively eroded the older sediments. Volcanism at this time in the Blue Mountains was limited to small eruptions of lava along the southern and western margins of the province as local lavas mixed with showers of ash from emerging Cascade volcanoes to the west. Sediments and lava flows from these events were deposited directly atop the Columbia River basalts. Of these, the Mascall Formation of volcanic tuffs, carried and deposited by streams, preserved a variety of middle Miocene fossils. A rich mammal assemblage of horses, antelope, camels, deer, and oreodons along with predators such as dogs, bears, weasels, and racoon suggest an open woodland or savanah in association with cool, temperate broadleaf plants dominated by angiosperms.

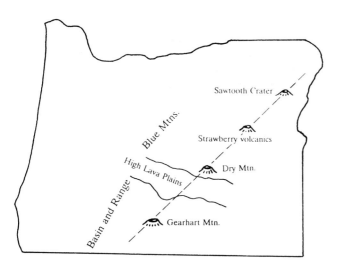

Lineation of Miocene andesitic volcanic centers in eastern Oregon (after Robyn and Hoover, 1982)

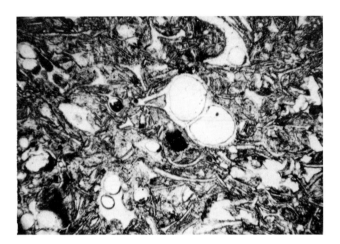

Thin section of Rattlesnake ignimbrite ash-flow reveals a complex mass of broken shards (photo courtesy of Oregon Dept. of Geology and Mineral Industries)

The last major widespread eruption in eastern Oregon took place during the late Miocene. Streams deposited up to 1,000 feet of sands, gravels, and silts of the Rattlesnake Formation which was capped by an ash flow that erupted from vents 90 miles to the south in Harney County. Today the distinctive Rattlesnake ignimbrite forms the skyline of much of Wheeler and western Grant counties. Horses, camels, rhinoceroses, mastodon, antelope, and sloths were supported by the high plains grasslands of this time. Fossils of these animals along with those of weasels, cats, coyotes, bears, and a variety of rodents are preserved within Rattlesnake sediments.

Pleistocene

Glaciation came to eastern Oregon with the Pleistocene or Ice Ages beginning a little less than 2 million years ago and winding down about 11,000 years ago after several ice advances and retreats that lasted about 150 to 200 thousand years each. With Ice Age glaciation, moving ice masses took over valleys originally cut by streams. After sculpting the valley, the glaciers melted, and the streams reoccupied the canyons to resume fluvial erosion. Continental glaciation, where large ice sheets expanded south across the landscape during the Pleistocene, was more typical of the Great Lakes and upper Mississippi Valley. In the western part of North America continental glaciers only spread as far south as Washington, Idaho, and Montana.

The Wallowa Mountains have been called the Oregon Alps because of their glaciated, picturesque slopes. With the exception of the Ochoco Mountains, all of the individual ranges in the Blue Mountains province display the unmistakable marks and deposits of valley glaciation ranging from glacial striations on bedrock to thick layers of till and moraines. Most of the higher valleys are U-shaped in cross-section marking canyons previously occupied by ice masses. Along valley floors, many lakes are glacial tarns or low spots cut by a previous glacier.

Great symmetrical piles of glacial till called moraines are easily identified by the poorly sorted mixture of large and small pebbles and clay in the same exposure. Perhaps the most dramatic glacial terrains of the range are the lateral and end moraines adjacent to the small community of Joseph. These moraines have created a natural dam for Wallowa Lake that occupies the toe of the valley. Joseph itself is built upon the ancient outwash plain, characterized by irregular hummocky ground, that spread out in front of a melting glacier.

Covering about 280 square miles, nine large glaciers, each more than 10 miles long, occupied the Wallowas during the Pleistocene. The Lostine glacier was 22 miles long, the Minam 21, and the Inmaha extended 20 miles. Averaging 1,000 feet in thickness, the ice masses reached down the valleys but failed to cover the ridges above 8,500 feet so that a true ice cap did not occur. The longest of the Wallowa glaciers in the Lostine River basin developed an ice mass up to 2,500 feet thick before melting and retreating.

As late as 1929 the last small glacier existing in the Wallowa Mountains was recorded on the ridge of Glacier Lake in Union County. The glacier was 800 feet long, 60 feet wide, and 24 feet thick. Today small stagnant masses of ice, but no moving glaciers, occupy a few of the cirque basins.

Structure

The Blue Mountains are caught between two massive earth blocks that have been in motion for millions of years. Stresses built up by the movement of these huge blocks initially created folds then faults within the Blue Mountains area. Due to changes in plate motion, direction, and speed, folding and faulting has occured in stages approximately every 5 million years. During Tertiary time, the Blue Mountains block was rotated 65 degrees in a clockwise direction. This rotation appears to have resulted from what is called dextral shear where the block was caught between large-scale, north-south faults that dragged it clockwise.

The Wallowa Mountains were carved by 9 major glaciers during the Pleistocene epoch. Reconstructed here the glaciers followed a radial pattern out of the range.

The east-west orientation of the Blue Mountains contrasts sharply with the rest of Oregon's north-south ranges because they are part of a much larger structural system that starts in California and trends northward through the Klamaths to turn sharply northeast before passing under the Cascades. At the Snake River, the system swings abruptly northwest to merge with the Cascades in Washington, where it again angles northward. Known as an "orocline", this major S-shaped bend in the older Mesozoic rocks is called the Columbia Arc.

Extending from north of Powell Buttes in Crook County almost to Lewiston, Idaho, one of the most obvious structures in the northern part of the province is the Blue Mountains anticline, which began

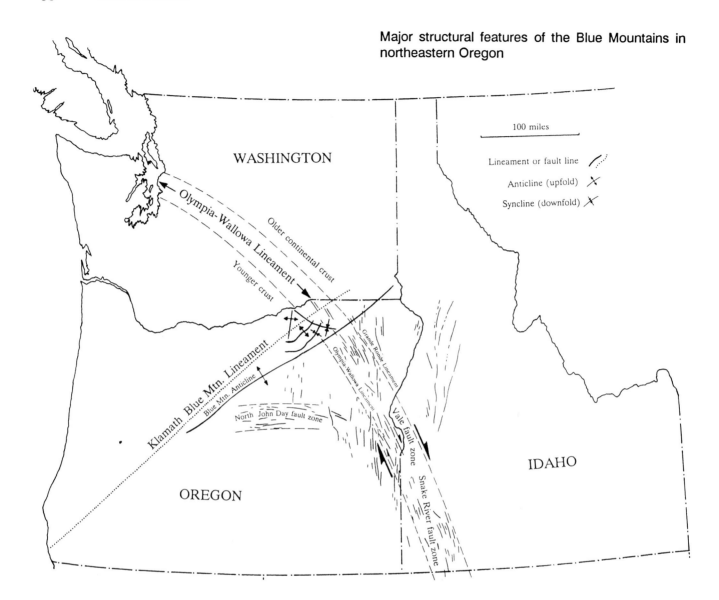

Major structural features of the Blue Mountains in northeastern Oregon

to fold upward around 35 million years ago in early John Day time. The development of this structure correlates with the breakup of the Farallon plate between the Murray and Mendocino fracture zones in the Pacific Ocean and records a strong north-south compression. Running parallel to and only slightly southeast of the Klamath-Blue Mountains lineament, the Blue Mountains anticline may mark a deeply buried suture between two microplates in the crust below.

A number of extensive large-scale lines or lineaments across Oregon are of such magnitude that they are seen best on high altitude aerial photographs. The cause of the lineaments is not understood, although they apparently represent a boundary or structure buried deep in the crust. These lineaments are traced along interconnected faults and folds or more subtle features such as the alignment of straight stream valleys. In the Blue Mountains, two lineaments

running across the northeast corner of the province are the Grande Ronde lineament and the Olympic-Wallowa lineament. The Grande Ronde alignment parallels Eagle Creek and the upper Grande Ronde valley for miles across eastern Oregon. To the southwest the Olympic-Wallowa lineament, a very straight structural trend extending across the Columbia plateau and northern Blue Mountains, connects with the Vale and Snake River fault zones.

Mining and Mineral Industry

Gold

The region of the Blue Mountains has produced about three-fourths of all of Oregon's gold. These deposits occur in the gold belt of the Blue Mountains, so named by Waldemar Lindgren. The strip

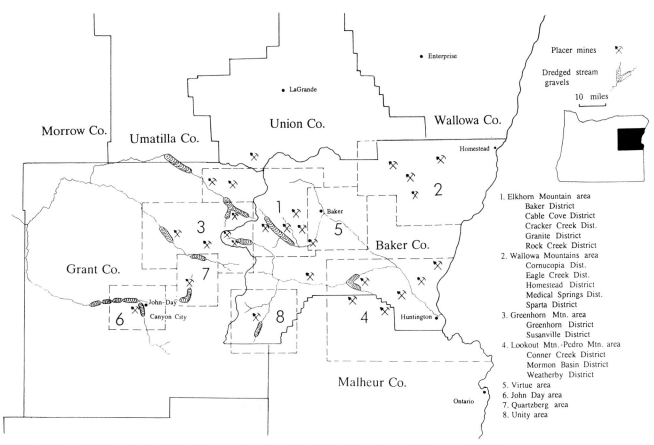

Gold mining districts in the Blue Mountains (modified after Brooks and Ramp, 1968)

is 50 miles wide and 100 miles long from John Day to the Snake River and Idaho.

Gold in the Blue Mountains province is of two types, lode gold and placer gold. Lode gold occurs as veins and is mined by tunnelling into the rocks to follow tabular bodies rich in the metal. Virtually all of the Blue Mountains vein gold is found along the margins of batholiths which were intruded during the Cretaceous and Jurassic periods. Found in the shattered wall rocks of the batholith, gold was formed along with quartz and a host of other low temperature minerals that were last to crystallize from the original magma. Placer mining exploits areas where older gold veins have been exposed by erosive action of present and ancient streams. Gold is seven times heavier than quartz and almost twenty times heavier than water. It can be carried and moved by only the fastest flowing streams where it quickly settles in the deepest part of the channel with the slightest waning of the water velocity. Most placer mining tactics such as panning and sluicing take advantage of gold's heavy weight.

Gold was first reported in the Blue Mountains region by immigrants in 1845. Under the leadership of Stephen Meek, a wagon train on its way to the Willamette Valley became lost when it struck across the desert of central Oregon. Looking for straying cattle, several members of the party picked up pieces of dull yellow metal from the bed of a small stream. One of the nuggets was hammered flat on a wagon tire then discarded into a tool kit. Eventually the pioneers found their way to the Willamette Valley, the piece of gold forgotten. Several years later, the discovery of gold in California and Idaho triggered a memory of this nugget, and several men of the wagon party thought they could find what was now called the Blue Bucket Mine. This designation has several imaginative origins. One story goes that children on the wagon train filled a blue bucket with water from a stream which had gold nuggets in the bottom, while in another verison an immigrant said he would have filled his blue bucket with the nuggets had he known what they were. Since that time, many unsuccessful efforts have been made to find the creek which people claim must be located in the vicinity of Canyon City in Grant County.

Although gold finds were reported during the 1850s, it wasn't until 1861 that prospectors were willing to face hazards from Indians and a harsh climate to open eastern Oregon to the excitement of a gold rush. Henry Griffin from Portland made the original discov-

The gold dredge "City of Sumpter" operated from 1913 until 1954 in Baker County (photo courtesy of Oregon Dept. Geology and Mineral Industries).

ery of gold just west of Baker in Griffin Gulch. A second group of miners, harassed by Indians, found gold in Canyon Creek near John Day in Grant County. The gold here was so plentiful that the men searched no farther, and by July 5, 1862, a tent city of 1,000 men had grown up along the stream. Within a few years, prospectors spread out to most of the districts in the Blue Mountains and adjoining Idaho.

Water for working the placers was scarce in the desert, and long, extensive ditches were dug to carry it to the workings. The Auburn ditch, parts of which are still used by the municipality of Baker, was completed in 1863. The Rye Valley ditch was completed in 1864, and the 100 mile long El Dorado ditch, finished in 1873, carried water from the head of Burnt River to the Malheur placers. The water supply stimulated production for 10 to 15 years until the placer gravels were exhausted. Bucket line dredges, which were first used in Sumpter in 1913, also increased gold production for a time on the creeks and rivers of eastern Oregon.

Lode mining began shortly after placers with the development of the Virtue Mine near Baker in 1862 where a 10-stamp mill was constructed to process the ore. Stamp mills, employing heavy steel rods, pulverized the gold ore to extract the fine gold. Virtue and Sanger mines in the Wallowa Mountains and Connor Creek mine in the Lookout Mountain region dominated lode gold production until the early 1900s when production of vein gold from the Cornucopia mines began. Ranking as the top lode producer, Cornucopia mines were active for over 60 years, frequently yielding more than half of the total lode gold of Oregon for several consecutive years before closing in October, 1941.

As late as June, 1913, a large nugget weighing 80.4 ounces was picked up in the gravels of a placer mine near Susanville in Grant County. George Armstrong and his partner found the nugget while mining by hydraulic pressure. Today the gold specimen worth between $30 and $40 thousand dollars is exhibited by the U.S. Bank of Oregon in Baker.

Early gold production for the Blue Mountains can only be estimated because exact figures were never recorded, and much of it was sent to the mint at San Francisco where it was credited to California. Placer mines in the region west of Baker had an approximate yield of $5 million dollars, while placers elsewhere in Grant and Baker counties had an estimated total output of $18 million dollars by 1900. By way of comparison, lode mining operations totaled $5 million dollars by 1900. Mines in the North Pole-Columbia

lode, extending 5 miles southwest from Elkhorn Ridge, were the most lucrative in Oregon. Production figures from 1895 to 1908, the most active period, were $2,485,000 for 100,450 ounces of gold.

After 1900 activity increased. Peak years were from 1900 to 1908, from 1913 to 1942, and from 1946 to 1954 with the estimated dollar value between 1900 and 1965 placed at $47 million. After that, production dropped drastically. For example, during 1954 there were 5,175 troy ounces of gold from both mines and placers in Baker County. In 1955 only 299 ounces of gold were recorded from the same sources. Since 1945, most lode gold has come from the Buffalo Mine in the Elkhorn Mountain area, while the Bonnanza Mine north of Halfway is currently the state's major placer gold producer. Today data on individual companies is no longer available to the public.

Now residing with the U.S. Bank of Oregon in Baker, the Armstrong nugget was among the largest found in Oregon during the gold rush days (Oregon Dept. Geology and Mineral Industries).

Silver is typically recovered with gold as part of the extraction process since gold and silver are usually associated in nature. Production of silver in the Blue Mountains province reached just over $2 million dollars from 1900 to 1965, the most lucrative period.

Southwest and northeast Oregon are prime targets for gold exploration. Large placer mining operations, recreational mines using small portable dredges, and the installation of gold-saving devices at private sand and gravel plants have increased the amount of gold recovered. The use of a process called heap-leaching, where sodium cyanide or other chemicals are employed to extract the gold from piles of low-grade ores, has brought vigorous protests from environmentalists. Large open pits, created in the extraction process along with lagoons which contain toxic materials would eventually be abandoned.

Copper

Deposits of copper are often found with gold and silver in the Wallowa Mountains and along the Snake River canyon. Copper prospecting began in the late 1800s, and the bulk of the copper came from the Iron Dyke Mine near Homestead on the Idaho border. Copper here occurs in sheared, faulted zones within Permian and Triassic volcanic and sedimentary rocks. The mineralization or emplacement by hydrothermal fluids along fractures in the host rock probably came later in the Jurassic period. Copper occurs chiefly as chalcopyrite with lesser amounts of other minerals present.

The Iron Dyke was discovered in 1897 although serious production didn't begin until 1915. A "rush" to Homestead occurred in 1890 with the promise of major copper discoveries, however, output from most of the deposits remained small. From 1900 to 1965 approximately $3 million dollars worth of copper was extracted. The peak years were from 1913 to 1920.

Mining ceased in 1928, although operations sporadically continued for several years after that. There was even limited gold and copper output from the Iron Dyke in the 1980s.

Mercury

Mercury ore or cinnabar in north-central Oregon occurs in the Clarno Formation where it is associated with intrusive volcanic plugs and adjacent faults. The Horse Heaven district in Jefferson County is responsible for the highest production. Discovered in 1933, the Horse Heaven Mine operated from 1934 to 1944 and from 1955 to 1958 when the reserves here were exhausted. During this period the mine yielded 17,214 flasks of quicksilver ranking among the top 5 in the state. The Bonanza Mine in Douglas County is the state's largest producer, while three other top mines in the southeastern corner of the state averaged 15,000 flasks each. No other Oregon mines have produced more than one thousand flasks each.

Thundereggs

Of Oregon's gem-like minerals, the thunder-egg, a spherical nodule or geode, is the most famous, and in 1965 it was designated as the official state rock by the legislature. In actuality, the thunderegg is not a rock but a structure created in rhyolite, tuff, or perlitic

The Iron Dyke Mine included a main tunnel, eight levels from four adits, and a 650 foot vertical shaft. The mine is no longer in production.

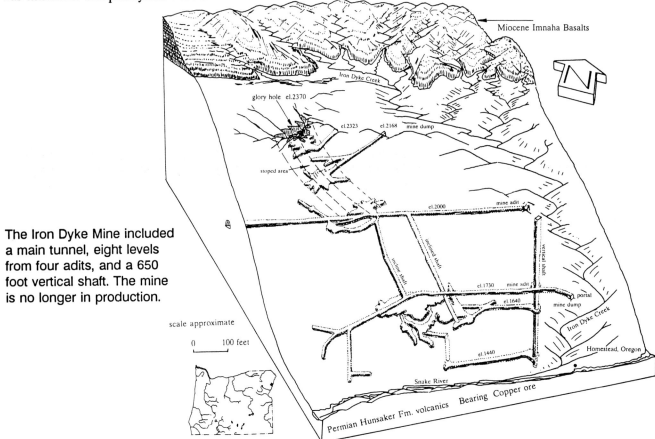

rocks. How thundereggs form isn't certain, but the process probably begins with an internal cavity produced by gasses expanding in highly viscous lavas. Once the lavas cooled, the chamber is filled in with agate or opal. Thundereggs range in size from less than one inch to over 4 feet in diameter, with the average geode about the size of a baseball. Within a rough, knobby and drab exterior, the thunderegg reveals a colorful, star-shaped interior when cut open. Thundereggs can be found in a number of localities in eastern Oregon, but they are best-known from rocks of the John Day Formation. According to Indian legends, angry thunder spirits who lived in the craters of Mt. Hood and Mt. Jefferson once hurled these spherical rocks at each other accompanied by thunder and lightning.

Geothermal Resources

The potential for the commercial exploitation of geothermal energy in the Blue Mountains is only moderate compared to that in the High Lava Plateau or the Basin and Range provinces to the south. Initially thermal waters in the Blue Mountains were utilized by pioneers, and a health spa was erected in 1880 at Hot Lake near La Grande. At nearby Cove and Union, public and commercial facilities were built to exploit thermal springs and wells. Here water averages less than 100 degrees Fahrenheit, while waters from Hot Lake at 185 degrees Fahrenheit are the warmest.

Thermal waters follow narrow conduits along fault zones by which they reach the surface. This has the effect of localizing the hot waters. At Hot Lake, for example, a cold water well is located less than 150 feet from the hot springs.

Features of Geologic Interest

Wallowa Mountains

Formerly known by a number of names, the rugged, glaciated Wallowa Mountains were called the Granite Mountains, the Powder Mountains, or the Eagle Mountains before an official decision was made to use the current name. "Wallowa" is a Nez Perce Indian name meaning a series of stakes arranged in a triangular pattern in the river for catching fish. Traps of this sort were used in the Wallowa River. Extending over the northeast Oregon border, the Wallowa Mountains are called the Seven Devils Mountains in Idaho. Between these two ranges, Hells Canyon of the Snake River has carved a gorge, 1,000 feet deeper than the Grand Canyon of the Colorado River.

During the Ice Ages, glaciers atop Wallowa Mountains radiated outward from the center of the

Thundereggs, a popular term for agate-filled geodes (photo courtesy of Oregon Dept. Geology and Mineral Industries)

range. Spectacular, high triangular peaks with steep walls are known as matterhorns and are cut out by the headward erosion of three or more glaciers back to back. Headward erosion by a glacier is primarily a freezing-thawing process where rocks are plucked or torn directly off valley walls by the ice. This style of erosional attack leaves bare, vertical rock faces and dramatic scenery at the heads of valleys.

A diversity of peaks in the Wallowas rise to heights of nearly 2 miles above sea level, producing both a sharply serrated as well as a rounded skyline. Southwest of Wallowa Lake, Matterhorn Mountain, composed of blue-white marble, rises to 10,004 feet just opposite Sacajawea Peak at 9,880 feet. Other mountains, as Aneroid, so named because an aneroid barometer was used to determine the elevation of the nearby lake, is more rounded and gives a different appearance to the landscape.

The Wallowa Mountains contain some of the finest scenery and recreational facilities in Oregon. Three separate episodes of Ice Age glaciation have shaped ridges, carved valleys into U-shaped troughs, and left crystal clear glacial lakes to be enjoyed whether driving, camping. or hiking here. There are close to 100 lakes of all shapes hollowed out of the granitic

Looking south at the Wallowa Mountains and Wallowa Lake surrounded by glacial moraines. The community of Joseph lies just at the bottom of the picture on the irregular glacial outwash plain (photo courtesy David A. Rahm).

valley floors with water spilling over the hard rock lip to create waterfalls. Minam, Moccasin, Jewett, Wallowa, Aneroid and Echo lakes are just a few of these. One of the most dramatic of these, Wallowa Lake, at the foot of Chief Joseph Mountain, is 3 1/2 miles long, 3/4 of a miles wide, and nearly 4,440 feet in elevation. Fed by the Wallowa River, the lake is hemmed in by moraines made up of boulders and mud left behind by glaciers around the edge of the lake when the ice retreated. The moraines were deposited during different glacial periods dating back only a few thousand of years. Unfortunately, plans are currently being made to carpet the tops of the Wallowa Lake moraines with a large housing development. It is difficult to understand the mentality of a developer that would call for destroying a natural wonder with housing in the midst of the vastness of eastern Oregon.

Steins Pillar

About 20 miles east of Prineville in Crook County, picturesque crags have been eroded from layers of welded tuff. One of the most imposing columns is named Steins Pillar. The original spelling was Steens

Pillar after Major Enoch Steen of the U.S. Army who explored this region for road development in the 1860s. However, the name was misspelled so frequently the incorrect version became official. The pillar stands 350 feet high and 120 feet in diameter.

The pillar had its beginnings almost 44 million years ago during the Clarno episode when hot avalanches of ash and pumice accompanied by clouds of volcanic dust spewed out from local volcanic centers filling an ancient valley. Ensuing flows of the John Day covered one another before the first had time to cool. At least three separate successive volcanic deposits make up Steins Pillar. Upon cooling, the deposit split into long vertical joints, and erosion over millions of years progressed along these cracks to isolate the steep columns seen today. Oxidizing iron from the ash has stained the pillar's surface yellow-brown.

Officer's Cave

Although the Blue Mountains province is not known for its caves, among of the most interesting geologic features is a cave approximately 11 miles south of Kimberly in Grant County. Named Officer's

Cave after the Floyd Officer family which homesteaded here, the cave is one of the largest developed in non karst terrain in North America. The cave was formed by an unusual subsurface process where tunnels, sinks, natural bridges, hills, and valleys are eroded in soft clays and silts instead of in limestone. Caves of this type can develop quickly as a result of a major storm, when rushing water scours steep gullies or sometimes causes the entire roof of the subterranean cave to collapse. In the case of Officer's Cave, large, broken blocks of the John Day Formation, fallen and tipped in all directions, are flushed out by underground streams. The John Day Formation is particularly susceptible to erosion because of its soft, poorly consolidated nature, a characteristic frequently leading to landslides and rockslides.

Officer's Cave is a maze of various levels originally over 700 feet long when first explored by geologists in 1914. The cave has increased in volume 5 to 10 times since then and when examined in 1975 was

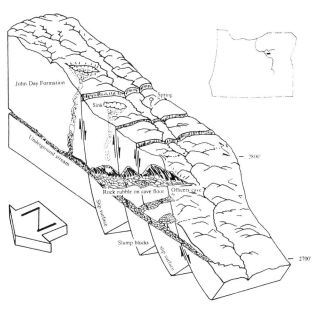

Officers Cave is developed in the John Day Formation where soft clays have been washed out by subterranean streams.

Steins Pillar in Crook County is an erosional remnant of volcanic ash-flow tuffs of the Clarno and John Day formations (photo courtesy of Oregon State Highway Department).

over 1,500 feet long. Underground channels follow the east-west direction of nearby Officer's Cave Ridge. As the cave erodes, large slabs drop from the roof creating an unusual display of rockfalls.

John Day Fossil Beds National Monument

The 14,012 acres of the John Day Fossil Beds National Monument record the Cenozoic history of the southwestern edge of the Blue Mountains. Travelling to three different national park locations in Grant and Wheeler counties, the visitor is treated to a fascinating view of extinct animals and plants as well as a colorful eroded landscape of volcanic and sedimentary deposits of both the Clarno and John Day formations. Oxidation of layers of ancient soils gives them a highly varigated or multi-colored appearance.

The story told in the rocks here began when volcanic eruptions of the Clarno entombed plants from a tropical environment 44 million years ago. Only a few large animals have been found from this period, however. Beginning 36 million years ago, renewed, explosive volcanic activity of the John Day episode released great volumes of ash which covered and preserved both plants and animals living in the ancient grasslands and along waterways. The John Day beds of the National Monument are famous for the many types of cats, dogs, camels, rodents, and cow-sized rhinoceroses. Among these were the small, 3-toed horse and the common oreodon. This was a sheep-like animal whose skulls were collected by early Oregon pioneers. The Metasequoia, or dawn redwood, was one of the most prevalent plants encountered during the temper-

The pig-like entelodont, the delicate and deer-like Hypertragulid, and the oreodons, the size of a sheep, are among the fossil mammals found in the John Day beds of eastern Oregon.

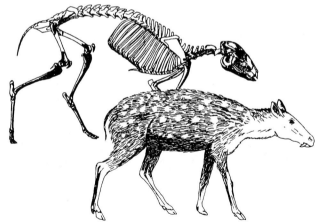
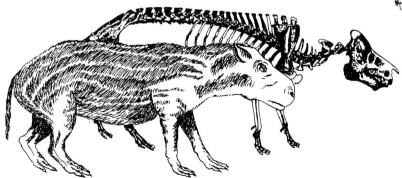

Lava-topped Sheep Rock is one of the landmarks in the heart of the fossil beds, where pre-historic animals roamed in a tropical climate 25 million years ago (photo courtesy Oregon State Highway Department).

At Castle Rock north of Picture Gorge, crenellated spires and columns typically result from erosion of the John Day sediments (photo courtesy of Oregon State Highway Department).

Painted Hills in eastern Wheeler County is a succession of ancient soil profiles developed in the John Day Formation (photo courtesy Oregon State Highway Department).

ate John Day period. Once thought to be extinct, Metasequoia was re-discovered in China and brought to Berkeley by paleobotanist, Ralph Chaney in the 1930s.

The display areas of the National Monument are in three different locations beginning with the Palisades at Clarno where Eocene mudflows of shales, siltstones, conglomerates, and breccias are exposed. These mudflows or lahars are particularly distinctive, eroding into steep cliffs, spires, and columns along vertical fracture lines in the softer sediments. This is the oldest section of rocks in the Monument.

At the Painted Hills north of Mitchell a badlands physiography reveals colorful exposures of rust red soils alternating with yellow, and light brown tuffs of the John Day Formation. Evidence of plant and animal life 30 million years ago can be seen in the few fossil remains found. The Bridge Creek fossil plant locality, famous for its Metasequoia leaves, is nearby.

The third section of the Monument at Picture Gorge and Sheep Rock 38 miles west of John Day tells the story of life here 25 million years ago. The narrow gorge was so named because of ancient Indian pictographs painted onto the basalt. Probably one of the most scenic localities in Oregon, the gorge cuts through 1,500 feet of basalt in numerous flows separated by occasional soil layers. A small cap of hard Miocene basalt protects the softer, buff-colored volcanic tuffs of the upper John Day beds making up Sheep Rock, one of the most familiar sights of the National Monument.

Grand Canyon of the Snake River

Beginning at the oxbow on Oregon's eastern border, the deep, narrow Grand Canyon carved by the Snake River cuts between the Wallowa Mountains of Oregon and the Seven Devils Mountains of Idaho as it journeys northward. From the summit of He Devil Peak in Idaho to the depths of the Snake River bed, Hells Canyon measures nearly 8,000 feet deep, which is 1,000 feet deeper than the Grand Canyon of the Colorado. For more than 50 miles, Hells Canyon averages 5,500 feet in depth. At Hat Point in Oregon the depth is 5,620 feet. The 1,300 foot drop in river elevation from Huntington to Lewiston inspired the Idaho Power Company to construct three dams in this short interval.

The 250 to 75 million year old Permian to Cretaceous rocks exposed in the canyon walls were deposited long before the Snake River canyon existed. These rocks were broken and shattered by folding and faulting only to be later intruded by granitic batholiths 160 to 75 million years ago. A 60 million year long period of erosion stripped off the sediments and exposed the deep batholiths only to have the surface covered over once again during the Miocene with flow after flow of lavas. The varied scenery of the canyon is the product of rapid erosion. As the area rose, tributary rivers from the mountains became deeply incised cutting down through the successive layers of rock to create a deep gorge. For almost two-thirds of its route, the Snake River dissects the dark-colored Miocene

basalts of the Columbia River group which are up to 6,000 feet thick in places. Vistas from such elevations as Hat Point on the western rim or Kinney Point on the eastern side show a panorama of vertical cliffs, sharply cut canyons, and narrow ledges.

Despite its size, the Snake is regarded as a relatively youthful stream. Along its course the river makes a major directional change 50 miles north of Huntington at the oxbow. Here the river begins to cut deeply into adjacent rocks as the oxbow marks the beginning of Hells Canyon. There is evidence that during the Pleistocene the headward limit of the Snake River was at the oxbow, "capturing" the drainage of a much larger south flowing stream, which entered the Pacific somewhere to the south. The Snake eroded downward at the oxbow, breaking through the divide which separated it from the other river. Capture of the upper Snake may have been aided by a large body of water, Lake Idaho, which had filled the older canyon as far south as Hagerman before spilling over into the gorge.

Layered Miocene Imnaha flood basalts exposed in the Grand Canyon of the Snake River in eastern Wallowa County (photo courtesy Oregon State Highway Department).

Additional Readings:

Ash, Sidney R., 1991. A new Jurassic flora from the Wallowa terrane in Hells Canyon, Oregon and Idaho. Oregon Geology, v.53, no.2, pp.27-33.

Ave Lallemant, Hans G., Schmidt, W.J., and Kraft, L.J., 1985. Major late-Triassic strike-slip displacement in the Seven Devils terrane, Oregon and Idaho: a result of left-oblique plate convergence? Tectonophysics, v.119, pp.299-238.

Bentley, Elton B., 1974. The glacial morphology of eastern Oregon uplands. PhD., Univ. of Oregon, 250p.

Blome, Charles D., and Nestell, M.K., 1991. Evolution of a Permo-Triassic sedimentary mellange, Grindstone terrane, east-central Oregon. Geol. Soc. Amer., Bull., v.103, pp.1280-1296.

Brooks, Howard C., 1979. Plate tectonics and the geologic history of the Blue Mountains. Oregon Geology, v.41, no.5, pp.71-80.

-----Ramp, Len, 1968. Gold and silver in Oregon. Oregon Dept. Geol. and Mineral Indus., Bull.61, 337p.

-----and Vallier, Tracy L., 1978. Mesozoic rocks and tectonic evolution of eastern Oregon and western Idaho. In: Howell, D., and McDougall, K., eds., Mesozoic paleogeography of the western United States. Pacific Coast Paleogeography Symp. 2, Soc. Econ. Paleont. and Mineralogists, Pacific Sect., pp.133-145.

Dickinson, William R., 1979. Mesozoic forearc basin in central Oregon. Geology (Boulder), v.7, no.4, pp.166-170.

-----and Vigrass, Laurence W., 1965. Geology of the Sulpee-Izee area Crook, Grant, and Harney counties, Oregon. Oregon Dept. Geol. and Mineral Indus., Bull.58, 109p.

Fisher, R.V., and Rensberger, J.M., 1972. Physical stratigraphy of the John Day Formation, central Oregon. University of California, Publ. Geol. Sci., v.101, 45p.

Getahun, Aberra, and Retallack, Gregory J., 1991. Early Oligocene paleoenvironment of a paleosol from the lower part of the John Day Formation near Clarno, Oregon. Oregon Geology, v.53, no.6, pp.131-136.

Gerlach, David C., Ave Lallemant, H.G., and Leeman, William P., 1981. An island arc origin for the Canyon Mountain ophiolite complex, eastern Oregon, U.S.A. Earth and Planet.

Sci. Letters, 53, pp.255-265.

Jones, David L., Silberling, N.J., and Hillhouse, John, 1977. Wrangellia - A displaced terrane in northwestern North America. Canadian Jour. Earth Sci., 14, p.2565-2577.

Mullen, Ellen D., 1983. Paleozoic and Triassic terranes of the Blue Mountains, northeast Oregon: Discussion and field trip guide. Part I. A new consideration of old problems. Oregon Geology, v.45, no.6, pp.65-68.

Noblett, Jeffrey B., 1981. Subduction-related origin of the volcanic rocks of the Eocene Clarno Formation near Cherry Creek, Oregon. Oregon Geology, v.43, no.7, 1981.

Parker, Garald G., Shown, Lynn M., and Ratzlaff, K.W., 1964. Officer's Cave, a pseudokarst feature in altered tuff and volcanic ash of the John Day Formation in eastern Oregon. Geol. Soc. Amer., Bull. v.75, pp.393-402.

Retallack, Gregory J., 1991. A field guide to mid-Tertiary paleosols and paleoclimatic changes in the high desert of central Oregon - Part 1. Oregon Geology, v.53, no.3, p.51-59.

Robinson, Paul T., Brem, G.F., and McKee, E.H., 1984. John Day Formation of Oregon: A distal record of early Cascade volcanism. Geology, v.12, p.229-232.

-----Walker, G.W., and McKee, E.H., 1990. Eocene (?), Oligocene, and lower Miocene rocks of the Blue Mountains region. U.S. Geol. Survey, Prof. Paper 1437, pp.29-61.

Sarewitz, Daniel, 1983. Seven Devils terrane: Is it really a piece of Wrangellia? Geology, v.11, p.634-637.

Silberling, N.J., and Jones, D.L., eds, 1984. Lithotectonic terrane maps of the North American Cordillera. U.S. Geol. Survey, Open File Report 84-523, various pagings.

Stanley, George D., 1986. Travels of an ancient reef. Natural History, v.87, no.11, pp.36-42.

-----and Beauvais, Louise, 1990. Middle Jurassic corals from the Wallowa terrane, west-central Idaho. Jour. Paleo., v.64, no.3, pp.352-362.

Vallier, Tracy L., and Brooks, Howard C., eds., 1986. Geology of the Blue Mountains region of Oregon, Idaho, and Washington. U.S. Geol. Survey, Prof. Paper 1435, 93p.

-----Brooks, H.C., and Thayer, T.P., 1977. Paleozoic rocks of eastern Oregon and western Idaho. In: Stewart, J.H., et al., eds. Paleozoic paleogeography of the western United States, Soc. Econ. Paleont. and Mineralogists, Pacific Coast Paleogeogr. Symp.1, pp.455-466.

Coast Paleogeogr. Symp.1, pp.455-466.

Walker, George W., 1990. Miocene and younger rocks of the Blue Mountains region, exclusive of the Columbia River Basalt Group and associated mafic lava flows. U.S. Geol. Survey, Prof. Paper 1437, pp.101-118.

-----ed., 1990. Geology of the Blue Mountains region of Oregon, Idaho, and Washington: Cenozoic geology of the Blue Mountains. U.S. Geol. Survey, Prof. Paper 1437, 135p.

-----and Robinson, Paul T., 1990. Paleocene(?), Eocene, and Oligocene(?) rocks of the Blue Mountains region. U.S. Geol. Survey, Prof. Paper 1437, pp.13-27.

Wardlaw, Bruce R., Nestell, M.K, and Dutro, J.T., 1982. Biostratigraphy and structural setting of the Permian Coyote Butte Formation of central Oregon. Geology, v.10, p.13-16.

Waters, Aaron, 1966. Stein's Pillar area, central Oregon. Ore Bin, v.28, no.8, pp.137-144.

Wheeler, Harry E., and Cook, Earl F., 1954. Structural and stratigraphic significance of the Snake River capture, Idaho-Oregon. Jour. Geol., v.62, pp.525-536.

Wilson, Douglas and Cox, Allan, 1980. Paleomagnetic evidence for tectonic rotation of Jurassic plutons in Blue Mountains, eastern Oregon. Jour. Geophys. Res., v.85, no.B7, pp.3681-3689.

Klamath Mountains

Physiography

The Klamath Mountains physiographic province is an elongate north-south trending area of approximately 12,000 square miles, three-fifths of which is in northern California and the remainder in southwestern Oregon. In Oregon the province is bordered on the north by the Coast Range and on the west by the Pacific Ocean. The broad Bear Creek Valley separates the Klamaths from the Western Cascade mountains to the east, and the state border marks the southern limit of the Oregon Klamaths.

Although the region boasts deep, narrow canyons and mountain peaks reaching over 7,000 feet, for the most part the province exhibits an even relief throughout that represents the old land surface before it was worn down then uplifted for renewed erosion. Mt. Ashland at 7,530 feet is the highest peak in the Oregon Klamaths, while the Siskiyou Mountains, forming the southern end of the province, display the greatest relief. West of the Siskiyou summits, the tilted upland surface drops to 2,500 feet before the land abruptly breaks off to a narrow coastal plain with steep headlands.

The Oregon Klamath Mountains province is drained by the Rogue River, its main tributaries the Illinois and Applegate, as well as by coastal streams, the largest of which is the Chetco River. The watershed of the Rogue begins on the western slopes of the Cascade Mountains at Boundary Springs near Crater Lake. Flowing southwesterly to Medford and Grants Pass, the river drops 5,000 feet, making several major turns before entering the Pacific Ocean at Gold Beach after a distance of 215 miles.

Geologic Overview

The Klamath Mountains are made up of pieces of exotic terranes that were once parts of ocean crust or island archipelago environments spanning the early Paleozoic to Jurassic. Each of the terranes have distinct rock layers and fossils by which they are identified. Formed in an ocean setting, the tectonic slices were carried eastward toward the North American landmass where they collided with the existing continent. As successive terranes were added, one after the other, the mosaic of terranes arranged themselves like fallen dominoes. After being accreted, the terranes were

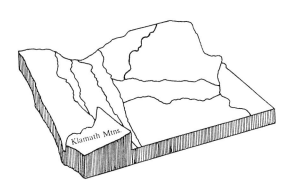

securely welded to the mainland by granitic instrusives before being rotated as much as 100 degrees clockwise by the early Cretaceous.

Today seven separate terranes are recognizable as making up the Klamath Mountain province. While not extending into Oregon, the Eastern Klamath terrane of California forms the oldest rock mass within the province, to which the later terranes were affixed. Because they were fabricated elsewhere, rocks of the Klamaths are much older than those in any other part of western Oregon, and the area may contain some of the oldest formations in the state. While the oldest known rocks in the Oregon Klamaths are Triassic, Klamath province rocks as old as Ordovician at 450 million years are recorded in California.

The intrusion of Klamath Mountains terranes took place in several waves of granitic rocks during the middle and late Mesozoic. Occurring in four northeast trending belts, most of the granite magmas making up the plutons were emplaced after the terranes were amalgamated. About 140 million years ago the Cretaceous ocean that covered much of Oregon deposited sediments in a broad basin extending northward from the upper corner of the Klamath Mountains. Material deposited in this seaway was derived from volcanoes to the east. Only a scattering of Tertiary sediments are represented in the province by sands and silts carried

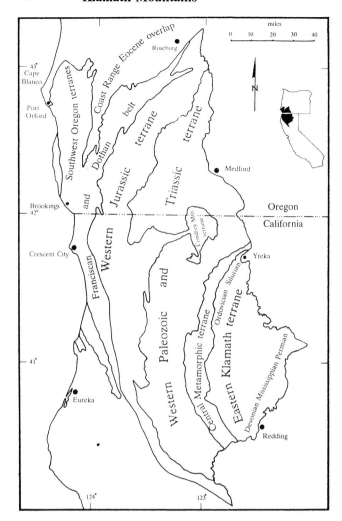

Distribution of terranes in southwest Oregon and northern California (after Irwin, 1985)

westward by a well-developed riverine system. Draining the Idaho batholith and Klamaths, the rivers deposited the sediments in the forearc basin along the margin of the mountains. Uplift in the Miocene resulted in extensive erosion of the region. The Pleistocene brought only small glaciers to these mountains, while continuing uplift along with rising and falling sea level produced coastal terraces.

In their long history as part of the sea floor as well as during the emplacement of granitic intrusions, Klamath terrane rocks were enriched with a diversity of economic minerals including gold, copper, nickel, and chromite.

Geology

Because the Klamath Mountains are made up of composite belts of rocks formerly part of an ocean environment, the concept of displaced accreted terranes is fundamental to understanding geology here. Terranes

are separate groups of rocks formed in an open oceanic or coastal environment, each group with its own layered sequence of distinctive rocks and fossils by which it is recognized. These slabs of oceanic rock were rafted toward North America from the west where they were bent, folded, and broken upon collision. As they were accreted to North America, the succeeding terranes were thrust beneath each other like shingles on a roof with the oldest to the east and the youngest to the west. All of the terranes are separated from each other by fault zones.

During the Paleozoic and Mesozoic more than 250 million years ago, the Klamaths began as an ocean chain or island archipelago that extended in a northwest line down from British Columbia and Washington into Idaho and California. Assembled very close to the North American West Coast, terranes were accreted to North America in the middle and later Jurassic. In middle Jurassic time, a subduction zone between two tectonic plates generated a series of volcanoes atop these older, accreted terranes. As the volcanic arc separated from the landmass and migrated westward away from North America, a backarc basin developed between the older volcanic chain and the eruptive centers situated above the subduction zone. During the late Jurassic, sills and dikes were intruded into the basin. In the final stage, the arc, basin, and remnants of the older volcanic chain migrated toward North America where they were accreted and imbricated in thrust sheets over each other.

From south to north the provinces today include the Sierra Nevada in California, the Klamath and Blue Mountains in Oregon, Idaho, and Washington, and the Cache Creek area in central British Columbia. Terrane rocks of the Klamath province are linked to those of the Sierras of California and the Blue Mountains in northeast Oregon by strikingly similar fossils and rock layers. Even though the relationship between these regions is still not well understood, all three form a discontinuous belt of Paleozoic and Mesozoic rocks. This 1,000 mile long ancient mountain chain bends in a northeast direction across the state beneath the Cascades from the Klamaths to the Blue Mountains before turning back toward the north in the Blue Mountains to continue into northwestern Washington. This kink is thought to have developed during early Tertiary with rotation and westward displacement of the Klamath and Blue Mountains provinces.

After initial contact of the exotic blocks to the North American continent, the Klamath block was rotated into a final position. It is particularly important to fix the time the terrane was accreted to the main

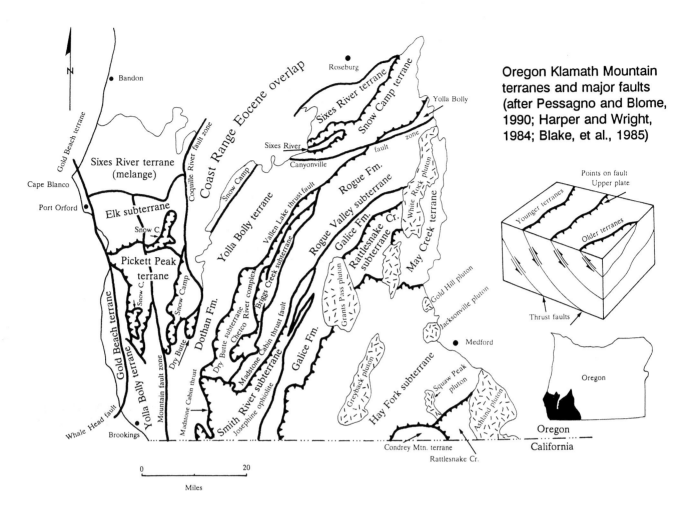

Oregon Klamath Mountain
terranes and major faults
(after Pessagno and Blome,
1990; Harper and Wright,
1984; Blake, et al., 1985)

continent and when the same terrane was rotated into its present orientation. The timing determines whether the plate rotation occurred when the terrane was still an independent part of an oceanic plate or whether it was moved after it had become welded to the main continent. Present evidence suggests that all Klamath terranes had completed their swinging clockwise motion by the early Cretaceous. This established the early Cretaceous as the time by which the multiple, composite Klamath terranes became joined to the stable North American continent.

Individual Terranes

After delineating and mapping these terranes, the process of grouping the many isolated pieces together is the next step in reconstructing their origins and history. In the Oregon Klamath province there are seven recognized terranes which are further subdivided into multiple subterranes. Two terranes in the Klamaths of northern California, the Eastern Klamath terrane and the Central Metamorphic terrane, do not extend into Oregon. From east to west, or from the oldest to youngest, Oregon Klamath terranes include the Western Paleozoic and Triassic belt, the Western Klamath terrane, also known as the Western Jurassic belt, the

Snow Camp, Pickett Peak, Yolla Bolly, Gold Beach, and Sixes River terranes. The overall grain of the Klamath terranes curves northeast by southwest with the convex side to the northwest.

Among the Klamath Mountain terranes, the Western Paleozoic and Triassic terrane, which is significantly older than subterranes of the Western Klamath belt, was accreted in middle late Jurassic then rotated to its current northwest facing configuration in Mesozoic and Tertiary time. Also known as the Applegate, this terrane has been subdivided into three subterranes. The Rattlesnake Creek subterrane originated as an ocean crust or "ophiolite". The Hayfork subterrane, which represents a volcanic island archipelago, is laminated between the Rattlesnake and May Creek subterranes. The May Creek, which is also ophiolitic, has been heavily distorted and altered by heat and pressure to high grade metamorphic rocks. The Western Paleozoic and Triassic terrane has tropical fossil faunas similar to those found in the Cache Creek terrane in British Columbia and the Baker terrane in the Blue Mountains.

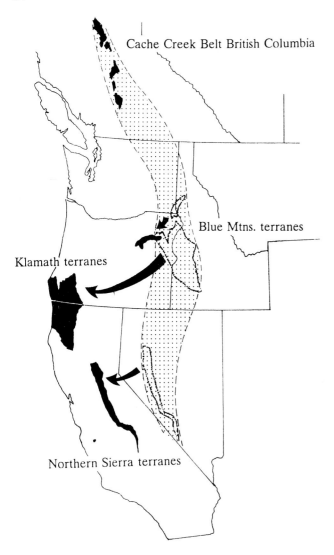

Model for pre-rotation configuration of West Coast tectonic terranes (after Miller, 1987)

ocean crust as well as upper mantle rocks from more than three miles below the seafloor.

The upper pillow basalt layers of an ophiolite sequence are highly porous, and during the ocean spreading process the convection of seawater in through the ophiolitic rocks and out near the ridge crest contributes to the precipitatation of massive sulfide deposits. In this way the upper portion of the ophiolite is richly mineralized with copper, lead, and zinc with smaller amounts of gold, silver and platinum. Nickel and chromium ores are primarily associated with the deep periodite layer of the ophiolite sequence. Basalts, ultramafics, and gabbros in the Klamath ophiolites have been altered to form the low-grade metamorphic rocks, greenstone and serpentine. The word, "ophiolite", meaning snake, refers to the rock serpentine which is completely fractured and broken up by faults giving it a smooth surface.

The Western Paleozoic and Triassic belt is to the southeast of the Western Klamath terrane. These two terranes are separated by a fault marking the surface along which the Western Klamath terrane was pushed beneath the Western Paleozoic and Triassic terrane. Studies of the magnetic alignments of mineral crystals in the rocks of the Western Klamath terrane suggest that it has been rotated less than 100 degrees in a clockwise direction since its origin in the late Jurassic to arrive at its present orientation of northeast by southwest. Presently the Klamath Mountains are in approximate alignment with the Blue Mountains. Both provinces display extension and clockwise rotation suggesting that the Klamaths may have connected with the Blue Mountains beneath the Cascades.

The Western Klamath terrane has been subdivided into six subterranes. From east to west, they are the Condrey Mountain subterrane, the Smith River subterrane, the Rogue Valley subterrane, the Briggs Creek subterrane, and the Dry Butte and Elk subterranes. Two of these, the Condrey Mountain and Smith River subterranes, have been the focus of considerable attention because of their geologic environment and economic minerals.

Southwest of Ashland, the Condrey Mountain subterrane, 90% of which projects across the border into California, is an inverted drop-shaped exposure formed during the late Jurassic between 146 to 148 million years ago. A hot oceanic slab and cooler ocean sediments, brought together by thrusting plates, were altered to the metamorphic rock, schist. Within the body of the Condrey Mountain, the schists range from the low-grade greenschist on the outside, to the middle layer of graphite-rich blackschist, and the core of high-grade blueschist, all of which have been folded with tight crenulations. Intensive erosion of the dome today

Commonly occurring in Klamath terrane rocks, ophiolites are layered rock sequences up to 3 miles in thickness that develop in the deep ocean floors between two spreading tectonic plates. At the bottom of an ophiolitic series, dark-colored ultramafic rocks of peridotite are overlain by gabbros that form the base of the ocean crust. These in turn grade upward into a complex of sheeted dikes without an apparent host rock. Layered over the intruded dikes are pillow basalts which are lavas extruded underwater onto the sea floor. These pillow-shaped blobs of lava are capped by deep sea cherts and pelagic clays with fossils of radiolaria and foraminifera typically found today in the open ocean under thousands of feet of water. Ophiolites are significant as they offer a chance to examine first hand

Major tectonic terranes of the Oregon and California Klamath Mountains

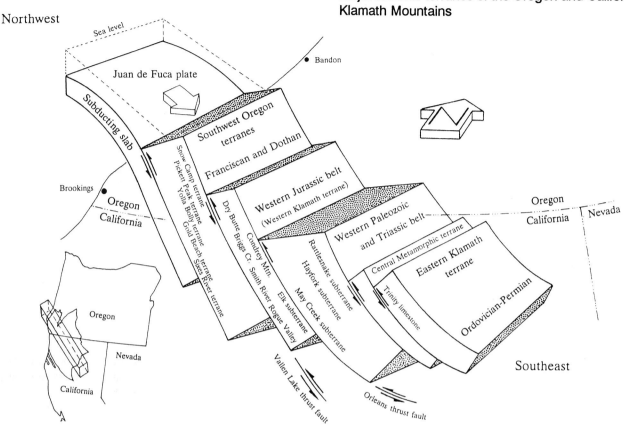

creates a deep window into the blueschists. The advanced metamorphic state of the Condrey Mountain subterrane makes it difficult to correlate with rocks of the Klamaths, although it may be derived from shales of the Galice Formation exposed further west.

An intriguing aspect of the Condrey Mountain exposure is that it displays a distinct doming believed to have developed in the Neogene. It was probably during the Miocene epoch as the Juan de Fuca plate was actively being subducted beneath the larger North American plate that lateral pressure between the two converging plates caused the bowing or doming, compressing the rocks by as much as 5% and raising them over four miles in elevation. Compression for this doming may have come from a slight change in the subduction direction of the Juan de Fuca plate where the forearc basin, containing the Condrey Mountain dome out in front of the Cascade volcanic arc, was pushed upward. Although the broad uplift of the Klamath Mountains in the later Tertiary was recognized early, the dome configuration was only identified in 1982.

With its mineral wealth, the Smith River is, perhaps, the most important subterrane within the Western Klamath block. Made up of two parts, this subterr-

ane has an underlying, deep ocean crust suite of rocks known as the Josephine ophiolite and a three mile thick overlying sandstone and slaty shale called the Galice Formation. The Josephine ophiolite was originally a slab of ocean floor, approximately 163 million years ago, lying in a spreading backarc basin between the mainland and an active volcanic archipelago to the west. Microfossils from the ophiolite sediments suggest that the terrane developed at a tropical latitude lying well to the south of the present Klamath Mountains. In addition to thick sequences of volcanic flows and ash, the Galice contains turbidites which grade into shale at the top. Structural evidence suggests that softer Galice rocks were literally scraped off the ocean floor and piled against the proto-Klamaths as two crustal slabs collided and slid past each other.

One of the largest and most complete ophiolite sequences in the world, the Josephine ophiolite is an assemblage of rocks representing fragments from an ancient ocean crust and upper mantle that are rich in magnesium, iron, and serpentine. The Josephine ophiolite is famous for its massive sulfide deposits which are similar to those on the island of Cyprus in the eastern Mediterranean. Within the Josephine

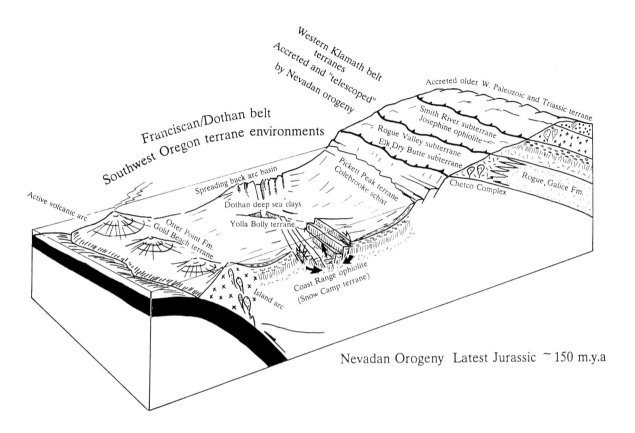

Nevadan Orogeny Latest Jurassic ~150 m.y.a

Reconstructed environments of southwest Oregon
Klamath terranes (after Saleeby, et al., 1982; Harper
and Wright, 1984; Harper, Saleeby, and Norman,
1985)

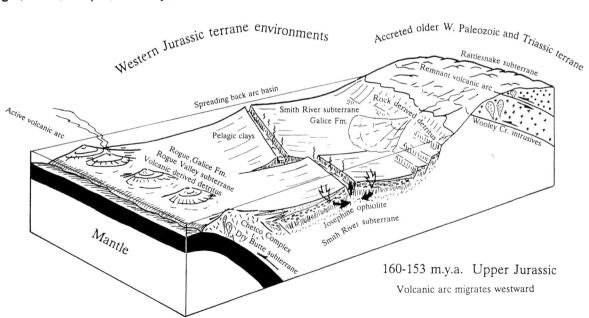

160-153 m.y.a. Upper Jurassic

Volcanic arc migrates westward

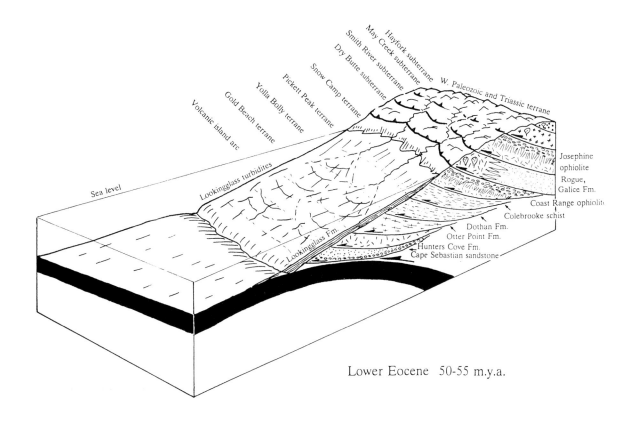

Lower Eocene 50-55 m.y.a.

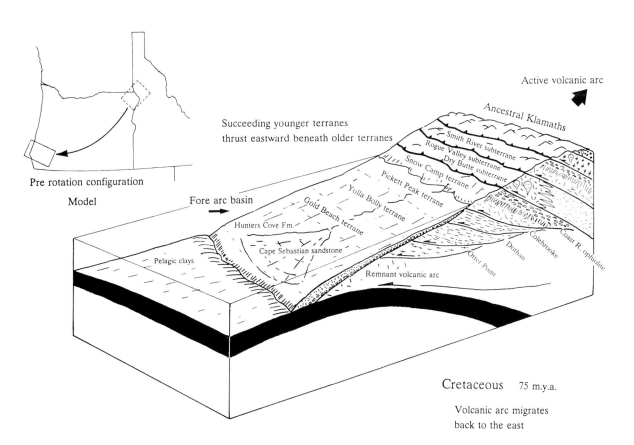

Model

Pre rotation configuration

Succeeding younger terranes
thrust eastward beneath older terranes

Active volcanic arc

Cretaceous 75 m.y.a.

Volcanic arc migrates
back to the east

ophiolite, one of the better exposures of a massive sulfide deposit is in the Turner-Albright mining district on the southern Oregon border. Here hydrothermal deposits yield gold, silver, copper, zinc, and cobalt from what has been interpreted as an ancient ocean spreading ridge where minerals precipitate from submarine hot springs at temperatures up to 650 degrees Fahrenheit.

A second important mineral-bearing ophiolite sequence is the Jurassic Preston Peak terrane of the Klamaths. Best exposed in California, the Preston Peak has been thrust into a position over the Josephine ophiolite. Originally the Preston Peak lay a distance to the east and was either a poorly developed volcanic archipelago or a piece of the Josephine interarc basin.

The Rogue Valley subterrane forms a five-mile wide strip just northwest of the Smith River terrane. The Rogue Valley consists of volcanic flows and ash of the Rogue Formation covered by shales and sands of the Galice Formation. The volcanics have been altered to greenstone which in turn contain gold and other base metals. The Briggs Creek subterrane, west of the Rogue Valley, consists of folded and altered garnet-bearing amphibolites. The Dry Butte subterrane includes the Chetco River complex of plutonic rocks, locally known as the Illinois River gabbro, and is considered to have been the roots of the volcanic arc.

Over 20 miles west of the main exposures of other Western Klamath terrane rocks, the Elk subterrane is included within the Western Klamath because it contains the unmistakable sandy turbidites, shales, and andesite volcanic flows of the Galice Formation. Atop the Galice and within the Elk subterrane coarse marine gravels of the lower Cretaceous Humbug Mountain Conglomerate and Rocky Point Formation are found.

Five slightly younger terranes, the Snow Camp, Pickett Peak, Yolla Bolly, Gold Beach, and Sixes River are separated from the overlying Western Klamath terrane by a series of faults of early Cretaceous age. The Snow Camp terrane is a disjunct group of rocks including the Coast Range ophiolite which has been overlain by the lower Cretaceous Riddle and fossilifer-

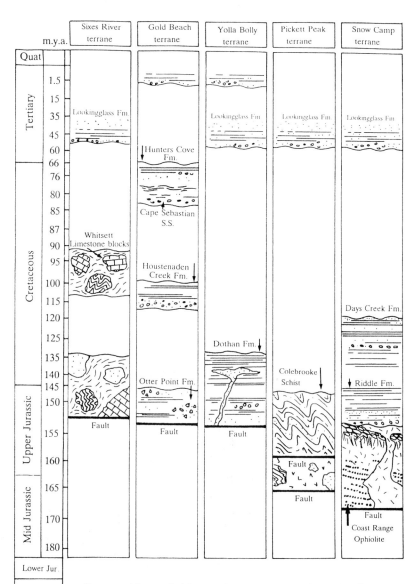

WEST ◄——

Oregon Klamath Mountain terrane stratigraphy (after Blake, et al., 1985)

ous marine Days Creek formations of conglomerates, silts, and sands. The highly folded Colebrooke Schist of late Jurassic age makes up the Pickett Peak terrane. Metamorphosed in early Cretaceous, the Colebrooke is a blueschist that was originally a mixture of tuffs, cherts, and pillow lavas in a deep sea, oceanic setting. The Yolla Bolly terrane of upper Jurassic-lower Cretaceous age has been divided into east and west sections, both of which include the distinctive Dothan Formation. Deriving its sand from continental and

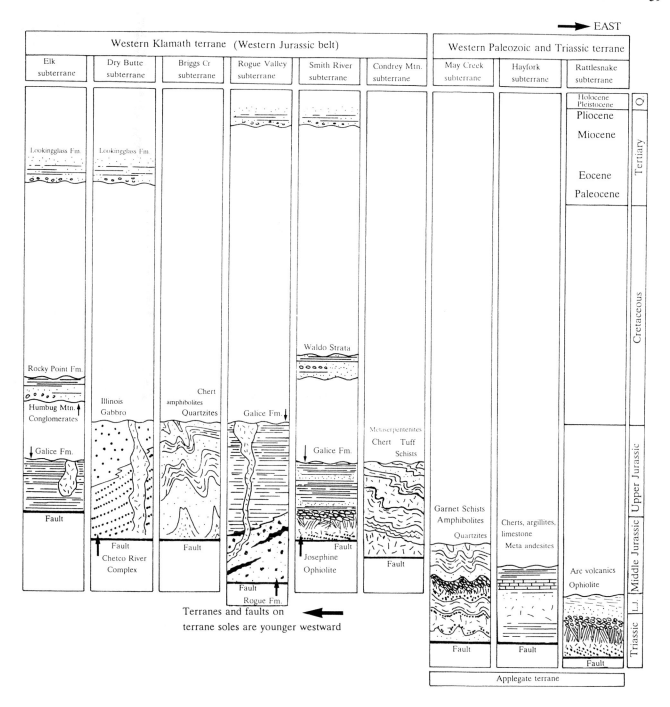

volcanic arc sources, the Dothan is composed of oceanic continental slope rocks of turbidite sands and muds with some deep water cherts. The Gold Beach terrane is a "melange" or mixture of upper Jurassic Otter Point Formation silts and sands mixed with lavas, breccias, chert, and even schists. Overlying that terrane, the upper Cretaceous Cape Sebastian sandstone and Hunters Cove siltstone reflect storm wave conditions with turbidite sands and shales. Bounded by a fault, the cohesive block of the Gold Beach terrane may have been moved north out of California in Tertiary time.

The Sixes terrane is exposed just north of Cape Blanco and in a small sheet south of Roseburg. It consists of Jurassic and Cretaceous mudstones, sandstones, and conglomerates chopped up by faults and studded with huge separate blocks of blueschist and the distinctive high-grade metamorphic rock eclogite. Near Roseburg, blocks of the middle Cretaceous Whitsett limestones with corals and deep water shales along with pillow lavas are found in the same terrane. The Sixes terrane has been compared to the Central Melange terrane of northern California and, like the Gold Beach terrane, may have been displaced northward into southern Oregon by faulting in late Cretaceous or lower Tertiary time.

Mesozoic Plutons

During an interval near the end of the Jurassic period intensive folding and faulting in the Klamath Mountains was accompanied by the intrusion of plutons. This mountain building episode was historically called the Nevadan Orogeny because it was thought to be synchronous with intrusions of the Sierra Nevada region further to the south. In actuality, the tectonic activity in the Sierras took place millions of years later in the middle Cretaceous. However, the phrase, late Jurassic and early Cretaceous Nevadan Orogeny, was so firmly entrenched in West Coast tectonic geology that it is still a useful milepost to understanding the history of the region.

Except for the coastal terranes, the Klamath Mountain province has been intruded by igneous plutonic bodies which cooled to coarsely crystalline rocks at depths beneath the earth. The plutons vary in size from exposures covering less than one square mile

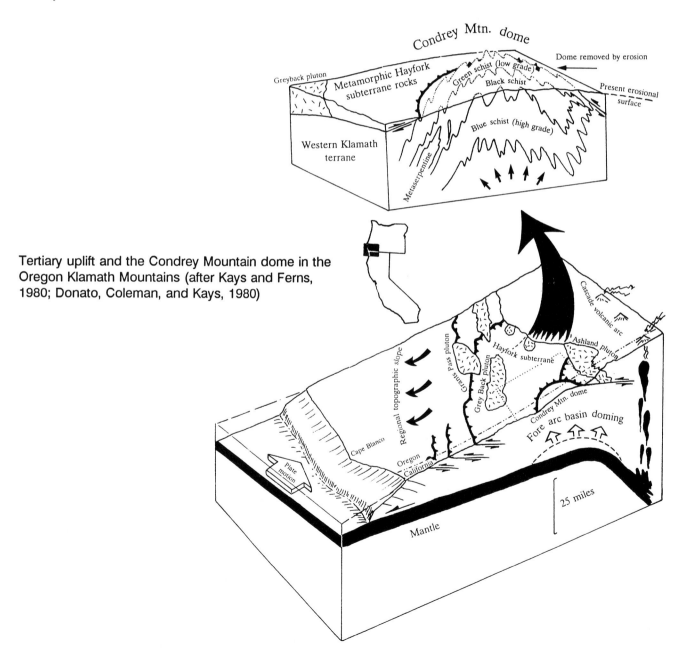

Tertiary uplift and the Condrey Mountain dome in the Oregon Klamath Mountains (after Kays and Ferns, 1980; Donato, Coleman, and Kays, 1980)

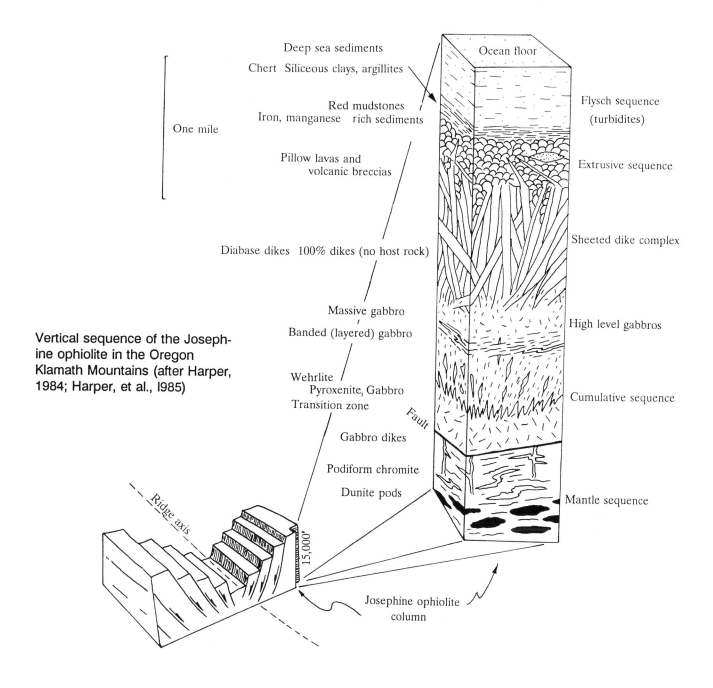

Deep sea sediments — Ocean floor

Chert Siliceous clays, argillites

Flysch sequence
(turbidites)

Red mudstones
Iron, manganese   rich sediments

One mile

Pillow lavas and
volcanic breccias

Extrusive sequence

Diabase dikes 100% dikes (no host rock)

Sheeted dike complex

Massive gabbro
Banded (layered) gabbro

High level gabbros

**Vertical sequence of the Joseph-
ine ophiolite in the Oregon
Klamath Mountains (after Harper,
1984; Harper, et al., 1985)**

Wehrlite
Pyroxenite, Gabbro
Transition zone

Cumulative sequence

Gabbro dikes

Fault

Podiform chromite
Dunite pods

Mantle sequence

Ridge axis

15,000'

Josephine ophiolite
column

to those covering over 100 square miles. Intrusives are situated in four major northeast trending belts with the oldest in the southeast and the youngest in the northwest. The oldest of these, the Wooley Creek belt of late middle Jurassic age at 155 million years ago includes the Ashland plutonic complex of quartz monzonite and granodiorite. About 150 square miles in total size, approximately two-thirds of the Ashland pluton is in Oregon where Mt. Ashland is its major exposure. The Wooley Creek pluton belt intruded the Rattlesnake and Hayfork subterranes of the Western Paleozoic and Triassic belt terrane after the subterranes had been amalgamated.

Northwest of that region, the Greyback belt of late Jurassic age includes the large Greyback pluton as well as the smaller Gold Hill and Jacksonville plutons. This belt, dating back to 153 million years ago intruded the Hayfork, Rattlesnake Creek, and May Creek subterranes of the Western Paleozoic and Triassic belt. Just northwest of the Grayback belt, the Grants Pass early Cretaceous plutonic belt at 140 million years includes the large Grants Pass and White Rocks plutons. Displaying 85 degrees of clockwise rotation, the Grants Pass plutons intruded the Smith River subterrane on the south and the Rattlesnake Creek and Rogue Valley subterranes to the north after amalgam-

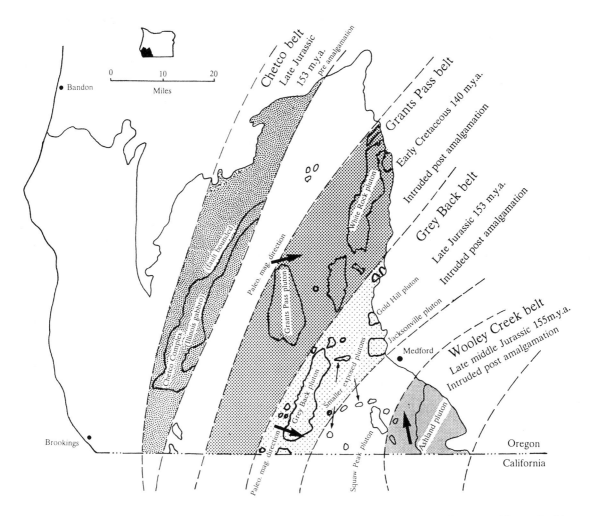

Plutonic rocks of the Oregon Klamath Mountains
(after Holtz, 1971; Irwin, 1985)

ation had taken place. The Chetco belt of late Jurassic age at 153 million years ago includes the Illinois River gabbro and is synchronous with the Greyback. In contrast to other intrusive belts in southwest Oregon, the Chetco belt, intruding the Dry Butte subterrane before it was amalgamated, formed the roots of a volcanic island arc.

Cretaceous

Covering the tectonically emplaced plutons of the Jurassic period, sands, shales, and conglomerates between 3,000 to 4,000 feet thick were part of a transgressive seaway during the Cretaceous. As the seaway advanced or transgressed toward the east, it became progressively deeper over most of Oregon. Broken fossil shells and hummocky, uneven stratification of both the Cape Sebastian and Hornbrook formations indicate a turbulent stormy ocean. The steady deepening of the sea in Oregon correlated with a similar late Cretaceous world-wide transgression.

Marine sands, shales, and conglomerates of the Hornbrook Formation were deposited in a broad basin located along the northeast corner of the Klamath Mountains. The basin probably extended much farther to the northeast where it connected with the Ochoco basin near Mitchell and Suplee as well as to the south where it merged with the Great Valley near Redding, California. The interrelated basinal structure would have formed a complex forearc basin to the west of the later Cretaceous volcanic archipelago. Rare Cretaceous floras from the Port Orford area include ferns, cycads, and ginkgos that grew at higher elevations in the warm moist climate.

Tertiary

A rapid northwestward withdrawal of oceanic waters at the end of the Cretaceous brought the southern end of the shoreline up against the northern edge of the Klamath Mountains by early Eocene time. Within the beginning of Tertiary time, the Klamaths lay at the southern tip of the newly formed volcanic

Coast Range block, an extensive volcanic archipelago that extended northwest by southeast across eastern Washington and into Idaho. Only a very thin veneer of Tertiary sediments are found in discrete areas of the Klamath Mountains along the coastal margin, from Powers to Agness, along the Rogue River above Illahe, and at Eden Ridge and Bone Mountain. Micaceous sands and silts of the Lookingglass, Flournoy, and Tyee formations were initially carried westward by a well-developed riverine system that drained the Klamaths and the Idaho batholith to east to be deposited in the forearc basin. Accumulations of plant fragments intermixed with marine fossils indicate an interfingering of near-shore and nonmarine sediments. Late Eocene nearshore environments are represented by conglomerates and sands of the Payne Cliffs Formation exposed in a northwest trending belt along the Bear Creek Valley near Ashland. Grading upward into ash, tuffs, and lava flows, the Payne Cliffs also record the earliest volcanic activity of the Western Cascades, and deposition was primarily by an extensive braided river which flowed to the north. This formation is overlain by nonmarine volcanic sediments of the Colestin Formation. With the continued uplift, the shoreline moved northward, and intensive erosion and leveling of the Klamath Mountains took place.

Pleistocene

Small glaciers formed in the Oregon Klamaths during the Ice Ages, although this range of mountains didn't have the extensive glacial system of the Blue Mountains, Cascade Range, or of the substantially higher California portion of the Klamaths. Glaciation is most evident in the U-shaped valleys, ampitheater-like cirques, and lakes carved in Chetco Peak near the California border at altitudes between 4,000 to 6,000 feet.

Present in patches along the western edge of the Klamath Mountains province, high marine terraces at different levels are the combined effect of uplift and fluctuating global sea levels during glacial and interglacial periods. Although extensive terraces, for the most part, are missing between Port Orford and the mouth of the Chetco River, south of the Chetco a broad marine terrace is present with an occasional raised sea stack projecting above the plain.

Structure

Large-scale lines or lineaments across Oregon and Washington, best seen on aerial photos, are poorly understood. Extending in a northeasterly direction, the Klamath-Blue Mountains lineament runs over 400 miles from the northern boundary of the Klamath Mountains to the northern boundary of the Blue Mountains. The lineament is displayed on gravity maps of Oregon where it reflects a crustal thickening from 20 miles thick northwest of the line to a crust 30 miles thick southeast of the lineament. The meaning of such a zone is unclear, but it may represent a preTertiary continental margin which was oriented north-south

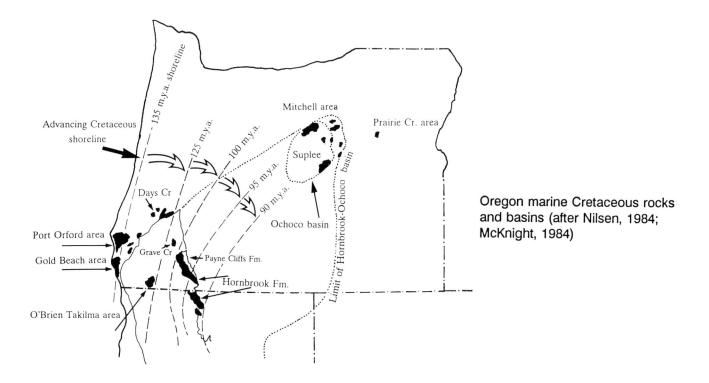

Oregon marine Cretaceous rocks and basins (after Nilsen, 1984; McKnight, 1984)

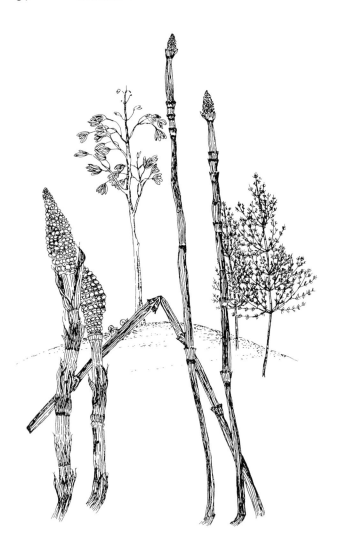

Equisetum or horsetails and ginkgo grew as early as Jurassic in southwestern Oregon.

during the early Cretaceous and which later rotated 45 to 50 degrees clockwise to its present position. It is worthwhile to note that foundation or basement rocks north of the lineament are of oceanic crustal composition, while basement rocks south of the lineament are of continental crustal composition.

Mining and Mineral Industry

Long exploited for their mineral wealth, the Klamath Mountains contain rich deposits of gold, silver, copper, nickel, and chromite, as well as smaller amounts of other minerals. The presence of precious metals in southwestern Oregon relates primarily to mineralization as part of the sea floor spreading processes and secondarily to the intrusion of plutonic

rocks situated throughout the Klamath Mountain province. Within the Klamaths, mining has focused on what is called the Josephine ophiolite. This is a belt of distinctive rocks that are part of the larger Western Klamath terrane forming an arcuate outcrop from southwest to the northeast across the province. What was once crust beneath a deep ocean floor, ophiolites have been raised up above sea level by continental collision to make the mineral deposits in them accessible. The Josephine ophiolite closely resembles mineral-rich seafloor deposits elsewhere in Cyprus, Turkey, Newfoundland, and Italy that are referred to collectively as "Cyprus type" massive, iron-copper sulfide deposits.

The Josephine ophiolite is believed to have developed originally in a backarc spreading environment, where the sea floor is pulled apart on the continental side of an island archipelago between the volcanic chain and the mainland. With the spreading process, tensional step faults form parallel to the axis of spreading. Sea water begins to circulate by convection along the faults and through porous undersea pillow basalts to emerge along the axis itself as sea floor hydrothermal vents. During cooling, minerals dissolved in the super-heated sea water precipitate in faults as well as within the volcanics and sediments to create metalliferous muds. In this way a host of elements as iron, copper, cobalt, zinc, nickel, gold, and silver, with lesser amounts of other minerals accumulate along the deep ocean spreading ridge to form ore bodies near the base of the ophiolite sequence. Chrome and nickel precipitate in pods associated with periodite in the upper mantle.

In another mineralization process, smaller amounts of ores are precipitated when mineral-laden, hydrothermal fluids from a cooling pluton invade and enrich the enclosing rock. Typical intrusions with associated gold and silver can be found in the Chetco complex of western Josephine County and in the Ashland pluton in southern Jackson County.

Gold

The search for gold stimulated the development of southern Oregon which, in turn, led to the discovery of other economically viable minerals as silver, copper, and zinc from the mining regions of the Klamath Mountains. Early miners examined river sands and gravels for placer gold which had been eroded from lode deposits. The discovery of placer gold along the Illinois River in 1850 by a party of miners, most of whom were from Illinois, set the stage for the gold rush of 1851 in southern Oregon. In December of that year thousands of men arrived in Jacksonville following reports of gold at this campsite along a wagon route to

California. Soon Jackson was the most populous county in Oregon with mining camps springing up in adjacent Josephine County as well.

Rich stream deposits of placer gold were washed out of gravels by hydraulic methods that necessitated the construction of ditches and flumes to bring water from miles away to the diggings. Many of the ditches were dug by Chinese laborers, and the famous 23- mile long Sterling Creek ditch constructed in 1877 was fed with water from the Little Applegate River. Most of the old placer workings are still visible. With over 75% of the gold produced in the Klamaths from placers, the Sterling Creek find proved to be one of the most productive with a total output of more than $3,000,000.

As the placers were fully exploited, gold was traced along the streams back to its lode sources. Lode gold mining began during the 1860s, and in the peak years from 1879-1908 over 150 lode mines were operating in the Klamath province. Of these, 6 mines supplied the bulk of the gold estimated at $7 million, while 14 other mines produced $600,000. Aside from the North Pole Mine in Baker County, the Greenback Mine located on Grave Creek in Josephine County was the most productive in Oregon at $3.5 million.

Six leading mineral-rich regions in the Klamath province are found near Galice and Canyonville, in the Greenback and Grants Pass district, near Gold Hill and Jacksonville, in the Ashland district, in the Applegate River valley and Takilma area, and in the upper Illinois and Chetco river watersheds.

The Galice-Silver Peak mining district, a narrow zone about 5 miles wide and 15 miles long through the rugged Rogue River valley, began with placer mining about 1854. By the 1880s, Chinese were working small placers on Grave and Galice creeks for an estimated total of $8,000 a year. A 5-stamp mill at the Gold Bug Mine in 1886 and a 100 ton furnace at the Almeda Mine north of Galice in 1908 initiated lode mining. Rising and falling gold prices dictated the opening and closing of mines in this and other districts before World War II. Since then, only a few seasonal placers have been worked in the Galice district.

At Galice, the Almeda mine exploited the Big Yank lode, discovered in 1874. With more than 3,000 feet of underground workings, the extensive Almeda was only exceeded in length by the 5,000 feet of tunnels at the Benton mine southwest of Glendale, the largest underground operation in southern Oregon. The presence of massive copper sulfide in association with gold here was soon recognized. Once the Almeda

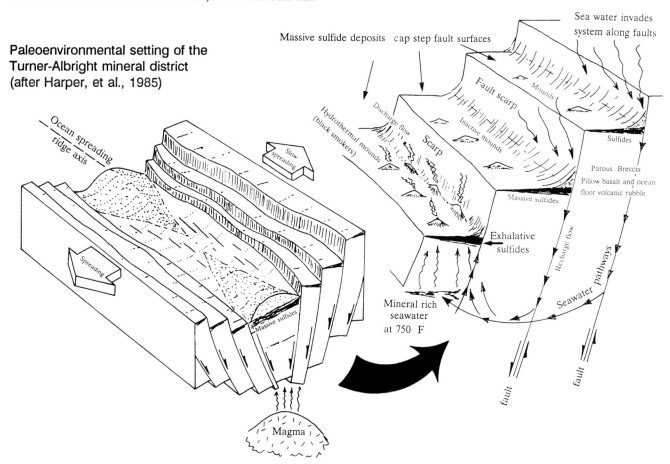

Paleoenvironmental setting of the Turner-Albright mineral district (after Harper, et al., 1985)

Mining Company had formed, a small copper ore smelter operated intermittently from 1911 to 1917 when 17,000 tons of ore yielded 260,000 pounds of copper, 7,000 pounds of lead, 1,500 troy ounces of gold, and 48,000 troy ounces of silver.

At the northeastern end of the Galice district near Canyonville in Douglas County, gold and silver mineralization in the Silver Peak region had a recorded production of 6,620 tons of ore between 1926 and 1937. Of this, 735,000 pounds were copper, 22,000 troy ounces were silver, and 500 troy ounces were gold. Minerals in the Silver Peak region were uncovered in 1919 on property owned by the Silver Peak mine, and 3,256 tons of ore were shipped out over the next ten years. A Swedish citizen is reported to have located the Gold Bluff mine here, but he was not allowed to claim the $7,000 in gold dust extracted because he was not a United States citizen. The Silver Peak area mainly produced copper, although a total of $216,000 in gold was taken from the mine prior to 1930. New operations here to extract copper, zinc, silver, and gold were begun in 1991. Minerals in the Galice-Silver Peak district are associated with island arc rocks of the Western Jurassic and Yolla Bolly terranes in which massive sulfide deposits occur in fragmented volcanics and sediments of the Rogue and Galice formations.

Rich placers along Jumpoff Joe Creek and the upper Grave creek watershed north of Grants Pass are responsible for most of the gold from the Greenback district, in operation since 1883. Occurring in sediments and volcanics of the Galice and Applegate formations, gold, copper, zinc, pyrite, and chalcopyrite are part of the Smith River and Applegate terranes that are ophiolitic in nature. Placer mining on Grave Creek produced $20,000 in gold, while the rich Columbia placer on Tom East Creek yielded more than $400,000. From September, 1935, to November, 1938, the Rogue River Gold Company operated the largest dredge in the history of Josephine County upstream from Leland in this district. Approximately 115 acres of gravels were worked before dredging ceased in 1939 only when the massive build up of loose gravel made it impossible to reach fresh bedrock. Equipped with sixty-five 7 1/2 cubic-foot buckets, this electrically powered behemoth could handle 5,000 cubic yards in 24 hours.

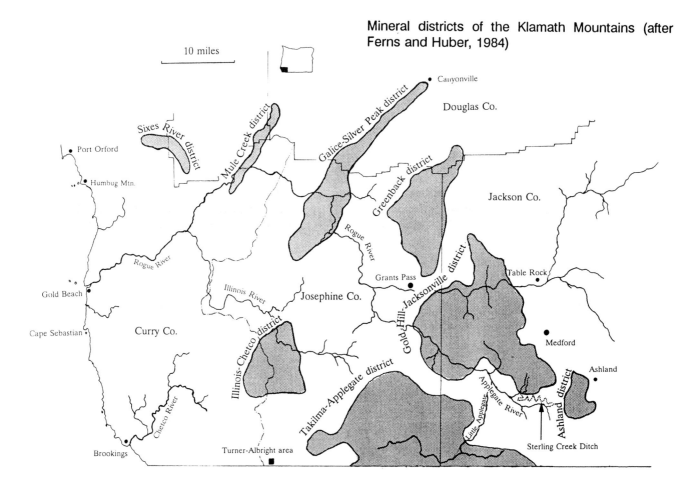

Mineral districts of the Klamath Mountains (after Ferns and Huber, 1984)

Almeda Gold Mine and smelter in Josephine County, 1913 (photo courtesy of Oregon Dept. Geology and Mineral Industries)

Gold discoveries in the Gold Hill and Jacksonville district originally attracted fortune hunters to southern Oregon in 1851. Several famous placers here occur in small, and shallow, but rich, concentrations of gold with few impurities or rock aggregates. The origin of the near-surface pocket gold deposits isn't well understood, but the gold occurs in sediments of the Applegate Group which have been leached by acidic groundwater. Once a pocket was found and worked out, the hole was abandoned. South of Jacksonville, gold at Sterling Creek was extracted by a large hydraulic system that yielded about $3 million by 1914. The Gold Hill, best known of the small gold pockets, was found in January, 1857, by a group of mining partners near the top of the hill now north of town. The miners discovered the outcropping rock was so full of gold that they had difficulty breaking it with hammers. The vein here only went 15 feet but produced $700,000 for its lucky discoverers.

A number of hydraulic and dredging activities took place on the Rogue and its tributaries near Gold Hill. Once the rich placer gold had been worked out, lode mining began. About 3 miles northeast of Gold Hill, the Sylvanite Mine on 80 acres started in 1916.

Owned by a succession of companies until 1939, the last was Imperial Gold Mines, Inc., which built a mill that could handle 140 tons of ore daily. The Sylvanite vein occurred in rocks of the Applegate Group that contained high grade ore. With more than 2,500 feet of underground workings, caving was prevalent. Caving along with wartime conditions forced closure after $700,000 in gold was extracted.

A small group of mines southwest of Ashland are in sedimentary rocks of the Applegate Group, part of the ophiolite or ocean crust making up the Rattlesnake Creek subterrane that has been intruded by the Ashland pluton. Gold here was extracted in placer operations for only a short time about the turn of the century. In this district, the Ashland vein, worked for its lode gold from 1886 to 1939, has been traced for more than a mile in length and to depths of 1,200 feet. Underground tunnels at the Ashland mine total 11,000 feet, and $1,500,000 was removed from this vein.

Gold seekers in the Applegate and Takilma district, which stretches along the southern Oregon border in Jackson and Josephine counties, began with placer activity on creeks and rivers in the 1850s. Operations were numerous, but limited, on bench gravels along the Little Applegate River, Carberry Creek, and Squaw Creek, but in the vicinity of Takilma (Waldo) gold was washed from gravels of Althouse and Sucker creeks. The largest hydraulic system was at the Llano de Oro or Esterly Mine northwest of Takilma that operated until 1945. Gravels here yielded close to $600,000. Pits, dug into an area of 30 acres on French Flat, were accessed by hydraulic elevators, and today these pits, below the water table, are called Esterly Lakes. At Takilma and Waldo, the rich shallow gravels of the celebrated Sailor Diggings paid as much as $2.00 per square yard of bedrock, yielding $4 million up to 1933.

Most copper mined commercially in southern Oregon has been from the Queen of Bronze Mine near Takilma in this district. Although discovered as early as 1860, significant amounts of copper weren't marketed until 1904 when a small smelter was constructed nearby. Over 25,000 tons of smelted ore was shipped from the Queen of Bronze and other mines during the main period of activity from 1903 to 1915. The Queen of Bronze was the second largest copper producer in Oregon after the famous Iron Dyke Mine in Baker County. Large-scale operations ceased in 1933 although exploration continues intermittently.

Copper deposits in the Queen of Bronze mine as well as gold from the Takilma district are developed within the Applegate Group of the Rattlesnake Creek subterrane within the Western Paleozoic and Triassic belt. Occurring within an ophiolite sequence, copper

Hanna Nickel Mine and smelter in Douglas County (Photo courtesy of Oregon Dept. Geology and Mineral Industries)

sulfide ores are found in veins, and the coarse crystals in the ores and adjacent rocks suggest that cooling of the hydrothermal fluids proceeded very slowly at considerable depth below the sea floor to deposit ores of pyrite, chalcopyrite, sphalerite, and pyrrotite.

Economic amounts of gold as well as silver, copper, zinc, and cobalt along the California border southwest of Takilma have been identified as part of the Turner-Albright sulfide deposit in the Western Jurassic belt. The Turner-Albright occurs in the Josephine ophiolite, a section of oceanic crust that was impregnated with minerals by superheated oceanic waters. No production of base metals has yet been reported, but serious efforts to examine the sulfide deposit began in 1954. Since then these mineral-rich deposits were the focus of intensive exploration in the 1970s and 1980s, and drilling programs indicate significant reserves of minerals exist here.

In the mining district along the upper Illinois and Chetco River watersheds, most of the gold was extracted from placers. After discovery in 1850, placer mining was most active on Josephine Creek and Briggs Creek in the upper Illinois River. Workings on these creeks are the principal producers, but incomplete records from both placer and lode mines make it almost impossible to accurately estimate how much gold was ultimately removed from this district. Gold in this district appears along shear zones within the

Galice Formation in association with the Josephine ophiolite.

Small quartz veins have been responsible for limited amounts of gold from northeast Curry and southeast Coos counties on Mule and Cow creeks where mineralization is associated with an ophiolite of the Snow Camp terrane. Although some of the veins are rich sources of lode gold, others contain only low grade ore. The Red River Gold Mining Company on Mule Creek, one of the largest placers in this district, had an hydraulic operation for a brief time in the late 1800s. Total production for the Mule Creek district was around $100,000.

Nickel

Along with gold and chromite, nickel is one of the most important economic minerals in southeast Oregon where it is mined from peridotites that have been deeply weathered leaving the isoluble nickel as a by-product in the soil. Formed deep within the ophiolitic sequence, peridotites containing the characteristic minerals olivine and pyroxene decompose quickly upon exposure to surface geologic processes. The soft laterite residue from deep weathering is relatively easy to excavate.

The only operating nickel mine in the United States is at Nickel Mountain, a few miles west of

Riddle in Douglas County, where the principal deposit is lateritic nickel in the form of the pale green mineral garnierite. Mining claims were first recorded in 1939, and extraction operations involve cutting parallel level benches around and down the mountain side 50 feet apart. Diesel shovels dig and load the ore, but in recent years only about half of the ore recovered is rated as acceptable for further processing. Claims were held by the Hanna Mining Company which operated here from 1954 until closure in 1987. Two years later the smelter was reactivated as the Glenbrook Nickel Company.

Other significant nickel deposits in the Klamath Mountains are on Woodcock Mountain in the Illinois River valley and at Eight Dollar Mountain near Cave Junction, both in Josephine County.

Chromite

Like nickel, chrome ore as chromite is found throughout the Klamaths deep in ophiolite rock sequences. Mines were begun in the Illinois River valley, Pearsoll Peak, Chrome Ridge, Takilma, and Vulcan Peak. Of these, the largest chromite producer was the Oregon Chrome Mine located in Josephine County about 20 miles west of Grants Pass in the central Illinois River area. Here individual layers of lode chromite 20 feet thick, occuring in a massive body of ore, produce up to 5,000 tons. In Josephine County a total of 48,941 tons of chromite ore was mined from 1917 to 1958. Only in times of national emergencies, during World War I, World War II, and between 1952 and 1958 when the U.S. government was stockpiling strategic materials, was Oregon chromite of economic interest.

Chrome ore-buying depot, Grants Pass, 1953. (photo courtesy Glunz Photo Studio)

Features of Geologic Interest

Oregon Caves National Monument

    In spite of the remote location 20 miles southeast of Cave Junction, the cavern at Oregon Caves National Monument was thoroughly explored by the turn of the century when the park was established. In actuality, Oregon Caves is really only one cave, approximately 1,600 feet from entrance to exit, with a variety of shapes, crystals, and graceful forms built up of calcium carbonate dripstone.

    The cave system was discovered in late 1874 by Elijah Davidson while hunting for a winter supply of meat. Following his dog through the thick underbrush, Davidson found a cave opening. The dog had been chasing a bear, and Davidson could hear the sounds of both animals fighting deep within the cave. Venturing inside, Davidson lit matches to see where he was going,

but was left in total darkness when his matches were expended. Fortunately he was able to follow a running stream back to the entrance, where his dog appeared shortly thereafter. The ill-fated bear was shot the next day.

    This cave has been developed in limestone lenses of the Applegate Formation, which is part of the Rattlesnake Creek subterrane. Here 190 million year old Triassic limestones were folded, fractured, and altered to marble by heat and pressure resulting from collision and accretion of the terrane. Percolating groundwater widened the cracks and joints to tunnels through the rock. In a final phase of cave formation, calcium carbonate carried in groundwater, precipitated icicle-like stalactites suspended from the ceiling and stalagmites rising off the cave floor.

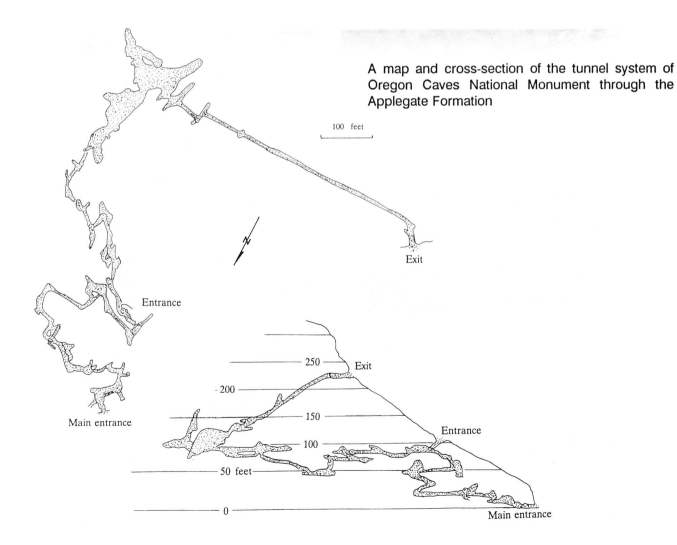

A map and cross-section of the tunnel system of Oregon Caves National Monument through the Applegate Formation

Dripstone stalactites and stalagmites of calcium carbonate in the Oregon Caves National Monument. Caves are dissolved into 200-million year old Triassic limestones (photo courtesy Oregon Department of Transportation).

The two basalt mesas, Upper and Lower Table Rocks are visible for a distance in the flat valley near Medford.

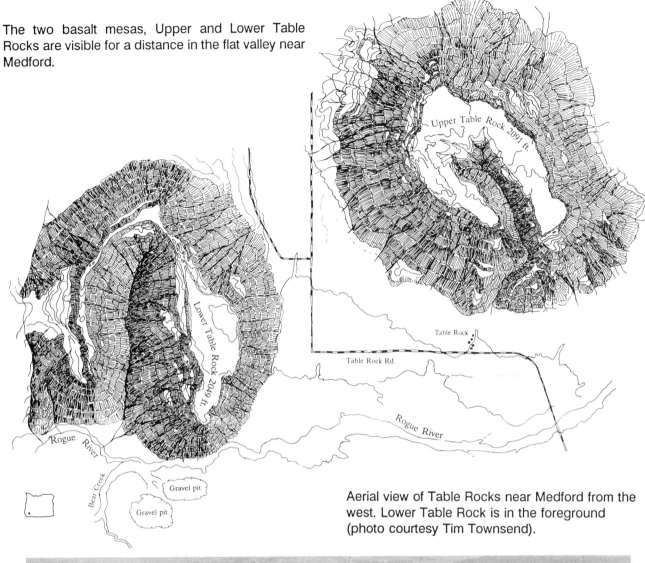

Aerial view of Table Rocks near Medford from the west. Lower Table Rock is in the foreground (photo courtesy Tim Townsend).

Table Rocks

Approximately 7 miles north of Medford, two flat mesas known as Upper and Lower Table rocks stand out 800 feet above the flat basin floor. With a distinctive horseshoe- shape when viewed from the air, Upper Table Rock is one mile square while Lower Table Rock is slightly smaller. Both are capped by an 125 foot thick layer of dark grey basalt.

One to 2 million years ago, lavas flowing from a

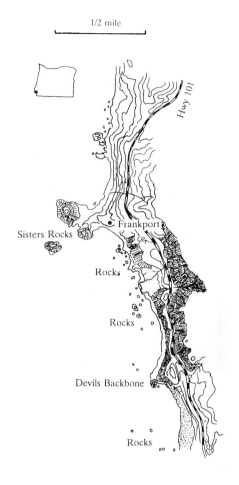

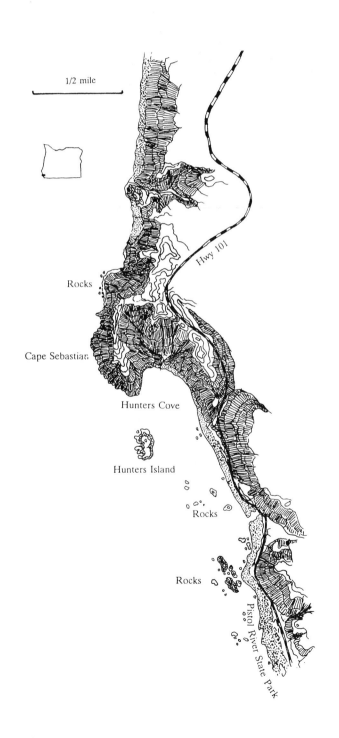

volcanic source in the upper Rogue River watershed were confined to the meandering canyon of the river at higher elevations but spread out upon reaching the broad Medford valley to cool and harden. Intensive erosion of the lava in later Pliocene and Pleistocene time left the two features as remnants of the once entensive flow. Reverse or inverted topography of this type occurs where stream valleys, cast in lava, later stand out in relief after erosion has planed off the local area. These two mesas near Medford are distinctive because their U-shape preserves the ancient paths of river meanders.

Coastal Region

Geologically part of the Klamath Mountains province, the Oregon coast from Cape Blanco south shows little resemblance to coastal regions to the north. Here resistant lower Jurassic and Cretaceous rocks have been eroded into an extremely rough coastline of headlands and offshore stacks, shoals, and rocky reefs, with few sandy terraces. The term, reef, refers to a chain of rocks or a ridge of sand near the surface of the water.

As viewed from the west, Humbug Mountain is composed of Cretaceous sandstones and conglomerates. (photo courtesy Oregon State Highway Department).

Along the southern coast, Cape Ferrelo, Crook Point, Cape Sebastian, Sisters Rocks, Humbug Mountain, and Nellies Point at The Heads project as the dominant headlands. Although most are composed of Jurassic Otter Point Formation rocks, the promontory at Cape Sebastian is geologically unique because it is one of the few coastal areas where late Cretaceous rocks occur. Here massive 900 feet thick sandstones and siltstones of the Cape Sebastian and Hunters Cove formations form the precipitous sea cliffs that extend around Hunters Cove on the southeast side of the cape. North of Cape Sebastian, the small headlands at Otter Point and Sisters Rocks as well as Nellies Point are braced by resistant Jurassic sandstones, mudstones, and volcanic rocks. The Heads mark the southern end of a Pleistocene terrace.

The highest point on the southern coast, Humbug Mountain, a rounded promontory overlooking the ocean, offers a sweeping view of the shoreline and ocean to anyone willing to climb the 3 mile long trail. A coarse bed of boulders at the base of the mountain grades upward into sandstones of the early Cretaceous Humbug Mountain Conglomerate. On the south side of the mountain, huge cohesive blocks of rock, sliding into the ocean, deposit rubble and massive piles of debris along the shore faster than the sea can carry them away.

The elongate Battle Rock, prominent at the cove south of Port Orford, was the site of a conflict between Europeans and Indians in 1851. A sea captain, William Tichenor, sailing into the harbor, left behind 9 men to begin a settlement. The men, attacked by the resentful Indians, took refuge on Battle Rock which projects above the water, until they could prudently escape to the settlement where Reedsport is located today. Composed of Mesozoic basalts, the rock is an obvious natural landmark in the area where a number of offshore stacks line the coast.

The most noticeable aspect of the southern Oregon coastline is the countless offshore stacks and shoals, many of which are part of the Oregon Islands National Wildlife Refuge. The age, composition, and arrangement of the rocks indicate they were originally connected to the mainland at a time that it extended further to the west. Once erosion shifts the shore landward, rock masses are cut off as isolated stacks, arches, and small knobs. Many are clustered in groups above the water level, while others are exposed only at low tide.

Along the southern-most section of the Oregon coast, Hastings Rock, rising 100 feet above the sands just south of Brookings, was once a sea stack when higher ocean waters covered the terrace. Nearby

Landslides into the sea just south of Humbug Mountain (photo courtesy of State Foresty Service)

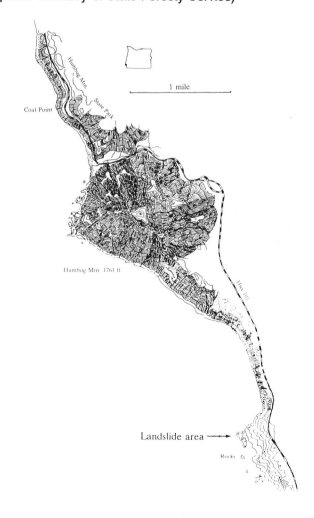

McVay Rock, an impressive stack also stranded ashore by a lowering sea level, has had its size reduced significantly by quarrying. Offshore here, Goat Island, with an area of 21 acres and a height of 184 feet, is Oregon's largest coastal island. All three structures are composed of 150 million year old dark gray sandstones, siltstones, and volcanics of the upper Jurassic Dothan Formation.

Near Crook Point, offshore rocks making up Mack Reef and Yellow Rock are a melange of resistant rocks that are part of the Jurassic Otter Point Formation. Mack Reef is a complex of parallel stacks running to the south and lined up with Crook Point. Offshore, the 325 foot high monolith, Mack Arch, is the remaining section of a reef which has had a tunnel cut through by persistent wave action, similar to Arch Rock Point in nearby Boardman State Park and Natural Bridge slightly north. From Mack Reef north to The Heads at Port Orford, most of the reefs and stacks scattered along the shoreline in a variety of extraordinary shapes are made up of mudstones, sandstones, conglomerate, and volcanic rock of this same heterogenous formation. An exception is found at Cape Sebastian, where the large Hunters Island has been eroded from the same massive Cretaceous sandstones that make up the cape.

A field of seastacks that make up Mack Reef south of Crook Point are composed of rocks of the Otter Point Formation (photo courtesy of E.M. Baldwin).

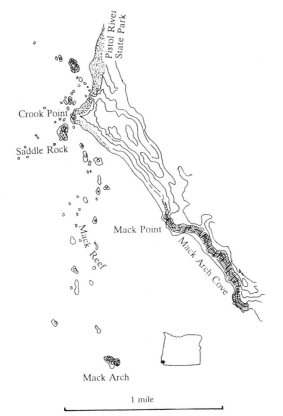

Groups of offshore stacks and submerged rocks making up a coastal belt of Cretaceous and Jurassic rocks from Gold Beach to Cape Blanco are part of an extensive reef complex that includes the Rogue River, Port Orford, and Cape Blanco reefs. Black Rock is the southern most projection of the Orford reef, and the larger Castle Rock above Cape Blanco is the northern most point. Sea stacks here with imaginative names as Arch Rock, Best Rock, Large Brown Rock, and Square White Rock are sandstones, volcanics, and conglomerates of these two reef systems. The third line of rocks, Rogue River reef offshore north of Gold Beach, is a complex jumble of conglomerates and sediments that project from the ocean as Pyramid Rock, Double Rock, and Needle Rock.

Suggested Readings:

Blake, M.C., jr., et al., 1985. Tectonostratigraphic terranes in southwest Oregon. In: Howell, D.G., ed. Tectonostratigraphic terranes of the circum-Pacific Region. Circum Pacific Council for Energy and Mineral Resources, Earth Science Series, v.1,pp.147-157.

Brooks, Howard C., and Ramp, Len, 1968. Gold and silver in Oregon. Oregon Dept. Geol. and Mineral Indus., Bull.61, 337p.

Ferns, Mark L., and Huber, Donald F., 1984. Mineral resources map of Oregon. Oregon Dept. Geol. and Mineral Indus., Geological Map Series, GMS 36.

Gray, Gary G., 1986. Native terranes of the central Klamath Mountains, California. Tectonics, v.5, no.7, p.1043-1054.

Harper, Gregory D., 1984. The Josephine ophiolite, northwestern California. Geol. Soc. Amer., Bull., v.95, p.1009-1026.

-----1984. Middle to late Jurassic tectonic evolution of the Klamath Mountains, California-Oregon. Tectonics, v.3, no.7, p.759-772.

-----and Saleeby, Jason, and Norman, Elizabeth A.S., 1985. Geometry and tectonic setting of seafloor spreading for the Josephine ophiolite, and implications for Jurassic accretionary events along the California margin. In: Howell, David G., Technostratigraphic terranes of the circum-Pacific region. Circum-Pacific Council for Energy and Mineral Resources, Earth Sci. Series no.l, pp.239-257.

Hotz, Preston E., 1971. Plutonic rocks of the Klamath Mountains, California and Oregon. U.S. Geol. Survey, Prof. Paper 648-B, pp.B1-B20.

Howard, J.K., and Dott, R.H., 1961. Geology of Cape Sebastian State Park and its regional relationships. Ore Bin, v.23, no.8, pp.75-81.

Howell, David G., Jones, David L., and Schermer, Elizabeth R., 1985. Tectonostratigraphic terranes of the circum-Pacific region. In: Howell, D.G., Tectonostratigraphic terranes of the circum-Pacific region. Circum-Pacific Council for Energy and Mineral Resources, Earth Science Series, no.1, pp.3-30.

Hunter, Ralph E., 1980. Depositional environments of some Pleistocene coastal terrace deposits, southwestern Oregon-case history of a progradational beach and dune sequence. Sedimentary geology, v.27, pp.241-262.

-----Clifton, H. Edward, and Phillips, R. Lawrence, 1970. Geology of the stacks and reefs off the southern Oregon coast. Ore Bin, v.32, no.10, pp.185-201.

Irwin, William P., 1964. Late Mesozoic orogenies in the ultramafic belts of northwestern California and southwestern Oregon. U.S. Geol. Survey, Prof. Paper 501-C, p.C1-C9.

--------1977. Review of Paleozoic rocks of the Klamath Mountains. In: Steward, J.H., Stevens, C.H., and Frische, A.E., ets., Paleozoic Paleogeography of the Western United States, Pacific Coast Paleogeogr. Symp.,v.1, pp.441-454.

Kays, M.A., and Ferns, M.L., 1980. Geologic field trip guide through the north-central Klamath Mountains. Oregon Geology, v.42, no.2, pp.23-35.

Koski, R.A., and Derkey, R.D., 1981. Massive sulfide deposits in oceanic-crust and island-arc terranes of southwestern Oregon. Oregon Geology, v.43, no.9, p.119-125.

Lund, Ernest, 1973. Landforms along the coast of southern Coos County, Oregon. Ore Bin, v.35, no.12, pp.189-210.

Mankinen, Edward, Irwin, W.P., and Sherman, G., 1984. Implications of paleomagnetism for the tectonic history of the eastern Klamath and related terranes in California and Oregon. In: Nilsen, Tor H., ed., Geology of the upper Cretaceous Hornbrook Formation, Oregon and California. Society Econ. Paleontologists and Mineralogists, v.42, p.221-229.

McKnight, Brian K., 1984. Stratigraphy and sedimentation of the Payne Cliffs Formation, southwestern Oregon. In: Nilsen, Tor, ed., Geology of the upper Cretaceous Hornbrook Formation, Oregon and California. Soc. Econ. Paleont. and Mineralogists, Pacific Sect., v.42, pp.187-194.

Nilsen, Tor H., 1984. Stratigraphy, sedimentology, and tectonic framework of the upper Cretaceous Hornbrook Formation, Oregon and California. In: Nilsen, Tor H., ed., Geology of the upper Cretaceous Hornbrook Formation, Oregon and California. Pacific Sect., Society Econ. Paleontolog. and Mineralogists, v.42, p.51-88.

Pessagno, E.A., and Blome, C.D., 1990. Implications of new Jurassic stratigraphic, geochronometric, and paleolatitudinal data from the western Klamath terrane (Smith River and Rogue Valley subterranes). Geology. v.18, pp.655-668.

Ramp, Len, and Peterson, Norman V., 1979. Geology and mineral resources of Josephine County,

Oregon. Oregon Dept. Geol. and Mineral Indus., Bull.100, 45p.

Roure, F., and Blanchet, R., 1983. A geological transect between the Klamath Mountains and the Pacific Ocean (southwestern Oregon): a model for paleosubductions. Tectonophysics, 91, p.53-72.

Walsh, F.K., and Halliday, W.R., 1976. Oregon caves; discovery & exploration. Grants Pass, Te-Cum-Tom Publ., 29p.

Basin and Range

Physiography

The elongate Basin and Range physiographic province can be traced into Utah, Nevada, Idaho, Oregon, Arizona, California, New Mexico, and Mexico encompassing an area of 300,000 square miles and comprising approximately 8% of the United States. Of this vast province, only the northern, butterfly-shaped area, primarily covering Nevada with small projections into Utah, Idaho, California, and Oregon, forms the Great Basin. The northwest extension of the Basin and Range province into southern Oregon is bordered on the west by the Cascade Mountains, on the north by the High Lava Plains, and on the east by the Idaho border, although the characteristic topography continues into Idaho. This region, which includes the stratigraphically similar Owyhee Uplands on the east, forms the southeast quadrant of the state and figures importantly in Oregon's geologic history.

As the name implies, the Basin and Range is a series of long and narrow, north-south trending fault-block mountain ranges alternating with broad basins. Prominent physiographic features from west to east across the province are Klamath Lake basin, Goose Lake valley, Winter Rim, Summer Lake, Chewaucan Basin, Abert Lake basin, Abert Rim, Warner Valley basin, Warner Peak and Poker Jim Ridge, Hart Mountain, Catlow Valley, the Steens and Pueblo mountains, the Alvord basin, and the broad Owyhee Valley. Most of the province is more than 4,000 feet in elevation. The crest of Steens Mountain, at 9,670 feet above sea level, is the highest and most scenic of Oregon's fault-block peaks. Extending northwesterly for 50 miles, the Steens merges to the south with the Pueblo Mountains. East of Warner Valley, Warner Peak rises to 8,065 feet, while Hart Mountain, the best defined fault-block mountain in the United States, is 7,710 feet high. To the west the sheer escarpments of Abert Rim and Winter Rim rise dramatically more than 2,000 feet above the valley floor.

The Great Basin, that portion of the Basin and Range which includes Oregon, was first named in 1844 by explorer John Fremont who recognized that the basin had no direct outlets to the sea, ending 20 years of speculation on the course of waterways in the region. In what is called the Oregon Great Basin, only

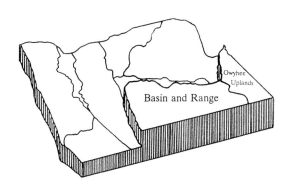

the Klamath and Owyhee rivers indirectly reach the sea, although Goose Lake has been known to overflow into the Pit River during historic times. In the western portion of the Great Basin, the Klamath River originates in Upper Klamath Lake to flow from the extreme southwestern edge of the province into California.

The Owyhee Uplands lies in the northwest corner of the Great Basin. This region differs from the rest of the province in that it is a flat deeply dissected plateau with little interior drainage where fault-block topography is less pronounced. The drainage basin of the Owyhee River encompasses the uplands. Originating in Nevada, the Owyhee River flows northerly through Idaho and Oregon to join the Snake River near Adrian, Oregon. In spite of low rainfall in the area, steep gradients give the the river and its tributaries well-defined drainage patterns and deep canyons. Cutting through the uplands over 6,000 feet above sea level, the river drops to approximately 2,000 feet where it joins the Snake. Small streams flowing in from the hills are largely intermittent.

The Great Basin portion of the Basin and Range province barely extends into southeast Oregon.

Geologic Overview

The Basin and Range is a tectonically youthful province with an anomalously high heat flow, thin outer crust, and a high regional elevation. The most pronounced structural phenomena effecting the Basin and Range province are the stretching or extension of the crust and the movement of large tectonic blocks. These forces are responsible for giving the basin its characteristic tilted, raised mountains and down-dropped basin structure as well as producing the volcanic activity of the late Tertiary. A complex network of faults and fissures resulted as the enormous crustal blocks were uplifted, tilted, or dropped while the basin was being stretched and distorted. Major faults are marked by spectacular scarps trending northward along the face of the mountain ranges.

Cenozoic volcanic activity in this province began in the Miocene and continued into the Pleistocene resulting in basalts, tuffs, and tuffaceous sediments totalling nearly 10,000 feet in thickness. As the crust thinned and cracks appeared, magma from below broke through to cover the area. Basalts and ash were extruded from a broad shield volcano in the region of Steens Mountain. Following this a number of large calderas in the southeastern part of the state were responsible for thick ash and stream-deposited sediments in the basin and Owyhee uplands.

During the Pliocene and Pleistocene increased rainfall was responsible for large lakes filling the basins. These lakes have since dried up or diminished considerably because of more arid conditions prevailing today. The colder climate of the Pleistocene also brought about a build-up of ice masses atop Steens Mountain. Relatively recent volcanism has created some unique topography at the lava fields of Diamond Craters, Saddle Butte, and Jordan Craters.

Geology

If Oregon has a foundation block upon which the rest of the state was constructed, it is buried under the Basin and Range province. In the Blue Mountains and Klamaths, the movement, amalgamation, and eventual accretion of late Paleozoic and Mesozoic exotic terranes to North America can be readily demonstrated. By contrast, the volcanic veneer of the Basin and Range region obscures its preCenozoic history, and older rocks here could predate the accretion of terranes of the Blue Mountains and Klamath provinces. However, recent geophysical work on the northern Basin and Range suggests that the crust is even thinner than previously thought. This means that the Miocene volcanics may, in fact, represent the basement or oldest rocks of the province.

Extensional Tectonics

Beginning in the Miocene and extending into the Pliocene, the forces of crustal stretching and the movement of tectonic plates triggered faulting, extensive volcanism, and the development of the basin and fault block mountain topography characteristic of this province. The most profound effect of tectonics in the Basin and Range is an east-west stretching, extensional phenomena that has expanded this province by as much as 100%

One interesting aspect of the extension in this province has been the clockwise shift in the stretching direction from northeast-southwest to a more northwest-southeast orientation. When the crust of the basin was pulled and drawn apart, it grew thinner like a piece of taffy being stretched. With additional tension, the thin, semi-brittle crust began to fail producing faults which provided a route for magma in the lower crust to escape to the surface as widespread lava flows and volcanoes. The thinning also brought the water table into contact with the hot crustal rocks below so that Nevada and southern Oregon today have scattered thermal springs and explosion craters or maars situated above ancient and modern faults.

Forces responsible for the stretching of the Basin and Range are still not well-understood. It is possible the basin may be above a spreading backarc that is near the western margin of North America. Alternately, the basin may represent a plate moving westward over a mantle plume or even an oceanic spreading center that has been overridden by the North American plate.

The extensional tectonics occurred in two distinct phases. An early event between 20 and 10

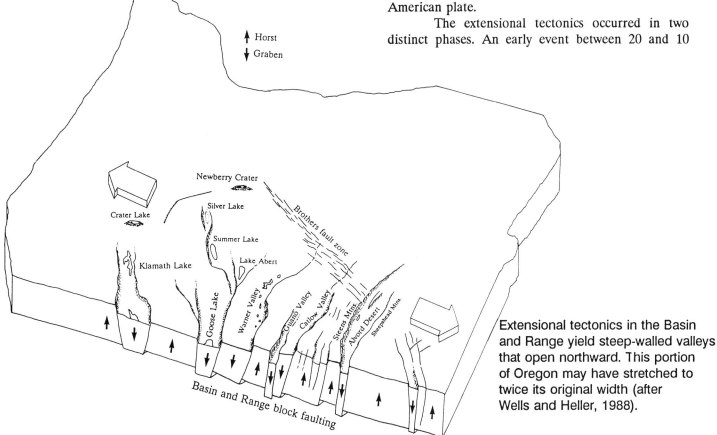

Extensional tectonics in the Basin and Range yield steep-walled valleys that open northward. This portion of Oregon may have stretched to twice its original width (after Wells and Heller, 1988).

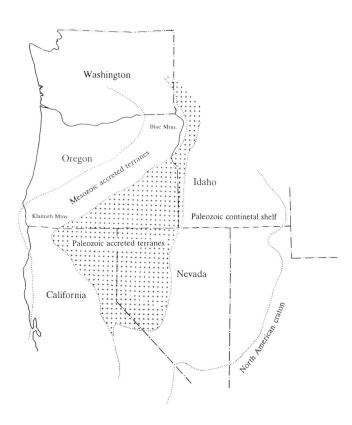

Possible Paleozoic accreted terranes in the Oregon Basin and Range are obscured by layers of younger Tertiary volcanics.

million years ago was responsible for opening the Nevada-Oregon rift, the western Snake River downwarp or graben, and crustal fractures in eastern Oregon which in turn gave rise to volcanic dike swarms that released Steens Mountain and Columbia River basalts.

A second sequence of Basin and Range extensional tectonics during the last 10 million years gave the basin its characteristic topography of fault-block mountains interspersed with depressions. The complex mosaic of cracks and faults here line up in two distinct trends, one to the northwest and the other to the north-northeast. The result is that basins formed by faults tend to be open at the north end and form a "V" at the southern end. Northwest trending faults in this system are relatively small showing only about 150 feet maximum displacement with spacing of about a mile. Faults running in a northeast lineup, by contrast, are several miles apart and show displacements of thousands of feet.

More important, in Oregon, this final extensional phase exerted differential pressure to spin or twist much of the state 60 degrees in a clockwise direction. One estimate is that extension of the basin is responsible for only 60% of the total rotation. The remaining 40% of rotation may be due to structural phenomena called "dextral shear" where the basin of Oregon is caught between two massive shifting blocks. As the west block moves north and the east block moves south, Oregon is rotated in a clockwise direction.

The spreading itself has created a distinctive pattern which has been recognized in Nevada with respect to gravity anomalies. The anomalies are mapped as high and low gravitational values across an area. This means that over a gravity low, for example, an object actually weighs less or there is less gravitational attraction compared to a gravity high. By carefully measuring and plotting the gravity map for the basin, it has been shown that the pattern in the region east of the 116 degree meridian is strikingly similar to that in the area west of the meridian. This near mirror image or twinning display of gravity data seems to support the thesis that a Basin and Range extension relates to some type of sub-crustal spreading mechanism.

Volcanism and Tectonics

In addition to the crustal stretching phenomenon, massive volcanism in this province was connected directly to movement of the crustal plates. Four tectonic plates, the Kula, Farallon, Pacific, and North American, converged at the North American West Coast during late Cretaceous and early Tertiary time. The Kula plate gradually disappeared northward under Alaska, while the Farallon plate was subducted eastward under North America. As the Farallon and Kula plates slowly sank beneath the North American plate, the Pacific plate expanded until today it is adjacent to the North American plate along the West Coast. Remaining fragments of the trailing edge of the old Farallon plate have been named the Juan de Fuca plate.

The angle of the subducting slab beneath the overriding plate is critical to the location of surface volcanic activity. Typically the volcanic archipelago is situated in a line on the overlying plate where the melting slab is 90 miles below within the crust. When the collision rate between two plates is slow, the angle of the subducting plate is steep, and the volcanic arc will be close to the subduction zone. As the convergence rate increases, the angle becomes very low and flat, and the archipelago of volcanoes will move toward the continent far from the subduction zone. As the rate of converging plates accelerated at the Mesozoic-Cenozoic boundary between 70 and 50 million years

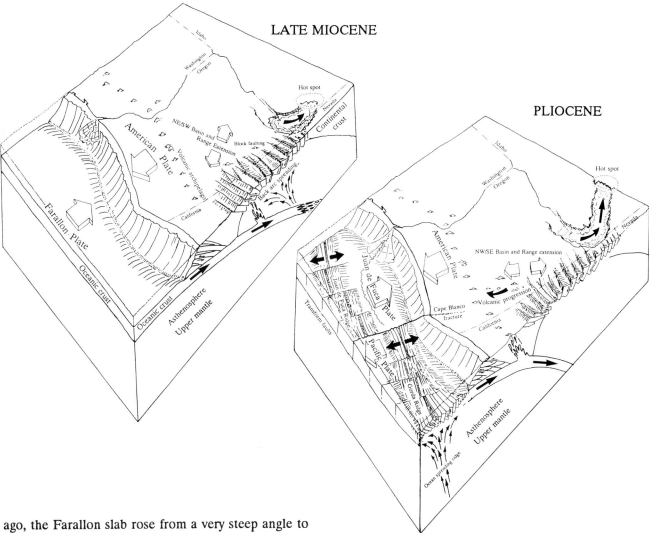

LATE MIOCENE

PLIOCENE

ago, the Farallon slab rose from a very steep angle to a relatively shallow position transferring volcanic activity in the Columbia Plateau and Basin and Range from west to east. About 30 million years ago, during the Oligocene, with decelerating convergence, the slab began to steepen. Then at 17 million years ago the subducting slab was detached and heat broke through at the rupture to initiate backarc spreading and issue as the Steens Mountain and Columbia River lavas. For the past 5 million years extensional spreading forces have intensified to create crustal thinning faults, and volcanism in the Basin and Range.

The most striking aspect of the Basin and Range volcanic activity is that eruptions of silica-rich rhyolitic lavas are progressively younger in a line running northwest across the province. In the eastern margin of the area during the late Miocene, volcanic activity migrated rapidly westward at a rate of 1 1/2 inches per year. However, the volcanic progression slowed to approximately 1/2 inch per year in the western section where volcanism is younger. While this migration of volcanic centers resembles a "hot spot", both the trend and direction are inconsistent with the steady westward movement of the North American

Stages in the extensional stretching of the northern Basin and Range as well as movement of the North American plate westward over the Yellowstone hot spot. In this model the subducting slab steepens sending the volcanic front back to the west. Simultaneously the direction of extension in the basin shifts clockwise from northeast-southwest to a more southeast-northwest orientation (after Zoback, Anderson, and Thompson, 1981).

plate during the last 10 million years of geologic time. It seems probable that after the subducting Farallon slab became detached, collision slowed considerably to steepen the slab and drive a volcanic wave gradually back to the west.

A similar pattern of volcanic progression extends from Steens Mountain in the southwest all the way across Idaho, through the Snake River downwarp,

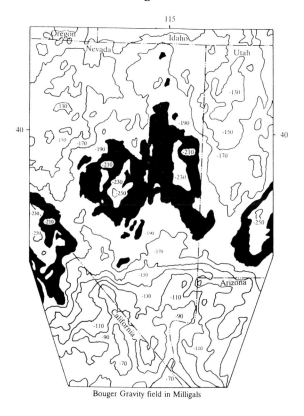

Bouger Gravity field in Milligals

A gravity map of Nevada and southern Oregon clearly displays a twinned image east and west of the 116 degree meridian which supports a spreading or stretching hypothesis of the Great Basin. Smaller map for comparison is a mirror-image of the eastern half of the area (after Eaton, 1984).

and up to very young volcanics at Yellowstone in northwest Wyoming. The rapid eastward movement of the volcanic centers, which is twice as fast as that across Oregon, is due to a combined effect of plate movement and extension of the Basin and Range. The hot spot presently beneath Yellowstone may have generated a large volume of Eocene basalt as it passed beneath the area of the Oregon Coast Range.

Volcanic "hot spots" are known from dozens of sites around the globe. Recognition of these volcanic centers is a simple matter of tracing a chain of volcanoes back to its source or youngest vent. As a continental or oceanic plate moves across such a hot spot, intermittent volcanic activity leaves a plume of volcanoes as is the case with the Hawaiian Islands. The more speculative geologists have suggested that these hot spots may represent locally weak places in the earth's upper mantle where incoming meteorites punched through millions of years in the past.

Mesozoic

Because of the two-mile thick mantle of late Tertiary volcanics and sediments, exposures of Mesozic and older rocks are very limited in the Basin and Range. It is necessary to step over into Nevada and compare rocks from there with those in Oregon in order to reach conclusions about the preCenozoic history of this province. Older Mesozoic rocks occur in the Pueblo and Trout Creek mountains on the extreme

southeastern Oregon border. Here metamorphic rocks, which include phyllites and greenstones as well as some granitic intrusions, are tentatively dated as Jurassic and older. Just over the border in Nevada identical rocks have been dated with some confidence as late Triassic and Paleozoic suggesting that the foundation of the Oregon Basin and Range could be part of a composite Paleozoic accreted terrane block that considerably predates terranes in the Klamaths and Blue Mountains.

Cenozoic

Early Cenozoic rocks are apparently missing, but the Miocene, Pliocene, and Pleistocene epochs are represented in the Basin and Range by basalts and tuffs from the extensive and fairly continuous volcanic activity. Coinciding with the initial stretching and extension of the Basin and Range 16 million years ago, the most productive volcanic event in the province was the eruption that centered in Steens Mountain when fluid basaltic lavas spread across southeast Oregon from a large, low profile shield volcano. The uneven topography of the surface covered by the liquid flows produced wide variations in thickness of the layers. Because of glacial erosion, faulting, and younger lava flows, the maximum thickness of the basalt can not be determined. However, the flows average 3,000 feet thick and cover 6,000 square miles for a total volume of over 3,000 cubic miles. Although the volume of the Steens basalt is less than 10% of that of the Columbia plateau lavas, Steens lavas are remarkable in that they

A view to the north of Abert Rim in central Lake County (photo by E.M. Baldwin).

were apparently extruded in an extremely short time span. One estimate of the duration of the eruption is only 50,000 years.

The basalt, making up the bulk of Steens Mountain, extends into the Pueblo Mountains to the south and to Abert Rim to the west. The upper 3,000 feet of the east scarp of the Steens as well as rock exposed in the glacial valleys is composed of andesitic basalt. In the vicinity of Alvord Creek, the "great flow" of Steens Mountain basalts is nearly 900 feet thick. Columnar joints, which mark this basalt layer, measure as much as 5 feet across and rise 300 feet to form a very steep scarp along the southern side of Alvord Creek. The columns are broad at the base and narrow toward the top. The similar Owyhee Basalts, that erupted onto the Owyhee Plateau 13 to 12 million years ago, attain a maximum thickness of about 1,500 feet in Owyhee Canyon but are much thinner elsewhere. These basalts do not contain large plagioclase feldspar crystals that are so common in certain areas of the Steens flows.

About the same time as the Steens Mountain eruption, extensive Miocene ash-flow tuffs covered southeast Oregon and northern Nevada, originating from more than a dozen different volcanic centers aligned in a northeastern direction across the basin and Owyhee Uplands. Eruptions from volcanic cones in the Lake Owyhee and McDermitt fields resulted in im-

mense calderas. The largest of these is McDermitt caldera which is bisected by the Oregon-Nevada line. Actually a series of overlapping calderas, the McDermitt complex has a combined diameter of 22 miles.

Three Fingers caldera west of Succor Creek, Mahogany Mountain caldera along the Owyhee River, Saddle Butte to the west, and Castle Peak to the north are the four volcanoes comprising the Lake Owyhee field in Oregon. With a north rim extending 1,000 feet above Leslie Gulch, Mahogany Mountain is the southeast rim of an ancient volcanic caldera 10 miles in diameter, whereas the Three Fingers caldera, just to the northeast, is a circular collapsed depression 8 miles in diameter. As rhyolitic magma erupted from these pre-caldera volcanoes, the subsequent ash, tuffs, and continuing eruption of lava produced the two calderas. Three Fingers rock, within the depression, is the erosional remnant of a rhyolite plug. Saddle Butte, a third caldera just to the southwest is 15 miles in diameter. This vast structure is almost completely buried beneath younger volcanic rocks of the Saddle Butte field. Castle Peak caldera, part of the same field, collapsed with the eruption of the Dinner Creek ash flow tuff on the southwest margin of the rim. By comparison, Crater Lake and Newberry calderas are 6 miles and 5 miles in diameter respectively.

Volcanic activity continued into the later Miocene with the eruption of Gearhart Mountain in southcentral Oregon just before major block faulting in the Basin and Range began. In contrast to other volcanoes of this province, Gearhart lacks the typical bimodal history where volcanoes initially erupt basaltic lavas to be followed later by rhyolites. The eruption of the Gearhart volcano began with basalt and ended with a veneer of andesitic lavas. Just north of Gearhart, the prominent volcanic peak, Yamsay Mountain, is significant because it erupted during the Pliocene coinciding with extensional stretching of the Basin and Range. Unlike the Gearhart eruption, Yamsay has a history of bimodal volcanism beginning with a low shield cone of basaltic then rhyolitic lavas. Near the end of its eruptive history, small amounts of basalts again extruded from its flanks. This kind of bimodal volcanic activity may have been due to crustal extension.

Pleistocene

Continental ice sheets advancing southward during the Pleistocene accompanied increased precipitation and valley glaciers in the mountains. Rainfall and mountain runoff filled depressions in southcentral Oregon, creating large pluvial lakes across the region. Pluvial lakes are dependant on rainfall, expanding or decreasing dramatically with changing climatic conditions. At the peak of the moist Pleistocene environment vast areas of Oregon were covered by freshwater lakes, whereas today, during an interglacial warm phase, lakes cover less than one percent of the state. During this period nine fair-sized pluvial lakes existed in the Basin and Range province of Oregon. The largest of these in the western section of the province was Lake Modoc covering 1,096 square miles, followed by Lake Chewaucan at 461 square miles in the central region, Lake Coleman, now Warner Lakes, spread over

Within the Owhyee Uplands of southeastern Oregon a complex series of overlapping volcanic calderas of immense size erupted during the Miocene (after Rytuba, et al., 1990).

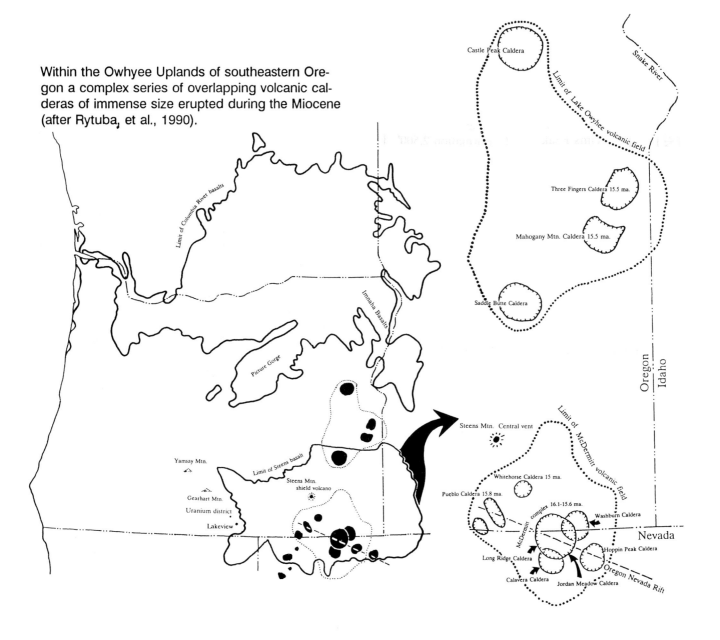

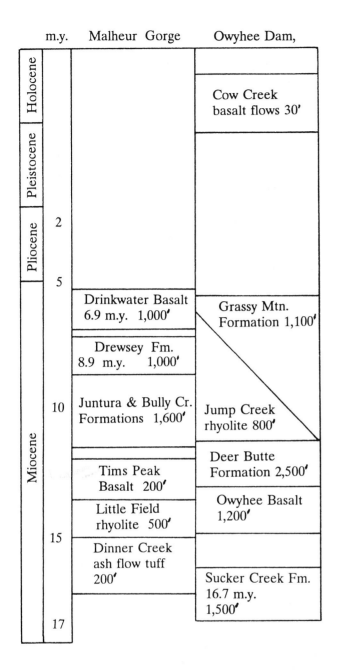

m.y.	Malheur Gorge	Owyhee Dam,
Holocene		Cow Creek basalt flows 30'
Pleistocene		
Pliocene 2 5		
Miocene 10 15 17	Drinkwater Basalt 6.9 m.y. 1,000'	Grassy Mtn. Formation 1,100'
	Drewsey Fm. 8.9 m.y. 1,000'	
	Juntura & Bully Cr. Formations 1,600'	Jump Creek rhyolite 800'
		Deer Butte Formation 2,500'
	Tims Peak Basalt 200'	
	Little Field rhyolite 500'	Owyhee Basalt 1,200'
	Dinner Creek ash flow tuff 200'	
		Sucker Creek Fm. 16.7 m.y. 1,500'

Great Basin Tertiary stratigraphy (after Walker, 1979)

483 square miles in Lake County, and Alvord Lake in Harney County extended for 491 square miles.

Ancient pluvial Lake Modoc, which once reached nearly 75 miles in length, has receeded to the Upper and Lower Klamath lakes and Tule Lake in California. With a climate change to dryer conditions at the end of the Pleistocene 11,000 years ago, Lake Modoc began to decrease in size. The lowering of water was uniform at first, but once the basins were no longer connected the water level fluctuated variously among the now separated and smaller 8 lakes. Eventually only four retained any size. Others as Yonna and Langell are now wide, flat valleys.

Pluvial Lake Chewaucan, whose remnants are Summer Lake, Lake Abert, and the Chewaucan Marshes, was almost 375 feet deep. The ancient lake, which occupied an enclosed basin, was bounded on the west by the Winter Rim fault and on the east by the Abert fault scarp. Former lake levels are shown by prominent beaches high above the present water line and by a broad sand and gravel delta built when high lake waters overflowed into the flat area just north of Paisley. The moist habitat around Lake Chewaucan provided water and vegetation for large herds of Ice Age bison, camels, horses, and even elephants whose skeletal remains are exposed along the old shorelines. Fossil bones of waterfowl, fish, and freshwater mollusc shells also litter the dry sands of the lake beds. The lakes today are fed by small amounts of water issuing from several springs and streams.

Two major events effected Lake Chewaucan. The first, about 10,000 years ago, began when the waters dried up to create the playa. The second occurrence during this warmer period was the catastrophic eruption of Mt. Mazama on the present site of Crater Lake about 6,900 years ago. Tremendous volumes of pumice, ash, and volcanic dust created a dune field in the northeast corner of Summer Lake and Lower Chewaucan Marsh, 60 miles away from the volcanic source.

Warner Valley and Alvord Desert occupy similar depressions where small lakes persist. Bluejoint, Campbell, Flagstaff, Anderson, Hart, and Crump lakes are remnants of what was a large pluvial lake occupying Warner Valley, a basin that extends in a north-south direction for 50 miles. Hart Mountain and Poker Jim Ridge form the eastern boundary of the valley. Alvord basin, bounded by a fault scarp along the base of Steens Mountain to the west and a second fault to the east, was once a lake 12 miles wide and 70 miles long. The basin today contains the Alvord Desert and a much reduced lake rimmed with a glistening white alkali crust resulting when lake waters evaporate in the summer leaving the residue.

Unfortunately during 1990 and 1991 an out-of-doors "artist" used a rototiller and volcanic cinders to mar the floor of Alvord Lake with a series of lines and circles. Because the grafitti is on such a grand scale that it can be seen from passing airplanes, the pristine lakebed may take decades to heal itself.

Cooling during the Pleistocene also signalled the outset of glaciation on Steens Mountain, the only region in the Basin and Range to support large ice

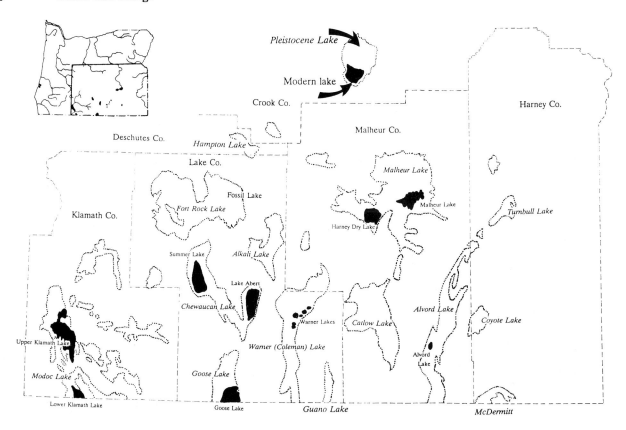

The small playas and pluvial lakes that dot southeast Oregon today are all that remain of much larger bodies of water here during the Ice Ages (after Allison, 1982).

masses. The highest part of Steens Mountain has been more thoroughly dissected by erosion, much of it glacial, than other scarps in southeastern Oregon. This is due to steeper gradients and the greater precipitation that accompanies higher altitudes. Glaciation on the western slope of the Steens sculpted lake basins and cut deep U-shaped valleys. The symmetrically curved

Kiger Gorge, the canyon of Little Blitzen River, Big Indian Creek, Little Indian Creek, Wildhorse Creek, and Little Wildhorse Creek were all gouged out by valley glaciers which filled the streambeds. Fish Lake occupies a shallow depression where Fish Creek was dammed by a glacial moraine. Other lakes, such as Wildhorse Lake, occupy glacially cut, saucer-shaped valleys.

Ducks, pelicans, herons, and cormorants used Fossil Lake as a stop-over during the Plesitocene.

A view southeast over Steens Mountain that shows glaciated valleys as well as the Alvord Lake playa (courtesy Oregon State Highway Department)

Well-developed glacial valleys along the front of Steens Mountain looking south (courtesy of Oregon State Highway Department)

The long glaciated face of Steens Mountain stretches northward across the southeastern margin of the Basin and Range in Oregon.

On the east side of the Steens, smaller glaciers extended only part of the way down the stream valleys. As the ice mass eroded the rim of the mountain, the crest moved westward. Glaciers working on both sides of the mountain rim cut the rock away leaving only a thin wall separating Little Blitzen and Big Indian canyons from the glacially cut valleys to the east. At Kiger Gorge, the east and west glacier eroded through the mountain crest to produce a gap called The Big Nick.

Ice masses advanced down Steens Mountain in two separate stages. The first formed a broad ice cap over approximately 115 square miles. On the western slope, the retreating ice left extensive debris, while on the east slope it reached about 2,500 feet above the Alvord Desert and cut large, bowl-shaped cirque valleys as the one at the head of Alvord Creek. The second glacial advance was much smaller, covering less than 50 square miles down in the canyons.

Structure

The nature of faulting in this province is still not well understood. Two phases of faulting are recognized. In an early phase horsts, or uplifted fault blocks, and grabens, or basins, developed from a series of tensional faults or tilted fault-blocks. In the Nevada area of the Basin and Range, the block faulting phase was succeeded by what is called "listric" faulting where the normal faults curve at depth to a nearly horizontal orientation as the crust is pulled and stretched to the extreme. Listric faulting typically yields a topographic pattern where all of the fault block surfaces slope away from a central axis. On the other hand, randomly tilted blocks suggest horst and graben topography. To understand the mode of faulting, the surface orientation of the fault-blocks is carefully mapped.

The age of faulting is particularly important to the geologic extensional history of the Great Basin, and, since stretching of the Great Basin was not uniform, tectonic activity shifted from area to area. During the later Teritary, faults became progressively younger northward so the Oregon Basin and Range province has many Pliocene events. Within the Quaternary, however, seismic activity shifted back to the south into Nevada where several large earthquakes have been recorded in historic time in the central Nevada seismic belt.

Of the uplifted blocks in this province, the Hart Mountain and Steens Mountain blocks are the most prominent. Hart Mountain and Poker Jim Ridge are part of a large complex of an uplifted block with a steep western face and a more gentle east-facing slope. The sheer scarp of Hart Mountain is more than 3,000 feet above Warner Valley, while the asymmetrical

The head of Little Blitzen Canyon in Steens Mountain is a cirque that nearly breached the mountain front (courtesy Oregon State Highway Department).

Kiger Gorge in Steens Mountain looking north displays the unmistakable U-shape of a glaciated valley (courtesy Oregon State Highway Department).

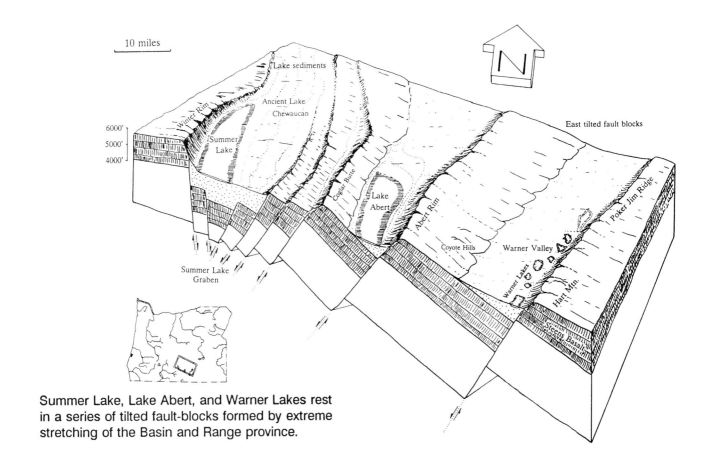

Summer Lake, Lake Abert, and Warner Lakes rest in a series of tilted fault-blocks formed by extreme stretching of the Basin and Range province.

Steens Mountains is an up-thrown fault-block that has been heavily glaciated on its top and sides by Pleistocene ice masses.

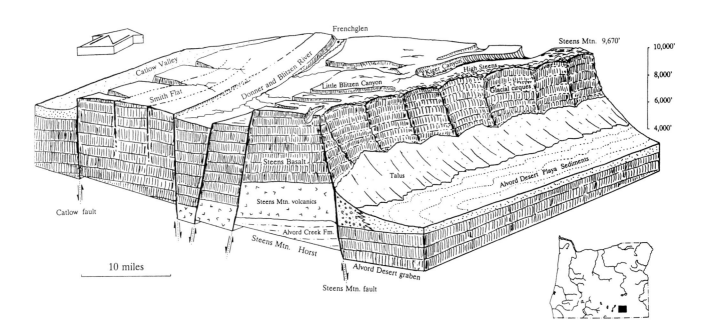

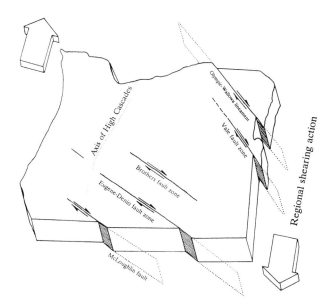

Caught between large-scale moving crustal masses, Oregon is being sheared by extensive faults running northwest to southeast across the state.

Poker Jim Ridge has an 1,800 foot precipitous scarp. Steens Mountain is a large horst where the block has been fractured into several pieces. The northern most portion tilted to the west as it rose giving that side the very gentle slope of Smith Flat. Extending for over 60 miles in a north-south direction, the steep east escarpment displays successive layers of Steens basalt in cross-section.

The southern and eastern margins of Oregon are cut by faults that run hundreds of miles across several physiographic provinces. These large-scale cracks, termed strike-slip faults, move laterally and are parallel to the southeast by northwest Olympic-Wallowa lineament. From south to north they are the McLoughlin fault, the Eugene-Denio fault, the Brothers fault, and the Vale fault. The faults are actually complex fault zones of smaller overlapping faults, but the trends are remarkably parallel. Faulting is most severe to the south and minimal to negligable along the short Vale fracture which runs northwest through the Owyhee Uplands. These fault systems relate to a shearing action where central and eastern Oregon are caught between two large-scale moving blocks. Action along the blocks is similar to movement of the San Andreas fault in California with the eastern block moving south and the western block moving north.

Mining and Mineral Industry

Uranium

Although deposits of lithium, antimony, gold, mercury, copper, and uranium have been known for several years in the Great Basin of Oregon, only uranium and cinnabar, or mercury, have been mined extensively. The Lakeview district in the southcentral part of the state as well as the McDermitt district in southern Malheur County have produced small amounts of uranium. Nearly 400,000 pounds of uranium have been recovered from two deposits in volcanic rocks just northwest of Lakeview. Most of the material was from the White King Mine, with smaller amounts from the Lucky Lass Mine where shipments in 1955, shortly after discovery, were the first uranium ores marketed from Oregon. A mill processing 210 tons per day was constructed at Lakeview in 1958, operating until 1965 when closure resulted after several years of only minimal production. Located in the McDermitt caldera, the Aurora and Bretz uranium prospects are the largest in total output of any yet found in Tertiary volcanic rocks of the United States. This potentially economic source of uranium ore, found as uraninite and coffinite, is associated with rhyolitic rocks and lake sediments. Once the thick, flat-lying lavas, covered by tuffaceous lake sediments, had been broken up by faulting, hot waters, containing minerals, followed fissures and cracks to deposit uranium and mercury in veins.

Mercury

Within the McDermitt caldera complex, the Opalite Mining district, which includes the McDermitt Mine, the Bretz Mine, and the Opalite Mine have had a total output of 270,000 flasks of mercury, the richest supply of mercury in the western hemisphere. Mercury exploitation in the Opalite district began in 1917 with the discovery of cinnabar by William Bretz who had prospected in this region for a number of years. By 1925 tunnels had been driven 80 feet below the ore body and the cinnabar brought to the surface where it was processed in a large rotary furnace completed here in 1926. Over the years production was sporadic, stimulated in 1940 by the rising price of quicksilver prior to and during World War II when large amounts of mercury were used as ballast in submarines. The opening of several new pits signalled renewed mining as late as 1957 when the ore was sent to Salt Lake City for processing.

Very small deposits of cinnabar, copper, and gold along the eastern edge of the Steens and Pueblo mountains are largely unproductive.

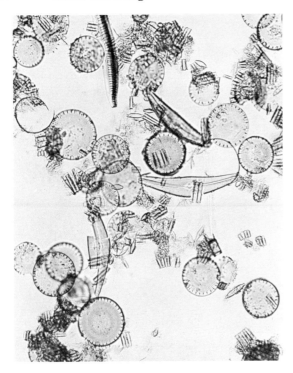

An assortment of freshwater diatoms from within
the Klamath Basin (photo Oregon Dept. Geology
and Mineral Industries)

Diatomite

Diatomite, bentonite, and zeolite minerals are
all exploited in small amounts from the eastern region
of the Great Basin province. Diatomite deposits in the
Juntura and Otis basins are the remains of microscopic
siliceous skeletons of diatoms, single-celled aquatic
plants that accumulated in a former freshwater lake.
Typically white in color, the diatomite beds here range
from a few inches to 20 feet in thickness within the
Miocene Juntura and Drewsey formations. First mined
between 1917 and 1934 and again commencing in 1986,
diatomite is used as an absorbent cat box filler and as
a filter for chemicals and drinking water.

To the southeast, bentonite (volcanic ash) and
zeolites are produced from localities in Malheur
County near the Idaho border. Mined from within the
Miocene Sucker Creek Formation, high-swelling
bentonies are utilized primarily as a sealant in dams,
landfills, and waste sites. Small amounts of potassium-
rich zeolites from the same formation were first
extracted in 1975, but commercial markets for this
mineral have developed only slowly. Currently zeolites
are used as carriers for agricultural chemicals.

Sunstone

One of the most outstanding minerals of semi-
precious quality is the state gem, the heliote or sunstone, a clear feldspar that varies in color from yellow
or red to green and blue. The color relates to the

amount of copper in the stone, ranging from a low of
20 parts per million for yellow to 200 parts per million
for red. Heliote is a calcium-rich variety of plagioclase
feldspar known as labradorite. It occurs as large crystals
in basalt from 4 small localities in Oregon, all of which
are privately owned except for one locality in Lake
County on Bureau of Land Management lands. Here
these gems are referred to as "Plush diamonds" because
of the nearby town of Plush. In Harney County near
Hines, the sunstones are being mined in commercial
amounts by private companies. Some of the stones are
priced over $1000 per caret.

Geothermal Resources

The Basin and Range is a region of high
temperature gradients, numerous faults, and volcanic
activity which make it a province of intense geothermal
phenomena. Because heat flow varies considerably from
place to place as it moves through the earth's interior,
higher temperatures are found in the mountains, while
lower temperatures characterize the valleys here.
Geothermal sources result when groundwater, flowing
along fault zones, encounters rocks which are abnor-
mally hot. Using the faults as conduits, the heated
waters often reach the surface as hot springs.

The area around around Vale, the Alvord
Desert, Lakeview, and Warner Valley have characteristically high heat flow records. The Vale hot springs
vicinity has a particularly high potential for geothermal
resources. In this zone of thermal activity, surface and
near surface springs and hot water wells have a temperature range as high as 200 degrees Fahrenheit. South
of Vale along the Owyhee River canyon, natural
mineral springs with waters up to 120 degrees Fahrenheit form a pool that is used by the public for bathing.

Along with the Vale region, the Alvord Valley
is thought to have significant geothermal potential.
Several thermal springs emerge south of Alvord Lake
along the north-south Steens fault zone. One of these,
Borax or Hot Lake, is a large pool 800 feet across
formed by discharging water with a temperature of 97
degrees Fahrenheit. The water contains borax salts
which have built up a low, mile-wide dome. It has been
estimated that water from two of the largest springs
here deposits enough salts to produce 150 tons of boric
acid a year. In the late 1800s the land and springs were
purchased for $7000, and the 20-Mule Team Borax
Company, later called the Rose Valley Borax plant,
began to mine and remove the borax. For 10 years
borax was taken out to the railhead at Winnemucca,
Nevada, with wagons drawn by teams of mules. Borax
operations ceased by 1902, but ruins of the old buildings can still be seen.

On the extreme northern edge of the Alvord

Diatomite processing mill near Christmas Valley in northern Lake County (photo Oregon Dept. Geology and Mineral Industries)

basin, Mickey Springs discharges boiling water through vents into several pools and a boiling mud pot. In the same region, the Alvord Hot Springs group has an average temperature of 169 degrees Fahrenheit and a flow rate of 132 gallons per minute. Many of the Alvord basin hot springs contain sodium bicarbonate and sodium chloride along with concentrations of fluoride, arsenic, and boron compounds.

One interesting feature seen on Alvord Lake and Guano Lake to the west are gigantic fissures in polygonal patterns ranging from 50 to 1,000 feet in size and up to 15 feet deep. These huge mud cracks develop in clay layers of the dry lake beds once the water level has dropped enough to cause shrinkage and contraction at depths. The fissures form irregular polygon shapes so large that they may not be obvious when standing nearby on the ground. The mud cracks on Guano Lake are up to 100 feet across making them the largest polygon structures of this type known in North America.

Hot springs are common a few miles north of Lakeview and in Warner Valley where a 50-foot wide zone of faults bordering Warner Peak connects to subterranean geothermal sources. Calcite, gypsum, and several other alkaline minerals precipitate as a white crust along fractures where hot springs reach the surface. Several of these warm mineral springs, where temperatures average 100 degrees Fahrenheit, are used as public facilities. Hunters Hot Spring north of Lakeview has been drilled and cased. It erupts in a geyser-like pulsating manner, spurting about every 30 seconds from levels only a few feet above the well casing to 40 or 50 feet in the air. Issuing from the ground at 180 degrees Fahrenheit, the waters and surroundings have been commercially developed.

The Crump Geyser, which first erupted in 1959, is located a short distance north of Adel in Warner Valley. It is a drilled well that formerly exploded with considerable violence.

A geyser at Hunter Hot Springs located north of the city of Lakeview began erupting in October, 1923, while a well was being drilled at this site (photo Oregon Dept. Geology and Mineral Industries).

Features of Geologic Interest

Jordan Craters

Although several areas of Malheur County have dramatic volcanic features, the Jordan Craters lava beds approximately 36 miles southwest of Adrian are of prime interest because of the recent volcanic events that took place here. The 28 square miles of Jordan Craters lava field has been designated a Natural Resource Area in order to preserve its unique character.

Before volcanism began, the flat plateau, now occupied by the lava field, was part of the Cow Creek drainage. Lying in a depression on the Owyhee Plateau, the first to erupt was a ridge of small spatter cones in the northwestern corner of the field, aligned along a fault trending 1,000 feet toward Coffeepot Crater. This activity between 4000 and 9000 years ago was followed by basalt flows from Coffeepot Crater, one of the youngest and largest of the Jordan Craters complex. Basalts covered about 28 square miles mainly southeast

of the main crater. The extremely fluid pahoehoe lava from the eruptions produced a lumpy, ropy surface of ridges and pits as well as deep fractures and channels in the lava crust. One large section of the crater wall was torn away by explosions, and after activity ceased volcanic debris filled much of the remaining spectacular 260 feet deep crater.

The flowing lava followed the drainage pattern of Cow Creek, developing a meandering lava tube system deep within the flow. Once the liquid magma had drained away, a hollow tunnel was left within the cooled lava crust. Two large pits, north of the main crater open into the lava tube system. Immediately south of the Jordan Craters, Clarkes Butte, Rocky Butte, and Three Mile Hill are small shield cones that sent forth lavas about the same time as the Jordan cones.

Saddle Butte Lava Tube Caves

Volcanism from Saddle Butte tuff cones that erupted just after the Steens basalt produced sinuous lava tube caves near Burns Junction in Malheur County, a short distance southwest of Jordan Craters. This group of caves can be traced for 8 1/2 miles through the Saddle Butte lava field. At one time the caves were a continuous long chain of interwoven tunnels before parts of the tube failed. Collapsed sections of the cave create winding deep trenches in the lava field, the longest measuring well over one-half mile in length.

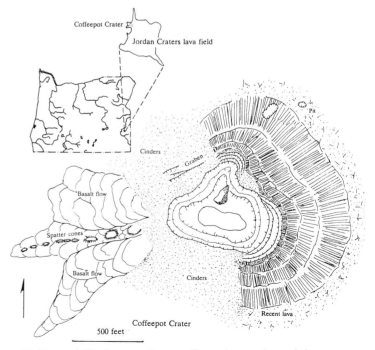

Coffeepot Crater, on the northwestern edge of Jordan Craters in Malheur County, produced most of the lava in the field (after Otto and Hutchinson, 1977).

Sucker Creek wends its way through the rugged canyon it has cut through the centuries in the rock of the Owyhee Mountains of southeastern Oregon (courtesy Oregon State Highway Department).

The sinuous lava tube system of the Saddle Butte
lava field includes caves and collapse trenches.

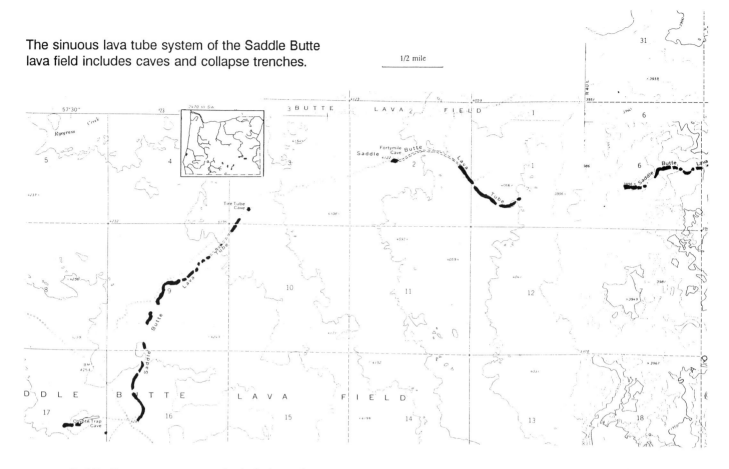

Saddle Butte caves are particularly hazardous
for exploration because of the tendency for roof
portions to fall. Walking across the thin lava shell over
a tube is dangerous, and none of the caves are safe to
enter.

Owyhee River and Succor Creek Canyons

The Owyhee plateau rises from 2,100 feet
above sea level where the Malheur River enters the
Snake to 6,500 feet at the top of Mahogany Mountain.
Dissecting the dry desert plateau of southeast Oregon,
canyons of the Owyhee, Malheur, and Snake rivers, as
well as the smaller creeks have been deeply cut into
some of the most scenic landscape in Oregon. Rock
formations exposed in walls of the river canyons tell
the geologic story of this region from Miocene volcanism that took place 15 million years ago through the
Pleistocene Ice Ages only a few thousand years in the
past. The oldest rocks in the Owyhee canyon are
colorful ash and lavas of the Sucker Creek Formation
violently expelled from fissures on the plateau during
the middle Miocene. Much of the ash was washed
down the hillsides to be deposited in stream valleys.
Ash and thick layers of basalt were covered over by
additional accumulations of lava. The fine-grained
layers of the tuffs are usually thin and nearly white,
whereas the coarse-grained, massive layers of fresh

unweathered tuffs are greenish. Weathered exposures
display shades of yellow, brown, or reddish brown
common on the badland topography. Particularly
striking along the river and at the reservoir are the
sheer 100-foot high cliffs of pink rhyolitic lava. Construction of the Owyhee Dam covered the vents from
which these lavas originated.

Between episodes of intermittent volcanic
eruptions, animals and plants flourished here throughout the Miocene. Fossil shells, bones, and plant remains, especially those of the famous Sucker Creek
flora, give clues to the Miocene environment. Beautifully preserved leaves of oak, birch, maple, willow, pine,
laurel, and sycamore signify a moist temperate climate
of more than 20 inches of rainfall annually in contrast
to the arid environment today. Ancestral horses, deer,
camel, and the pronghorn antelope fed on the extensive
grasslands and forests. Elephants lived on the wooded
slopes, and aquatic rhinoceroses along with giant
beaver inhabited the wetlands.

A brief period of erosion and uplift that produced low mountains was followed by thick sequences
of Owyhee basalts interbedded with red and brown
tuffs. This eruption blanketed more than 1,000 square
miles with hot volcanic material about 15 million years
ago. The main layers of Owyhee Basalt are visible near
Mitchell Butte, below the Owyhee Dam, at Hole-in-

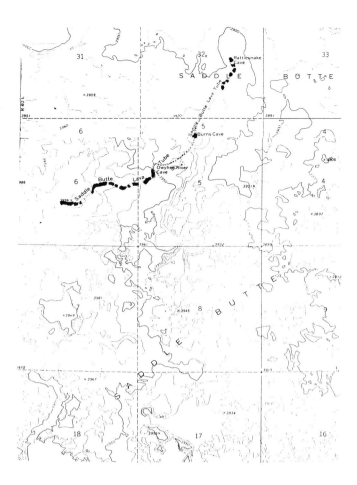

canyons.

The Owyhee was dammed in 1932 to back up the waters into a meandering 52 mile long lake. The top of the dam is 417 feet above the foundation, and the crest is 810 feet long. Developed as a state park in 1958, the surrounding 730 acres and river canyon with its colorful formations carved into a variety of erosional forms offer a geologic picture quite different from any other in Oregon.

A giant beaver, 7 1/2 feet long, lived in Oregon during the Pleistocene

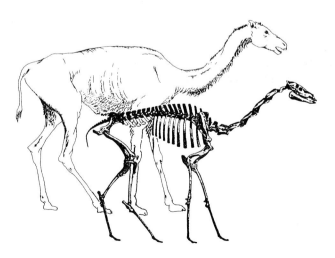

The long-limbed Miocene camel of eastern Oregon

the-Ground, a 5 mile wide basin just above the dam, and as flows forming the surface of Owyhee Ridge. Numerous dikes on the east wall at Hole-in-the-Ground are only a few feet wide where the fluid lava cut through the sediments of the Sucker Creek Formation to fill cracks and fissures.

Between the Miocene and Pliocene, great fault blocks developed creating basins where ash rich sediments, carried in by streams, accumulated with lava flows in a sequence up to 2,000 feet thick. Yellowish, orange, and brown beds of the Deer Butte Formation are exposed extensively in the western part of the Owyhee Reservoir and are responsible for the resistant knobs of Deer Butte, Pinnacle Point, and Mitchell Butte. Entombed in the conglomerates, sandstones, and siltstones in these small basins are fossilized skeletal parts of a wide variety of rodents as well as those of beaver, rhinoceroses, and the small three-toed horse Merychippus that lived on the eastern plains.

During the early Pliocene, 5 million years ago, the climate became dryer, and grasslands were interspersed with small ponds and a community of modern-looking mammals. By this time the Owyhee River had established its present channel. As the entire region was slowly raised to an elevation of over 4,500 feet above sea level, the ancestral streams continued downcutting their channels to produce deep winding

The Chalk Basin along the lower Owyhee River shows lake sediments covered by successive lava flows (courtesy Oregon State Highway Department).

Pillars of Rome and Leslie Gulch

Approximately 4 miles northwest of the small town of Rome in Malheur County, cliffs called the Pillars of Rome have been cut into the deep canyon where Jordan Creek merges with the Owyhee. Here 400 to 500 feet of Pliocene tuffaceous sediments filled the Owyhee River valley during an alluvial period. Since that time, erosion has cut through the beds to create a badland topography of high, vertical pinnacles and pillars that resemble Roman ruins.

Another outstanding feature produced by erosion in this dry plateau are the jagged peaks and columns at Leslie Gulch approximately 23 miles directly south of Owyhee Dam. The narrow 10 mile long canyon is traversed by a steep road that requires the frequent use of brakes when descending from an elevation of 4,600 feet to the Owyhee reservoir at 2,600 feet. Cut into Leslie Gulch ash-flow tuff, brown, cream, and white minarets and spires extend throughout the gorge. The stream-deposited tuffs are more than 2,000 feet thick in places and cover an area of about 100 square miles. The multi-colored volcanics were produced from eruptions of Mahogany Mountain, the oldest caldera in the Lake Owyhee volcanic field, 15.5 million years ago.

Suggested Readings:

Allison, Ira, 1982. Geology of pluvial Lake Chewaucan, Lake County, Oregon. Oregon State University, Studies in Geology, 11, 78p.

Blackwell, David, et al., 1978. Heat flow of Oregon. Jour. Geophys. Res., v.87, no.B10, pp.8735-8754.

Bowen, R.G., Peterson, N.V., and Riccio, J.F., 1978. Low-to intermediate-temperature thermal springs and wells in Oregon. Oregon Dept. Geol. and Mineral Indus., Geol. Map. Ser. GMS-10.

Castor, Stephen B., and Berry Michael R., 1981. Geology of the Lakeview uranium district, Oregon. In: Goodell, Philip, and Waters, Aaron, Uranium in volcanic and volcaniclastic rocks. Amer. Assoc. of Petroleum Geol., Studies in Geology, v.13, pp.55-62.

Christiansen, Robert L., and McKee, Edwin H.,1978. Late Cenozoic volcanic and tectonic evolution of the Great Basin and Columbia intermontane regions. In: Smith, Robert, and Eaton, Gordon P., eds. Cenozoic tectonics and regional geophysics of the western Cordillera. Geol. Soc. Amer., Memoir 152, pp.283-311.

Corcoran, R.E., 1965. Geology of Lake Owyhee State Park and vicinity, Malheur County. Ore Bin, v.27, no.5, pp.81-98.

Dicken, Samuel N., 1980. Pluvial Lake Modoc, Klamath County, Oregon, and Modoc and Siskiyou Counties, California. Oregon Geology, v.42, no.11, pp.179-187.

Eaton, Gordon P., 1984. The Miocene Great Basin of western North America as an extending back-arc region. Tectonophysics, v.102, pp. 275-295.

Kittleman, Laurence R., 1973. Guide to the geology of the Owyhee region of Oregon. University of Oregon, Museum of Natural Hist., Bull.21, 61p.

Lawrence, Robert D., 1976. Strike-slip faulting terminates the Basin and Range province in Oregon. Geol. Soc. of Amer., Bull., v.87, pp.846-850.

Mankinen, Edward A., et al., 1987. The Steens Mountain (Oregon) geomagnetic polarity transition 3. Its regional significance. Jour. Geophys. Res., v.92, no.B8, pp.8057-8076.

McKee, E.W., Duffield, W., and Stern, R., 1983. Late Miocene and early Pliocene basaltic rocks and their implications for crustal structure, northeastern California and southcentral Oregon. Geol. Soc. Amer., Bull., v.94, pp.292-304.

Otto, Bruce R., and Hutchinson, Dana A., 1977. The geology of Jordan Craters, Malheur County, Oregon. Ore Bin, v.39, no.8, pp.125-140.

Rodgers, David W., Hackett, William R., and Ore, H.T., 1990. Extension of the Yellowstone plateau, eastern Snake River plain, and Owyhee Plateau. Geology, v.18, pp.1138-1141.

Rytuba, James J., 1989. Volcanism, extensional tectonics, and epithermal mineralization in the northern Basin and Range province, California, Nevada, Oregon, and Idaho. U.S. Geol. Survey, Circular 1035, pp.59-61.

-----and McKee, Edwin H., 1984. Perkaline ash flow tuffs and calderas of the McDermitt volcanic field, southeast Oregon and north central Nevada. Jour. Geophys. Res., v.89, no.B10, pp.8616-8628.

-----et al., 1990. Field guide to hot-spring gold deposits in the Lake Owyhee volcanic field, eastern Oregon. Geol. Soc. Nevada and United States Geological Survey 1990, Spring field trip guidebook; field trip no.10, 15p.

-----Glanzman, Richard K., and Conrad, W.K., 1979. Uranium, thorium, and mercury distribution through the evolution of the McDermitt caldera complex. In: Newman, Gary W., et al., eds., Basin and Range Symposium and Great Basin field conference, Utah Geological Assoc., pp.405-412.

Smith, Robert, 1978. Seismicity, crustal structure, and intraplate tectonics of the interior of the western Cordillera. In: Smith, Robert, B., and Eaton, Gordon, eds., Cenozoic tectonics regional geophysics of the western Cordillera. Geol. Soc. Amer., Memoir 152, pp.111-142.

Walker, George W., 1979. Revisions to the Cenozoic stratigraphy of Harney Basin, southeastern Oregon. U.S. Geol. Survey, Bull.1475, 34p.

-----and Swanson, Donald A., 1968. Summary report on the geology and mineral resources of the Harney Lake and Malheur Lake areas of the Malheur National Wildlife Refuge north-central Harney County, Oregon. U.S. Geol. Survey, Bull.1260-L,M, 17p.

Wallace, Andy B., and Roper, Michael W., 1981. Geology and uranium deposits along the northeastern margin, McDermitt caldera complex, Oregon. In: Goodell, P.C., and Waters, Aaron C., Uranium in volcanic and volcaniclastic rocks. Amer. Assoc. Petroleum Geologists, Studies in Geology No.13, pp.73-79.

Wells, Ray E., and Heller, Paul L., 1988. The relative contribution of accretion, shear, and extension to Cenozoic tectonic rotation in the Pacific Northwest. Geol. Soc. Amer., Bull.110, pp.325-338.

Zoback, Mary Lou, Anderson, R.E., and Thompson, G.A., 1981. Cainozoic evolution of the state of stress and style of tectonism of the Basin and Range province of the western United States. Philos. Trans., Royal Soc. London, 300, pp.407-434.

In this view of the south rim of Newberry volcano, Big Obsidian Flow is accentuated by a recent snowfall [photo by Martin Miller].

High Lava Plains

Physiography

The High Lava Plains physiographic province, shaped roughly like a rectangle 50 miles wide and 150 miles long, is situated near the geographic center of Oregon. A high plateau averaging just over one mile above sea level, it is bordered by three other provinces, the Blue Mountains to the north, the Basin and Range and Owyhee Uplands to the south and east, and the Cascade Range on the west. Within the narrow rectangle, Newberry Crater and Harney Basin form the west and east boundaries respectively.

The overall topography, as the name implies, is smooth, with moderate relief. Elevation ranges from a high of 7,984 feet at Paulina Peak to 4,080 feet above sea level in Harney Basin. The lack of deep canyons and gullies as well as a poorly developed network of streams are due to the low rainfall. Precipitation of only 10 to 20 inches a year in the plains regions is the result of the Western Cascade rainshadow. This minimal moisture has created an area of little vegetation which makes the High Lava Plains an outstanding region to see evidence of recent volcanic and tectonic activity producing features that are still "fresh."

The headwaters of the Deschutes River are along the western edge of the province, and most tributary streams are seasonal with poorly defined channels. Malheur and Harney lakes, the largest in the plains, collect the drainage of the Silvies River and Silver Creek from the north and the Donner and Blitzen River from the Steens Mountain to the south. These undrained basins contain playa lakes at certain times of the year. At the center of the broad Harney Basin, Malheur and Harney lakes are now nearly dry.

Geologic Overview

The High Lava Plains is a province of remarkable volcanic features, most of which are the result of relatively young eruptions. A multitude of volcanic cones and buttes, lava flows, and lava tube caves are scattered throughout the province. Except for deposits of talus, lake sediments, and fluvial debris, most of the rocks in the province are volcanic, and thick flows are not unusual.

Volcanic eruptions of lava and ash beginning 10 million years ago continued into the Recent when

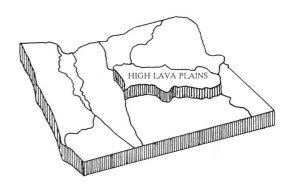

vents along a zone of faults erupted with pumice, ash, and cinders along with thick viscous lava. The volcanic activity here relates to a broad zone of faults and fractures running across the province that resulted from two enormous, underlying crustal blocks wrenching past each other. Molten lava reached the surface by way of the cracks to ooze onto the surface as large and small volcanoes which are the hallmark of the province. These volcanic events produced some of the most interesting features of the High Lava Plains, where the immense shield cone containing Newberry Crater, small cinder cones, tuff rings, and explosion craters stand out on the flat plateau. Many recent lava flows enclosing trees created the unique Lava Cast Forest, while lava caves formed in hollow tubes within the cooling lavas.

Lakes that characterized the Great Basin during the Pleistocene filled many of the large depressions of the Lava Plains. With the Ice Ages broad expanses of the flat plains were covered with continuous shallow-water lakes that served as a habitat for mammals and birds. Throughout the province, these basins received both fluvial sediments and ash from contemporaneous volcanism until drying conditions reduced them to the playas found today.

Geology

Structurally the High Lava Plains blends with the Blue Mountains to the north and the Basin and Range to the south, but its volcanic characteristics set it apart. The oldest rocks exposed in the High Lava Plains province are Miocene lavas. Five to 10 million years ago the landscape here was dotted with erupting volcanoes and slow moving thick lavas spreading over the flat surface. One eruption followed the other almost continuously for millions of years. Eruptions aligned themselves in a broad belt of overlapping faults, known as the Brothers fault zone, the dominant structural feature of the High Lava Plains and central Oregon. The zone runs for 130 miles from Steens Mountain in southeastern Oregon to Bend. Within the Brothers fault zone, individual faults are irregularly spaced a quarter to 2 miles apart with modest displacements of less than 50 feet. Over 100 separate rhyolite volcanic centers are located along the belt of faults where the silica-rich lavas have exploited the fractures and fissures as avenues to reach the surface.

The Brothers fault zone was generated by the same forces that twisted Oregon in a clockwise motion throughout the Cenozoic era. Large tectonic blocks share a zone of weakness running north-south through central Oregon. As the blocks move relative to each other, the eastern block moved south and the western block moved north. Caught in the middle, central Oregon was distorted by wrench faulting expressed on the surface as the wide zone of faults. Most of the faults along the zone are so recent that they are easily seen in aerial photographs. Similar large-scale wrench faults in Oregon following the same northwest-southeast trend are the Eugene-Denio and Mt. McLoughlin faults to the south in the Basin and Range province.

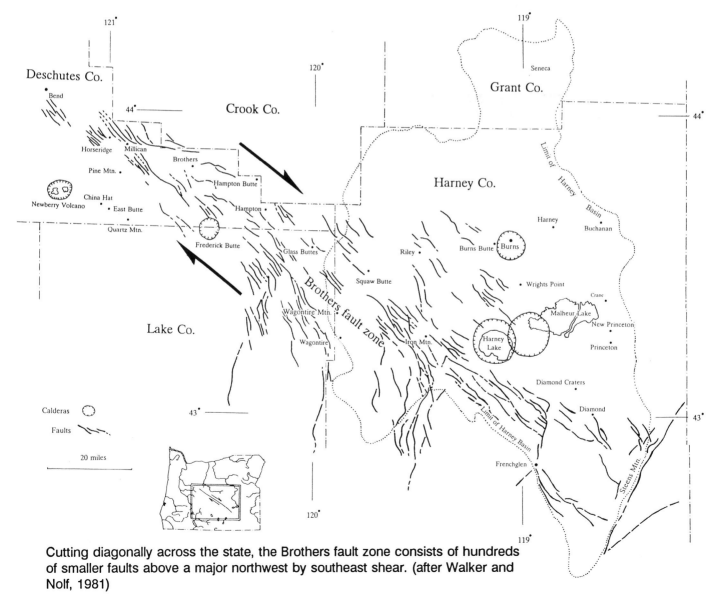

Cutting diagonally across the state, the Brothers fault zone consists of hundreds of smaller faults above a major northwest by southeast shear. (after Walker and Nolf, 1981)

Newberry Crater, at the western edge of the province, is situated at the apex of three converging fault zones, the easterly Brothers fault zone, the northerly Green Ridge and Sisters fault zones, and the southwesterly Walker Rim fault zone. Placed at the center of these fracture patterns, it is little wonder that Newberry volcano formed as a separate, younger eruptive site well to the east of the High Cascade vents. Today a small magma chamber may be less than two miles below the caldera. Although it is often suggested that Newberry Crater may again erupt, its blanket of rhyolitic rocks tends to diminish that possibility, because silica-rich rhyolitic lavas ordinarily appear very late in the life cycle of a volcano.

Lavas of the High Plains are distinctly bimodal. That is, they have strikingly different compositions varying from dark-colored basalt to light-colored rhyolite. Basaltic lavas tend to have a deeper source in the crust and are extremely hot, while the lower temperature rhyolitic lavas are from chambers at shallow depths. In addition, basalts are generally an early stage of eruption, while rhyolites appear late in

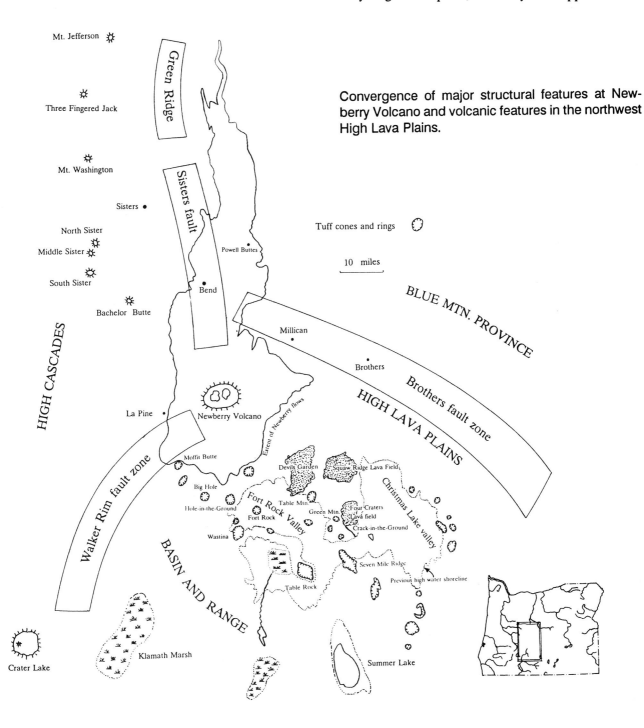

Convergence of major structural features at Newberry Volcano and volcanic features in the northwest High Lava Plains.

the cycle. In this province, the very fluid basaltic lavas predominate, but rhyolitic extrusions and domes are situated along major fracture zones. The association of basalt and rhyolite is is rare and usually occurs where the earth's crust thins because it is undergoing tension and being stretched.

Comparatively young eruptions and intrusions here between the upper Miocene and Recent occur in a broad northern belt of approximately 100 centers trending northwest across the Lava Plains and Owyhee Uplands. One of the most striking aspects of the High Lava Plains is the uniform decrease in the age of these volcanic eruptive centers geographically from east to west. Within the Harney Basin in the east, rhyolitic eruptions date back to 10 million years ago in the late Miocene, while near Newberry Crater in the west many lavas were extruded less than 1 million years ago. The eruptive zone moved steadily from the southeast toward the northwest at slightly more than one mile per 100,000 years. Such a progression of eruptions might be seen as an earth crustal plate moving over a hot spot, but two important aspects of local geology seem to preclude this notion.

First, it is well established that the North American plate, upon which this province rests, has moved progressively westward since well before Miocene time. The age progression should then be reversed with younger volcanics appearing in the east instead of in the west, as is the case with the Yellowstone hot spot. Additionally, another broad belt of rhyolitic and silicic domes to the south in the Basin and Range province between Beatys Butte in the southeast and

Age progression of silicic volcanic centers in the High Lava Plains and Basin and Range from the oldest (upper Miocene) in the southeast to the youngest (Pleistocene and Holocene) in the northwest. Contoured lines of simultaneous volcanic activity are in millions of years. (modified after Walker and Nolf, 1981).

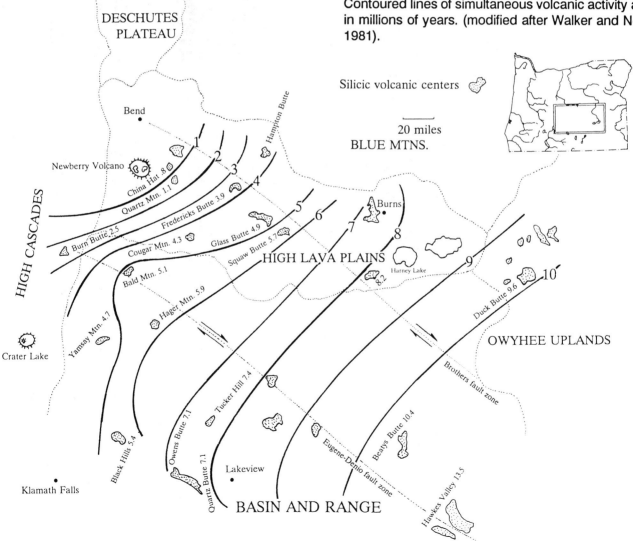

Yamsay Mountain to the northwest displays a similar age progression from east to west suggesting the deep magmatic source feeding these volcanic centers is a wide linear front. This age progression could relate to the clockwise rotational movement of central Oregon resulting from plate movement and Basin and Range extension. As the Basin and Range crust thins and stretches, differential movement with a center of rotation to the south could yield this age progression. More than likely, however, the age progression may be due to a steady steepening of the subduction zone known to be below eastern Oregon at this time. As the subduction of the Farallon plate slowed down, the descending slab angled more steeply below the North American plate. The eruptive front then moved westward which maintained the usual 90 mile distance between the melting rocks and eruptive volcanoes on the surface.

Pleistocene Lakes

During the late Pleistocene, up to 11,000 years ago, large and small lakes characterized the Great Basin in Oregon, Nevada, and Utah. Within the Great Basin, tensional faulting yields a basin and range topography, where today the shallow basins are either lakes or dry playas. The southern portion of the High Lava Plains merges gradually with the northern Great Basin and includes such lakes as those which once occupied the dry basins of Fort Rock Valley, Christmas Lake Valley, Silver Lake, and Fossil Lake situated south and southeast of Paulina Mountains. Prehistoric Fort Rock Lake, covering 1,400 square miles, was the largest of these followed by Malheur Lake at 900 square miles.

The sediments of ancient lakes in the High Lava Plains provide an outstanding record of Ice Age mammals as well as the activities of early man. Filled with broad expanses of water during the Pleistocene, the lakes and shores served as a habitat for mammals and waterfowl of every description. Pollen and plant fossils recovered from the former lake sediments reflect a lush Pleistocene vegetation that supported these large animals living along the margins of the lakes. Today these lake sands yield not only fossil vertebrate remains but worked flint tools and pottery from the Indians who hunted there. Early accounts mention that wagon loads of fossil bones were collected and removed from Fossil Lake where bird, fish, and mammal bones all have the distinctive, shiny black patina of desert varnish.

During highwater periods of the Pleistocene, Silver Lake, Christmas Lake, Fossil Lake, and Fort Rock Lake were interconnected, the contiguous broad lake reaching a maximum depth of over 200 feet. Some

of the higher elevations as Fort Rock were islands. During more arid intervals, islands became peninsulas, arms of the lake became bays, and with further drying bays became isolated marshes and swamps. Blowing winds made shallow depressions that became ponds during wetter, cooler climatic stages.

Old shorelines of Fort Rock Lake are easily recognized today by tracing gravels and beach erosion, however, 30,000 year old shorelines are obscured on the north by younger fresh lava flows in the vicinity of the Devils Garden. Postulated maximum shorelines are several hundred feet lower than river divides to the north. If the lake drained north to the Columbia, the former channel is probably covered by the younger lavas. Bones of the andromous salmon and the presence of a small snail, Limnaea, found at Fossil Lake, are known only in Columbia River drainage and point to the existence of a former outlet to the north. Dry River, that occupies a narrow canyon at the east end of Horse Ridge along Highway 20, may have been carved at the time some of the interior basins, perhaps Fort Rock Lake, overflowed making its way to Crooked River and then by way of the Deschutes River to the Columbia.

Harney Basin

The largest closed depression in the province, Harney Basin, is situated on the southeast corner of the High Lava Plains. Sitting directly upon the Brothers fault zone, the basin extends north and south well beyond the limits of the Lava Plains province. At 5,300 square miles, the depression is larger than the state of Connecticut. Harney Basin began to evolve as a downwarp assisted by large calderas which collapsed within the depression into an evacuated magma chamber. Large-scale eruptions of as much as 500 cubic miles of rhyolite accompanied the development. Today there is no surface outlet to the basin, and Harney, Malheur, and Mud lakes occupy the central southern part of the depression.

Throughout the later Cenozoic, the Harney Basin received lava flows, ash flow tuffs, and tuffaceous sediments derived from the surrounding volcanic activity. Miocene basalt from Steens Mountain was followed by air-borne pyroclastic material and ash-flows of the Danforth and Harney formations. Ash-flow tuffs of the former Danforth have been divided into the Devine Canyon, Prater Creek, and Rattlesnake members that form distinctive stratigraphic layers that can be traced and easily recognized over vast regions of southeast Oregon. The greenish-gray Devine Canyon extends from Steens Mountain in the southeast as far as Paulina Valley in the northwest, whereas the Prater Creek ash-flow tuff is limited to exposures in the

Harney Basin. The remarkable Rattlesnake ash-flow tuff, which extends to the vicinity of John Day 90 miles to the north, also has source vents in the Harney Basin. These formations are covered by alluvium and lake deposits eroded largely from the volcanic rocks of the adjacent uplands. At the outer margins of the basin, sand and gravel predominate, while silts and clays fill the center. Younger lavas from Diamond Craters along with some Pleistocene basalts cover these sediments. Within these deposits thin but well-defined layers of 6,900 year old Mazama ash lie three to six feet below the surface.

Shallow, ephemeral Malheur and Harney lakes, and smaller Mud lake which connects the two, are located in the south central portion of Harney Basin. During wetter conditions existing in the Pleistocene, a huge body of freshwater, ancient Lake Malheur, occupied the basin and extended all the way to Burns. The lake originally drained to the east along the south fork of the Malheur River and from there to the Snake River until the outlets near Princeton and Crane were dammed by a lava flow. As warmer, dryer conditions prevailed, about 10,000 years ago, the lake rapidly shrank in size creating the present series of smaller lakes, playas, and marshes. Lake Malheur, at its highest water level, was probably never more than 50 feet deep.

Even historically Malheur, Mud, and Harney lakes were not always joined. Explorer Peter Skene Ogden observed in 1826 that "*a small ridge of land, an acre in width, divides the freshwater from the salt lakes*". About 1880 Malheur Lake topped the ridge and overflowed into Harney Lake. Today these lakes frequently double or triple in size during wet spells or decrease to virtually nothing in a dry period. Flooding during three years of record rainfall in 1984 caused the three lakes, normally covering 125,000 acres, to expand to 175,000 acres. Periods of extreme drought are not abnormal to these lakes, and prior to 1930 they became virtually dry due to low rainfall.

Malheur, Harney, and Mud lakes have been designated as the federal Malheur National Wildlife Refuge, a home for thousands of wild birds, native plants, and animals.

Geothermal Resources

Percolating from deep within the crust through cracks and fractures, heated water reaches the surface as warm springs. Scattered over the Harney Basin, almost all thermal springs are aligned along faults. Although some springs record much higher readings, the average temperature of the waters is between 60 and 82 degrees Fahrenheit. Thermal waters, ranging from 64 degrees to 154 degrees Fahrenheit issue from

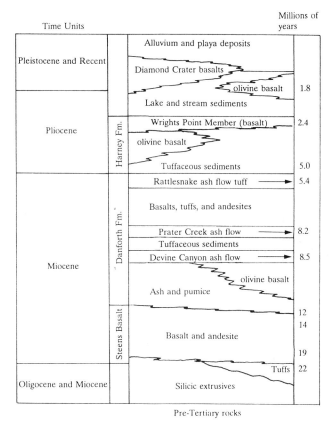

Correlation chart of High Lava Plains stratigraphic units

a number of perennial springs near Hines, Warm Springs valley east of Burns, and around Harney and Mud Lakes to the south. Water temperatures near Hines reach a high of 82 degrees Fahrenheit with temperatures diminishing outward from there. Some of the highest temperatures occur in spring waters southwest of Harney Lake where 90 to 154 degree Fahrenheit waters indicate the presence of a thermal system. Several springs here have commercial facilities, and one, Radium Hot Springs, popular since the 1890s, is the largest warm water pool in the state.

Features of Geologic Interest

Fort Rock, Hole-in-the-Ground, Big Hole

Distributed widely on the High Lava Plains, maars and tuff rings resulted when upward moving lava in the crust encountered underground water with devastating effects. Upon contact with the cooler water, the hot lava exploded catastrophically, creating a symmetrical, circular crater. Rocks and ash from the eruption, thrown into the air, settled close to the crater building up a high rim or tuff ring. Saucer-shaped maars are shallow explosion craters that are fed by a

In northern Lake County, Fort Rock is a tuff ring eroded by waves of a shallow Pleistocene lake that once occupied this basin (photo courtesy Oregon Dept. Geology and Mineral Industries).

"diatreme" or funnel-shaped vent which ultimately fills with angular pieces of volcanic breccia. Often these surface landforms are covered or eroded and not fully preserved. Three concentrations of tuff rings and maars in Oregon are located in northern and southern Lake County and in southern Klamath County. The best-known of these are Fort Rock, Hole-in-the-Ground, and Big Hole.

In Lake County, Fort Rock is a crescent-shaped tuff ring 1/3 mile across and 325 feet above the surrounding flat plains. Formed by a shattering explosion in early Pleistocene time, the crater has steep rock sides resembling a fort or castle. Once past the initial violent stage, the dying crater lacked the force to expell the ash, and much of it fell back within the ring where the yellowish and brown tuffs dip inward toward the center. Broken rock, ash and explosion tuffs that make up the crater walls are testimony to the extreme force of the Fort Rock event. The unusual shape of Fort Rock, with the wide breach in the south rim, was due to wave action from a former lake here eroding the thin walls of the ring.

Hole-in-the-Ground, a short distance north of Fort Rock in Lake County, is an explosion crater of remarkable symmetry. The pit is almost a mile in diameter, and the floor is over 300 feet below the surrounding land level. The bowl-shaped maar probably resulted from a single or very brief series of violent eruptions over a short period of time. Fine rock material expelled by the explosion forms the crater ring and floor. Drilling in the late 1960s has revealed buried massive blocks of broken rock which fell into the maar after the explosion.

To the northeast, Big Hole, a little over one mile in diameter is a circular maar created by a similar violent explosion. A wide ledge within the basin may be

Hole-in-the-Ground, near Fort Rock, is a remarkable explosion crater (photo courtesy of Oregon Dept. Geology and Mineral Industries).

the top of a large block which fell into the depression during the eruption. Unlike Fort Rock and Hole-in-the-Ground, Big Hole is heavily forested. The trees and its immense size make this volcanic feature difficult to visualize from the ground level.

Four Craters, Devils Garden, and Squaw Ridge Lava Fields

Continuing Pleistocene volcanic activity in the eastern High Lava Plains resulted in a number of unusual lava flows, domes, pumice and cinder cones, and lava tubes at three locations in Lake County. At Four Craters Lava Field on the northern edge of Christmas Lake Valley, four cinder cones with smaller mounds are surrounded by lava flows emitted from vents aligned along a fissure. Here blocky and broken aa lava from the cones covers about 12 square miles. Nearby Devils Garden and Squaw Ridge lava fields are

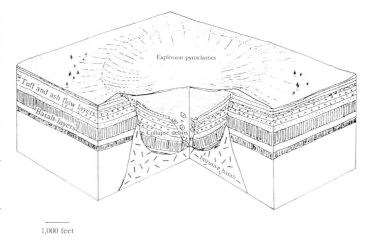

At Hole-in-the-Ground, the collapse blocks and crater rim are overlain by a thick layer of ash and tuff. (modified after Peterson and Groh, 1961)

large areas of clinkery chunks of broken aa lava and black swirly pahoehoe lava. Oozing from fissures, several large spatter cones were built up by cooling lava. Flows from the shallow Squaw Ridge cone covered over 200 square miles, while the adjacent Devils Garden is 45 square miles of lava, rubble, ash, basaltic bombs, and lava tube caves. Numerous small caves were formed by molten lava streams. In one of these, Derrick Cave, a collapsed roof forms the entrance. At spots along the tunnel of the cave, the roof rises as high as 50 feet. Benches inside the lava tube give the cave a keyhole-shape in cross-section to show that the lava drained out slowly enough to leave behind a lining along the wall.

The Devils Garden offers 45 square miles of rough black lava flows, spatter cones, and lava tubes in northern Lake County.

Diamond Craters

To the south and east of the lava fields near Bend, Diamond Craters in Harney County is one of the most dramatic of Oregon's volcanic attractions, displaying a varied landscape produced by late Miocene to Pleistocene eruptions. Named for the diamond-shaped cattle brand of an early settler, the craters are easy to reach. Within a 22 square mile natural area south of Malheur Lake, over 100 cinder cones and craters can be seen, 30 of which are located inside a 3,500 foot wide caldera which has collapsed to a depth of 200 feet.

About 9 million years ago vents here pumped out ash and lava to cover 7,000 square miles of southeast Oregon in layers up to 130 feet thick. This initial activity was followed by a series of explosions and intrusions 2,500 years ago forming small cinder cones accompanied by vast new lava flows. A surge of mag-

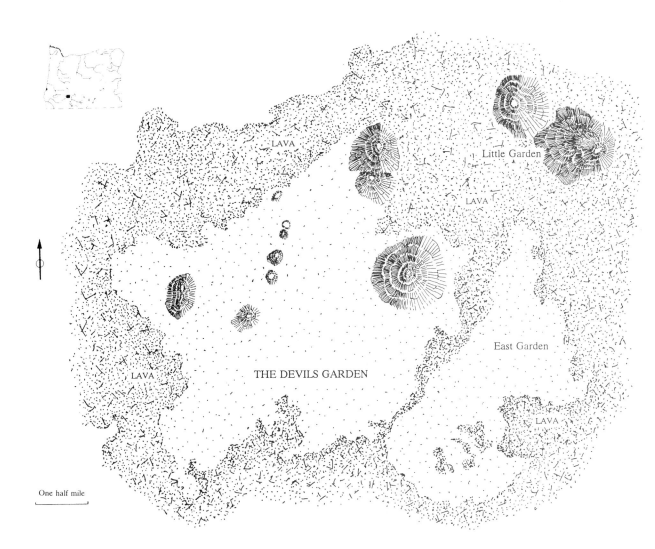

One half mile

LAVA

Little Garden

LAVA

East Garden

THE DEVILS GARDEN

LAVA

LAVA

The central crater complex about one mile in diameter is a moonscape of overlapping volcanic features (photo courtesy Oregon Dept. Geology and Mineral Industries).

matic intrusions bowed the ground surface into several low rounded domes. One of the eruptions produced spherical-shaped volcanic bombs along with flows of pahoehoe lava that created deep fissures and ridges. These bombs ranging from the size of a pea up to 2 feet in diameter were ejected as air borne blobs of liquid lava that cooled and cracked in flight. A more dramatic result of these volcanic events is the large depression or Graben Dome, 7,000 feet long, 1,250 feet wide, and 100 feet deep that developed as lava drained away underground causing the upper surface of the crater to collapse and create a basin.

Volcanic Buttes, Lava Cast Forest, Newberry Crater

Cinder and basalt cones stand out on the flat landscape near Bend and along the route from Bend to Burns. The most notable of these are Pilot Butte just east of the Bend city limits, Powell Buttes to the northeast, Newberry Crater and Lava Butte to the south, and Glass Buttes to the east.

Pilot Butte, a prominent feature in central Oregon, served as a landmark for pioneers travelling to the Willamette Valley. A view atop the lone cinder cone, 4,136 feet above sea level and 500 feet above the

Diamond Craters Lava Field south of Malheur Lake exhibits a wide variety of volcanic features.

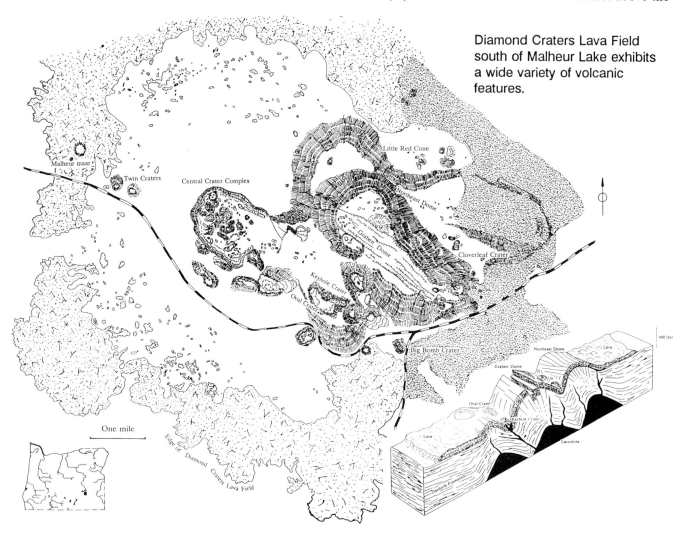

In the Lava Cast Forest on the northwest flank of Newberry Crater, lava has surrounded trees which burned to leave a hollow mold (photo courtesy of Oregon State Highway Department).

surrounding plains, offers an impressive panorama of the Cascades and surrounding buttes of the High Lava Plains, as well as a glimpse of the winding canyon of the Deschutes River. Volcanic debris, discharged from a vent into the air, settled as cinders and ash to build up the symmetrical cone. Pilot Butte has been a state park since 1927.

Nearby Powell Buttes complex of 11 separate volcanic domes represents a major silicic volcanic center that expelled large volumes of lava and ash-flow tuffs during the Miocene. Domes formed during the final stages of activity and subsequent erosion produced remnants known as "hat rocks". Exploratory geothermal tests have located warm water wells on the north side of Powell Buttes.

Lava Butte, familiar to anyone who has travelled south of Bend, is a classic basalt cinder cone projecting 500 feet above the surrounding jagged lava field. Because it is situated on the flank of Newberry Crater, Lava Butte is a "parasite cone" and is part of an almost continuous wide zone of faults and cracks extending from Newberry Crater to Lava Butte and beyond. More than 6,000 years ago, lava extruded from these vents in at least eight separate eruptive events, flowing northwesterly for 6 miles to dam and divert the Deschutes River. From the rim of the deep crater, built

up by cinders, a rugged volcanic panorama covering almost 10 square miles of lava field can be seen.

As successive, thick flows from this volcanic center between Newberry Crater and Lava Butte spread sluggishly across the landscape, pine forests much like those here today were engulfed by the slow moving lava. Some trees remained upright while others fell over after being buried by the molten lava. Once enclosed in the flow, the trees burned slowly leaving a mold the size and shape of the trunk. Contact with the trees chilled the lava in the flow into cylinders around trunks. Once the lava had receded, these cylinders were left standing above the surface flow. The Lava Cast Forest which resulted is encased in the rough black lava.

Newberry Crater, with a large caldera, lakes, obsidian flows and domes, and pumice and cinder cones, provides a view of spectacular volcanic geology in a concise area that is easily accessible. In the spring of 1991, Newberry Volcanoes National Monument was dedicated. This new monument which includes Newberry Crater, Lava Butte, Lava River Cave, the Lava Cast Forest, and Paulina Peak sets aside an area for viewing and studying a wide variety of volcanic features.

A low profile, shield volcano, Newberry is 40 miles long and 20 miles wide and contains Newberry

A cinder cone just south of Bend, Lava Butte has a well-developed crater on top and a vent which extruded lava on the south side (photo courtesy of Oregon Department of Geology and Mineral Industries).

Crater at the summit. Covering 500 square miles, the volcano is among the largest Quaternary volcanoes in the United States. The caldera was caused by collapse or sinking of the center of the mountain in somewhat the same manner that the caldera formed at Crater Lake. Today the central caldera is occupied by East and Paulina lakes which are separated by cinder cones and a massive obsidian flow.

With a comparatively short eruptive life lasting around 1/2 million years, the active period of Newberry volcano spans that of Crater Lake 70 miles to the southwest. About 500,000 years ago, a huge shield volcano, 4,000 feet above the surrounding plateau, was built by successive flows of basalt and silica-rich rhyolite interspersed with layers of volcanic ash, dust, and consolidated tuffs. The summit of the volcano may have collapsed at least twice after two large eruptions to create the caldera. Continued intermittent activity

produced small domes, lava flows, and obsidian 300,000 to 400,000 years ago that was followed by a huge ash eruption about 1600 years ago. Most recent activity began 2,000 years ago when lava discharged from a fracture on the south rim of the caldera spreading about 1 1/2 miles toward the center. This striking Big Obsidian flow is the largest of several young flows in the crater producing over 1/2 cubic miles of frothy pumice, ash, and obsidian as well as the dome marking the vent. A distinctive 6,900 year old layer of pumice and ash from ancient Mt. Mazama is laminated in among Newberry's own eruptive debris. Near the end of the cycle more than a half dozen separate obsidian flows, interlayered with a volcanic froth or pumice, spread out within the caldera.

The long outer slopes of the volcano are sprinkled with over 400 parasitic cinder cones, some in regular lines, indicating alignment along fractures. Nineteen rhyolitic domes, flows, small explosion

Newberry Volcano moument, covering 62,000 acres
south of Bend, provides a panorama of forests, lava
flows, volcanic cones, lakes, and craters.

craters, and other features on the flanks occur in zones
radial to the caldera. Basalt flows, forming most of the
north and south flanks, are younger than the ash flows
of the east and west flank which are covered by a thin
veneer of ash. Paulina Creek drains the caldera,
dropping over the west rim as Paulina Falls to flow
westward into the Deschutes River.

Over the past decade Newberry Volcano has been
the site of intensive investigation for its geothermal
potential, and drilling has encountered an active
hydrothermal system beneath the caldera. The recognition of the delicate ecological balance here should
eliminate any future exploitation at this site.

Glass Buttes east of Newberry Caldera is a 5
million year old silica-rhyolite volcanic center consisting of Glass Butte, Little Glass Butte, eastern Glass
Buttes, and Round Top Butte. The low domes aligned
in a west-northwest direction for about 12 miles built
up in a long eruptive process which may have taken as
much as 1 million years to complete. The Glass Buttes
group produced significant amounts of obsidian which
was gathered by Indians in prehistoric times and more
recently by rock collectors. Besides the common black
obsidian, there are mixtures of red and black as well as
iridescent silver obsidian. Obsidian or volcanic glass
has no internal crystalline structure. The failure to
form crystals may be due to the exceedingly high
viscosity prior to cooling. Very small amounts of
quicksilver have been mined from the volcanics of
Glass Buttes.

In Lava River Cave south of Bend the lava remained molten while the roof and sides solidified. Flowing out, the lava left flow lines, benches, and a keyhole-shape to the sides of the tunnel (photo courtesy of Oregon State Highway Department).

Lava Caves

A number of caves and long cavern systems exist in the Bend area as a result of volcanism which took place during the Pliocene and Pleistocene throughout the western High Lava Plains. Tubes and tunnels in basalt occur as molten lava cools to form a solid crust on the surface even as the hot lava stream continues to run out deep in the flow. As volcanic activity producing the lava ceases, the tube is drained leaving the cave or cavern. Tubes usually form when lava is thick and syrupy, and caves are most common in pahoehoe flows. Wide lava sheets often have several tubes within the flow, some of which may be interconnected by lateral passages. Pahoehoe lava, flowing through a narrow valley, will ordinarily produce only a single tunnel or a series of tubes which are stacked up over each other.

The pathway of lava caves are variously straight to meandering. Straight caves follow old fault and fracture patterns. More often, however, lava caves follow the twisting pathway of an ancient stream channel cut into the surface and filled in by lava. As the lava smoothed out the older topography, the thicker flows along stream channels remained liquid to drain out while the roof hardened. After the lava tube has formed, weathering processes attack the roof, leading to collapse, and producing cave openings.

Historically lava caves were used for shelter by Indians and early settlers. Eastern Oregon ranchers often kept meat chilled in caves which maintain a temperature of 40 degrees Fahrenheit year round. In the early part of the century, large blocks of ice were cut from Arnold Ice Cave to supply the nearby community of Bend on a commercial basis. One unusual advantage settlers found for the caves near Bend was as a place to brew whiskey. Federal agents conducting monthly raids, as the one in January, 1921, confiscated 2 stills, 3 quarts of whiskey, and several barrels of corn mash. Once raided, the caverns were sealed.

The Arnold Lava Cave system extends northeasterly from Paulina Mountains over 4 miles. In a number of places the roof has collapsed producing a series of smaller caves. This long lava channel system formed in basalt originating from fissures on the flanks of Newberry caldera. At the southwest end of the Arnold cave system is the well-known Arnold Ice Cave. Within the cave, water cascading down the walls freezes into silent waterfalls. Where the cave intersects the water table, ice forms. Since the water level is rising, ice now covers a stairway once in use, so that special climbing equipment is needed to enter the cavern.

Southwest of Arnold Ice Cave, the famous mile-long Lava River Cave, formerly known as Dillman Cave, is one of the longest, uncollapsed lava tubes in Oregon. Leander Dillman, trapper and rancher, discovered the southeast entrance about 1909, but when he was convicted of "gross immorality" in 1921 the name was abruptly changed. Entered where the roof has collapsed, the main passageway of the cave is 5,040 feet to the lower end which is blocked by sand. Delicate lava formations, vertically stacked lava tubes, ledges marking the levels of lava that flowed through the tube, and unusual sand castles where surface water drained down through the tube to carve out the sands into a miniature badlands make the cave of geologic interest.

Lavacicle Cave, south of Millican in Deschutes County, is best known for its unique spikes of lava hanging from the ceiling of the tunnel and projecting up from the cave floor. Discovered in 1959 and named for these unusual formations, the cave was once filled to the roof with molten lava. Once the lava had drained out, the material coating the ceiling of the cave dripped to the floor forming the pinnacles, called lavacicles. Lavacicles from the ceiling are only a few inches long, but those projecting from the floor are 2 to 6 feet in height. Because of the fragile formations, the cave entrace is barred and permission to enter must be obtained from the U.S. Forest Service.

On the extreme eastern edge of the Lava Plains, privately owned Malheur Cave is difficult to locate. At least one-half of the cave is filled with water creating a long underground lake where the lava tube intersects the ground water table. The cave must be explored by canoe or raft, although the lake waters and configuration of the tunnel prevent passage after a certain point.

Crack-in-the-Ground

A narrow northwest-southeast rift in the ground a few feet wide, 2 miles long, and up to 70 feet deep in places has remained open for several thousand years. Located in Lake County between prehistoric Christmas Lake to the south and the Four Craters lava field to

Crack-in-the-Ground from the south with Four Craters Lava Field in the background (photo courtesy of Oregon Department of Geology and Mineral Industries).

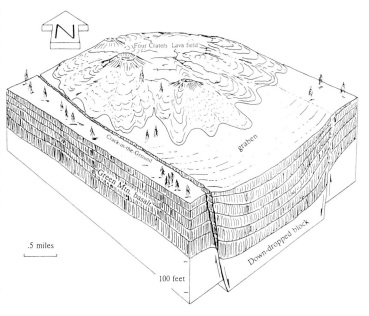

This narrow, two mile-long crack in Lake County opened in layers of lava when a block beneath the flow dropped down. (after Peterson and Groh, 1964)

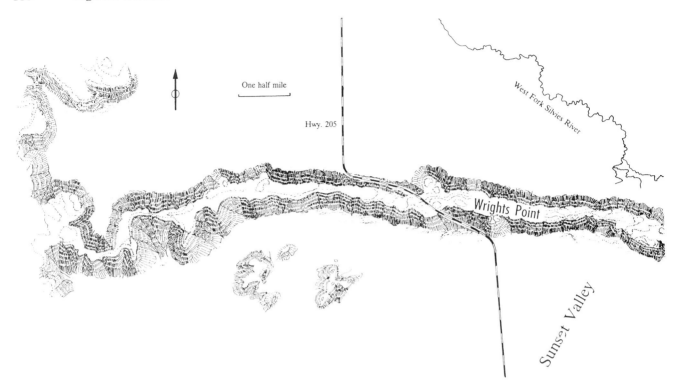

One half mile

Hwy. 205

West Fork Silvies River

Wrights Point

Sunset Valley

the north, this feature was referred to as "the crack" by homesteaders. Here picnics were held and ice cream made using ice deep from within the crevasses of the chasm.

During the Pleistocene successive flows of pahoehoe lava here piled up 70 feet thick in places, only to be covered later by streams of hummocky black aa lava. These basalts originated from Green Mountain and the Four Craters volcanoes. With the Four Craters eruption, a small tectonic depression or graben appeared about 2 miles wide extending south into the Christmas Lake basin. Crack-in-the-Ground opened in the lavas when the central block began to drop down to create the graben and trigger the crack seen on the surface. In a wetter environment than eastern Oregon, such geologic features would have been rapidly filled in with sediment and rubble or covered by vegetation.

Wrights Point

Lava flows following a prehistoric stream bed have created a 150 foot high sinuous ridge named Wrights Point 12 miles south of Burns. Extending a distance of 6 miles, the flat-topped ridge averages 400 yards wide before it merges with a broad mesa at its western end. Wrights Point is an excellent example of inverted topography. During late Pliocene time, a succession of fluid lavas followed an ancient, winding stream valley eventually filling in the canyon and hardening. As erosion took place, the softer rocks and soils of the adjacent valley walls were eroded away, leaving the streambed cast in lava projecting above the surrounding plain. Fossilized fish bones and freshwater

clams and snails found in the strata below the lava of Wrights Point are evidence of the stream environment.

Smith Rock

Smith Rock State Park, seven miles northeast of Redmond, is nationally renowned by rock climbers. The name for the park goes back to early Oregon history. In 1863, soldiers, in pursuit of local Indians, were camped near Smith Rock when one of the men named Smith climbed to the top of the monolith to get a better view. Unfortunately the boulder on which he was standing rolled out from under his feet.

On the very northwestern edge of the Lava Plains province at Smith Rock State Park successive colorful tuffs in the canyon of the Crooked River have been eroded to create the fantastic rock shapes visible in the cliffs. In the park, the river winds its way past massive, towering rock walls, picturesque pinnacles, crags, and great slabs of upright rocks. Smith Rock, one of these vertical pillars, is part of a volcanic center which erupted 18 to 10 million years ago. Included with Smith Rock are Gray Butte, Coyote Butte, Skull Hollow, and Sherwood Canyon, all of which are part of a larger Gray Butte complex of volcanoes. Rhyolitic lavas, emerging from vents near the Crooked River, marked the beginning of the Gray Butte events. Falling ash expelled from the volcanoes covered the region to be followed by mud flows and more volcanic debris ejected during subsequent eruptions. The land surface

Wright's Point, a sinuous 250-foot high ridge in Harney County, traces the channel of a meandering stream that was permanently cast in basalt.

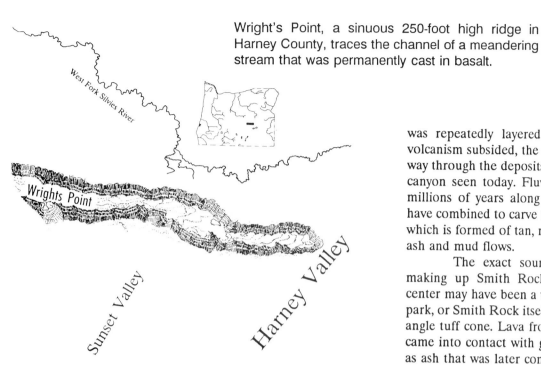

was repeatedly layered by this material. Once the volcanism subsided, the river slowly began to erode its way through the deposits to cut out the Crooked River canyon seen today. Fluvial erosion by the river over millions of years along with weathering of the rock have combined to carve the sheer walls of Smith Rock, which is formed of tan, red, and green tuffs of volcanic ash and mud flows.

The exact source of the volcanic material making up Smith Rock is uncertain. The eruptive center may have been a vent a short distance from the park, or Smith Rock itself may be the remnant of a low angle tuff cone. Lava from the cone exploded when it came into contact with groundwater and was expelled as ash that was later consolidated into tuffs.

Carved by the Crooked River into vertical pillars and sheer cliffs, the Miocene volcanic ash of Smith Rock is a favorite of rock climbers (photo courtesy of Oregon State Highway Department).

Suggested Readings:

Allison, Ira S., 1979. Pluvial Fort Rock Lake, Lake County, Oregon. Oregon, Dept. Geol. and Mineral Indus., Special Paper 7, 72p.

Bishop, Ellen Morris, 1989. Smith Rock and the Gray Butte complex. Oregon Geology, v.51, no.4, pp.75-80.

Catchings, R.D., and Mooney, W.D., 1988. Crustal structure of east central Oregon: relation between Newberry Volcano and the regional crustal structure. Jour. Geophys. Res., v.93, no.B9, pp.10,081-10,094.

Chitwood, Lawrence A., and McKee, Edwin H., 1981. Newberry Volcano, Oregon. U.S. Geol. Survey, Circular 838, pp.85-91.

Fitterman, David V., 1988. Overview of the structure and geothermal potential of Newberry Volcano, Oregon. Jour. Geophys. Res., v.93, no.B9, pp.10,059-10,066.

Greeley, Ronald, 1971. Geology of selected lava tubes in the Bend area, Oregon. Oregon Dept. Geology and Mineral Indus., Bull.71, 45p.

Gutmanis, J.C., 1989. Wrench faults, pull-apart basins, and volcanism in central Oregon: a new tectonic model based on image interpretation. Geological Jour., v.24, pp.183-192.

Jensen, Robert A., 1988. Roadside guide to the geology of Newberry Volcano. Bend, Cen-OreGeoPub, 75p.

Johnston, David A., 1981. Guides to some volcanic terranes in Washington, Idaho, Oregon, and northern California. U.S. Geol. Survey, Circular 838, 189p.

Niem, Alan R., 1974. Wright's Point, Harney County, Oregon. An example of inverted topography. Ore Bin, v.36, no.3, pp.33-49.

Peterson, N.V., and Groh, E.A., 1963. Recent volcanic landforms in central Oregon. Ore Bin, v.25, no.3, pp.33-42.

-----and Groh, E.A., 1964. Diamond Craters, Oregon. Ore Bin, v.26, no.2, pp.17-32.

Piper, A.M., , Robinson, T.W., and Park, C.F., jr., 1939. Geology and ground-water resources of the Harney Basin, Oregon. U.S. Geol. Survey, Water Supply Paper 841, 189p.

Walker, George W., 1979. Revisions to the Cenozoic stratigraphy of Harney Basin, southeastern Oregon. U.S. Geol. Survey, Bull.1475, 34p.

-----and Nolf, Bruce, 1981. High Lava Plains, Brothers faul zone to Harney Basin, Oregon. U.S. Geol. Survey, Circular 838, pp.105-111.

Weidenheim, Jan Peter, 1981. The petrography, structure, and stratigraphy of Powell Buttes, Crook County, central Oregon. Oregon State Univ., Ms., 95p.

Lambert Rocks, exposed in the Owyhee Canyon of Malheur County, are cut into fossiliferous Miocene lake deposits. Successive lava flows, that spread over the mud flats, are seen as darker layers [photo by Gary Tepfer].

Lava bomb (airborne blob of lava) atop McCarthy Rim, looking north toward Mt. Washington, Mt. Jefferson, and Mt. Hood [photo by Gary Tepfer].

Straightened and deepened by glacial activity, valleys now occupied by Horse Creek, Lightning Creek, and the Imnaha River display multiple lava flows of the Columbia River basalts in their walls [photo by Gary Tepfer].

Heceta head on the Oregon coast just north of Florence is the eroded remains of an Eocene volcanic edifice [photo by Gary Tepfer].

Mt. Washington, Three-Fingered Jack, Black Crater, Mt. Jefferson, Mt. Hood, Mt. Adams, and Black Butte are visible across the snowfield of McCarthy Rim [photo by Gary Tepfer].

In the foreground, Chambers Lakes occupy depressions in the irregular landscape of ash and cinders between the North and Middle Sisters. South Sister is in the distance [photo by Gary Tepfer].

View to the north in the Owyhee Canyon near Three Forks shows thick Miocene rhyolite layers being weathered into vertical columns [photo by Gary Tepfer].

Successive layers of iron-stained ash and lava within the crater of Broken Top in the High Cascades add bright colors to a rugged volcanic landscape [photo by Gary Tepfer].

Basin and Range topography of south-central Oregon shows the up-thrown fault block mountain of Winter Rim overlooking Summer Lake, which occupies a fault depression or graben [photo by Gary Tepfer].

Eastward dipping layers of the Eocene Coaledo Formation at Shore Acres State Park on the southwest coast are covered by iron-stained terrace deposits [photo by Gary Tepfer].

The rugged outline of Mt. Yoran in the Diamond Peak Wilderness of Lane County overlooks the Middle Fork of the Willamette River [photo by Gary Tepfer].

Viewed from Mt. Pisgah, the confluence of the Coast and Middle forks of the Williamette River can be seen with Eugene in the background. Skinner Butte is upper right in the distance [photo by Gary Tepfer].

The unmistakable irregular terrain of glacial deposits surrounds Dollar Lake in the Wallowa Mountains. Wallowa Lake and Joseph are in the distance [photo by Gary Tepfer].

Deschutes-Columbia Plateau

Physiography

The Deschutes-Columbia River Plateau is predominantly a volcanic province covering approximately 63,000 square miles in Oregon, Washington, and Idaho. The plateau is surrounded on all sides by mountains, the Okanogan Highlands to the north in Washington, the Cascade Range to the west and the Blue Mountains to the south in Oregon, and the Clearwater Mountains in Idaho to the east. Almost 200 miles long and 100 miles wide, the Columbia Plateau merges with the Deschutes basin lying between the High Cascades and Ochoco Mountains. The province slopes gently northward toward the Columbia River with elevations up to 3,000 feet along the south and west margins down to a few hundred feet along the river.

Primary rivers in the province are the west-flowing Columbia River and its tributaries, the northward flowing Deschutes, John Day, and Umatilla rivers, along with Willow and Butter creeks, all of which enter the Columbia between The Dalles and Wallula Gap. The third largest river in North America, the Columbia has the greatest impact on the plateau. Beginning in British Columbia on the slopes of the Canadian Rockies, the Columbia runs from the north to Wallula Gateway where it makes a sharp bend toward the west. Its watershed covers 259,000 square miles primarily in southwest Canada, southern Washington, and northern Oregon. Although smaller than the the Columbia, the Deschutes River crosses the province in a northerly direction to be joined by the Crooked and Metolius rivers south of Lake Simtustus. From its beginning as small streams near Mt. Bachelor, the Deschutes extends 250 miles before entering the Columbia just west of Biggs. These streams have cut intricate, deep canyons into the virtually horizontal lavas of the plateau. Divides between the canyons are dissected, but broad, flat uplands remain.

Geologic Overview

Geologic events in the Columbia-Deschutes province took place on a grand scale. Immense outpourings of lavas during the Miocene created one of the largest flood basalt provinces in the world, second only to the Deccan Plateau in India. Erupting from source vents in central and northeast Oregon as well as in southeast Washington and adjacent Idaho, flow after

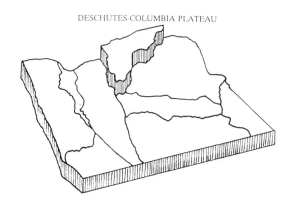

DESCHUTES-COLUMBIA PLATEAU

flow of basalts filled a subsiding basin to create a featureless plateau. This volcanic activity was followed by falling ash and lavas expelled from ancestral Cascade volcanoes aligned on the western border of the Deschutes Basin. Disrupted and blocked by the lavas, sediment-laden streams and the Deschutes River carried the material northward, depositing much of it along the channel filling the broad alluvial plain of the basin. With the gradual subsidence of the Cascade vents into a deep graben or depression and the resulting high Green Ridge escarpment, the accumulation of volcanic debris in the Deschutes basin shifted to lavas from small local cones during the Pliocene. In the Pliocene waters trapped behind structural ridges formed temporary lakes and ponds that filled with sediment. A broad uplift of the plateau triggered an aggressive new erosional phase where rivers carried away much of the unconsolidated clay, silt, and sand and cut deep channels. Volcanism here ceased after massive lavas from near Newberry Volcano entered the deep canyons, once again disrupting the flow of the rivers.

Vast lake waters, impounded by glacial ice, impacted the Columbia Plateau during the Pleistocene when they were released from Montana sending catastrophic floods across the landscape to scour southeastern Washington and the Columbia gorge.

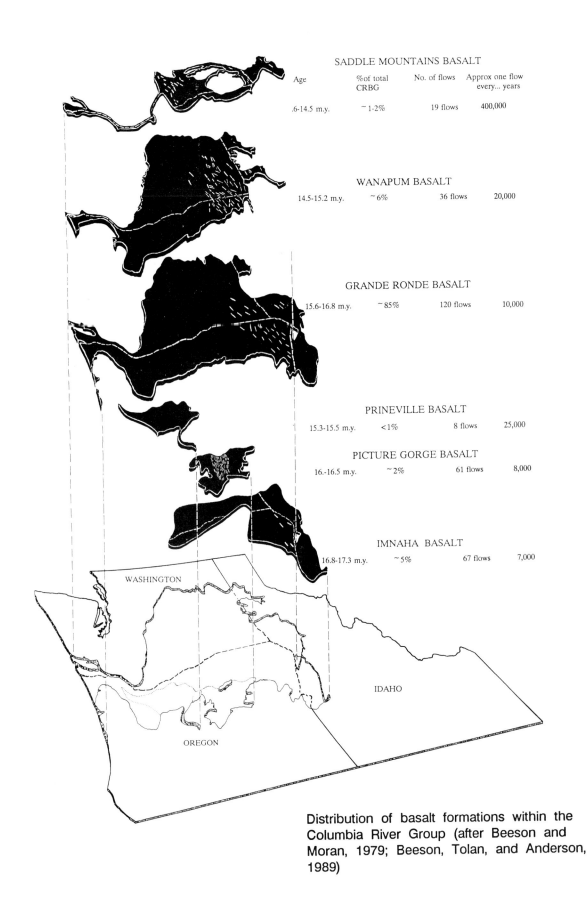

SADDLE MOUNTAINS BASALT

Age	%of total CRBG	No. of flows	Approx one flow every... years
.6-14.5 m.y.	~1-2%	19 flows	400,000

WANAPUM BASALT

| 14.5-15.2 m.y. | ~6% | 36 flows | 20,000 |

GRANDE RONDE BASALT

| 15.6-16.8 m.y. | ~85% | 120 flows | 10,000 |

PRINEVILLE BASALT

| 15.3-15.5 m.y. | <1% | 8 flows | 25,000 |

PICTURE GORGE BASALT

| 16.-16.5 m.y. | ~2% | 61 flows | 8,000 |

IMNAHA BASALT

| 16.8-17.3 m.y. | ~5% | 67 flows | 7,000 |

WASHINGTON

IDAHO

OREGON

Distribution of basalt formations within the Columbia River Group (after Beeson and Moran, 1979; Beeson, Tolan, and Anderson, 1989)

Geology

Small exposures of Paleozoic and Mesozoic rocks as well as parts of the Eocene, Oligocene, and Miocene Clarno and John Day formations that extend into the Columbia-Deschutes province from the south are treated with the Blue Mountains province.

The oldest rock known from the Columbia-Deschutes province is a single stone discovered near the base of Gray Butte in Jefferson County. Fusulinids, or fossil protozoa, extracted from the loose boulder were dated as late Permian, approximately 250 million years old. More remarkably, these fossils are of a variety typical only to the eastern Pacific regions of Japan, Timor, and China. Although the affinity of this rock was puzzling when it was discovered in the middle 1930 s, with the present knowledge, it can be related to one of the many exposures of displaced terranes found today in the eastern Blue Mountains. The stone was evidently eroded from one of the terranes before being transported by streams to the western part of the Columbia-Deschutes province.

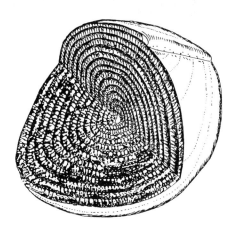

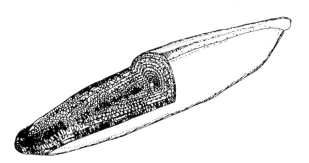

Comparison of Permian fossil fusulinids from the Orient (upper) and North America (lower) (specimens are 1/2 inch in length)

Miocene

Geologic study of the Columbia Plateau focuses upon the basalts that flooded this province, an episode that took place during the middle to late Miocene, 17 to 6 million years ago. Eruptions occurred on the average of one flow every 35,000 years with an output per flow of over 100 cubic miles of basalt. Larger eruptions, however, encompass over 500 cubic miles of basalt for a single flow. Thick, successive outpourings of lava spread over the landscape often moving at a speed of up to 30 miles an hour to form the Columbia River basalts, the primary group in the plateau today. Covering an area only slightly less than the entire state of Washington, the total volume of basalt, which was over 42,000 cubic miles, would be enough to construct a wall of lava one mile wide and almost 2 miles high around the earth. The thickness of individual flows is variable, but several flows up to 200 feet are known. Spreading across the Deschutes-Columbia River Plateau, the spoon-shaped mass of basalt is over 3 miles thick beneath Yakima and Pasco, Washington, thinning to the south from a mile thickness along the Columbia River to a feather edge within the Blue Mountains province. This suggests that the lava was extruded into a basin that subsided progressively with succeeding flows.

Some of the most extensive of the Columbia River flows may have taken decades to cool and harden. Cooling proceeds in a lava flow from the top down and from the bottom upward. Typical lava flows in cross-section have two distinct layers that have been named after parts of Greek temples. The lower portion or colonnade is so-called because the basalt has cooled and contracted to form columns perpendicular to the cooling surface below. The remaining section of the flow, which constitutes up to 4/5 of the entire mass, is the entablature with multiple directions to the columns. Near the top of the entablature, the basalt may be vesicular from gas which has come out of solution in the molten lava to form bubbles. Often these gas bubbles leave long vertical tubes or pipes through the lava to mark where the gas escaped.

There is disagreement with respect to exactly how lava in flood basalts flowed. As the lava moved, a thin surface skin may have hardened only to break up so that the new flow would look like a wall of broken rock rising up to 10 stories high. Inside the flow the lava was fluid, but at the base, along the top, and at the snout, it is armored by sharp, blocky pieces of broken, cooled basalt. The veneer of broken chunks atop the lava tends to roll over as the flow spreads, developing a steep, irregular front. As the lava moves farther from its source, the edge becomes progressively thinner until it is only a few stories high. At some point the lava,

The Columbia River Basalt Group develops its maximum thickness of nearly 15,000 feet under the Columbia Plateau near Yakima, Washington (after Reidel and Hooper, 1989)

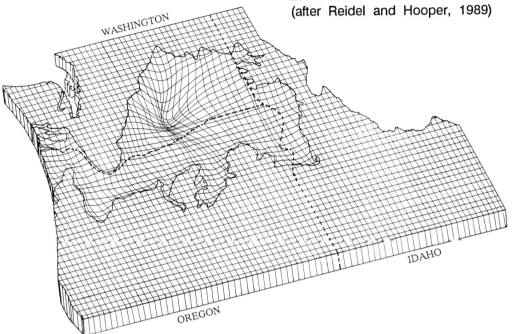

unable to push through the mound of hardened, rough rock along the edges, comes to a stop, essentially dammed up by its own cooled crust.

The relationship between the Columbia-Deschutes Plateau and the surrounding provinces is unclear because thick lavas obscure the deep crust which has not been well-studied. Crustal thickness in this province has been estimated between 15 to 20 miles. In the 1980s, a series of four wildcat wells were drilled through the lavas in the vicinity of Yakima, Washington, by Shell Oil Company. Targeted were Cretaceous and Tertiary rocks lying below the basalt along older anticlinal structures that were judged to have sufficient organic content and porosity to produce gas. Although the potential reservoir rocks were dry, rocks of the predicted formations were encountered by drilling deep beneath the basalt.

Of the four main formations making up the Columbia River lavas, the oldest is the Imnaha Basalt, followed by the Grande Ronde Basalt, the Wanapum Basalt, and finally the youngest Saddle Mountains Basalt, with the Grande Ronde composing more than 85% of the formations by volume. The Grande Ronde, Wanapum, and Saddle Mountains were formerly lumped together and designated the Yakima Basalts, but they are now distinguished by subtle differences in mineralogy and texture. Variations in magnesium oxide, titanium oxide, and phosphorus are also used to characterize individual flows.

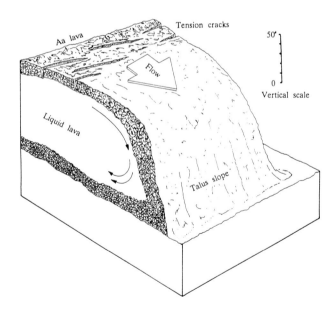

One model for the movement of lava in an Aa flow

SERIES	GROUP	FORMATION	MEMBER	ISOTOPIC AGE (m.y.)
MIOCENE — UPPER	COLUMBIA RIVER BASALT GROUP	SADDLE MOUNTAINS BASALT	LOWER MONUMENTAL MEMBER	6
			Erosional Unconformity	
			ICE HARBOR MEMBER	8.5
			Basalt of Goose Island	
			Basalt of Martindale	
			Basalt of Basin City	
			Erosional Unconformity	
			BUFORD MEMBER	
			ELEPHANT MOUNTAIN MEMBER	10.5
			Erosional Unconformity	
			POMONA MEMBER	12
			Erosional Unconformity	
			ESQUATZEL MEMBER	
			Erosional Unconformity	
			WEISSENFELS RIDGE MEMBER	
			Basalt of Slippery Creek	
			Basalt of Tenmile Creek	
			Basalt of Lewiston Orchards	
			Basalt of Cloverland	
			ASOTIN MEMBER	13
			Basalt of Huntzinger	
			Local Erosional Unconformity	
			WILBUR CREEK MEMBER	
			Basalt of Lapwai	
			Basalt of Wahluke	
			Local Erosional Unconformity	
			UMATILLA MEMBER	
			Basalt of Sillusi	
			Basalt of Umatilla	
			Local Erosional Unconformity	
MIDDLE		WANAPUM BASALT	PRIEST RAPIDS MEMBER	14.5
			Basalt of Lolo	
			Basalt of Rosalia	
			Local Erosional Unconformity	
			ROZA MEMBER	
			FRENCHMAN SPRINGS MEMBER	
			Basalt of Lyons Ferry	
			Basalt of Sentinel Gap	
			Basalt of Sand Hollow	15.3
			Basalt of Silver Falls	
			Basalt of Ginkgo	
			Basalt of Palouse Falls	
			ECKLER MOUNTAIN MEMBER	
			Basalt of Shumaker Creek	
			Basalt of Dodge	
			Basalt of Robinette Mountain	
			Local Erosional Unconformity	
		GRANDE RONDE BASALT (PRINEVILLE BASALT)	SENTINEL BLUFFS UNIT	15.6
			SLACK CANYON UNIT	
			FIELD SPRINGS UNIT	
			WINTER WATER UNIT	
			UMTANUM UNIT	
			ORTLEY UNIT	
			ARMSTRONG CANYON UNIT	
			MEYER RIDGE UNIT	
			GROUSE CREEK UNIT	
			WAPSHILLA RIDGE UNIT	
			MT. HORRIBLE UNIT	
LOWER		(PICTURE GORGE BASALT)	CHINA CREEK UNIT	
			DOWNEY GULCH UNIT	
			CENTER CREEK UNIT	
			ROGERSBURG UNIT	
			TEEPEE BUTTE UNIT	
			BUCKHORN SPRINGS UNIT	16.5
		IMNAHA BASALT	See Hooper and others (1984) for Imnaha Units	17.5

Basalt stratigraphy within the Columbia River Basalt Group (Beeson and Moran, 1979; Beeson, Tolan, and Anderson, 1989).

million years ago, large volumes of Grande Ronde basalts erupted with a frequency of one flow for every 10,000 years slowing to one flow for each 20,000 and 400,000 years for the Wanapum and Saddle Mountains lavas respectively. In a process that extended less than 1 million years, Grande Ronde basalts poured from fissures and cracks as much as 100 miles long in a complex network of openings in the crust known as the Chief Joseph dike swarm that extended from southeastern Washington and northeastern Oregon along a well-defined trend. Reaching westward all the way to the Pacific Ocean, many of the 120 individual Grande Ronde lavas spread out for as much as 460 miles and are among the most extensive on earth. Lava from these separate flows was so runny that it did not mound up to form volcanoes but rapidly filled in surface depressions to produce a flat topography. Following the Grande Ronde eruptions, 36 separate flows of the Wanapum lavas poured out over a 1 million year interval nearly covering the plateau, while the Saddle Mountains Basalt, with 19 flows, was mainly confined to the central part of the Columbia basin. To

Dikes of the Grande Ronde basalt (photo courtesy Oregon Department of Geology and Mineral Industries).

The oldest of the formations, the Imnaha Basalt, originated from vents and fissures along the Snake and Imnaha rivers in northeast Oregon, southeast Washington, and Idaho spreading along the Imnaha River in northeast Oregon. Following this 16.5

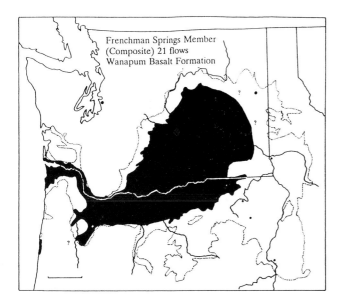
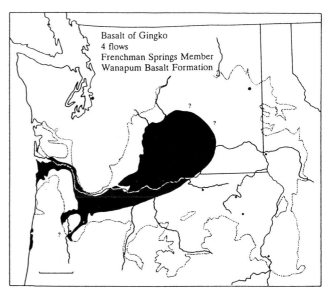

Geographic distribution of individual flows and members of the Columbia River Basalt Group (after Reidel and Tolan, 1989)

the south, widespread flows of the Prineville and Picture Gorge basalts in the Deschutes and John Day valleys correlate with the Grande Ronde Basalt, but differences in chemical composition of the three lavas distinguish them and indicate separate source magmas.

Prior to the extrusion of the Columbia River lavas, the channel of the ancestral Columbia River lay well south of its present course. Individual flows periodically swept into the canyon to plug and disrupt the drainage. As the frequent eruptions subsided, late Miocene and Pliocene compressional folding formed large-scale, east-west wrinkles that confined the river to its present course between Umatilla and The Dalles.

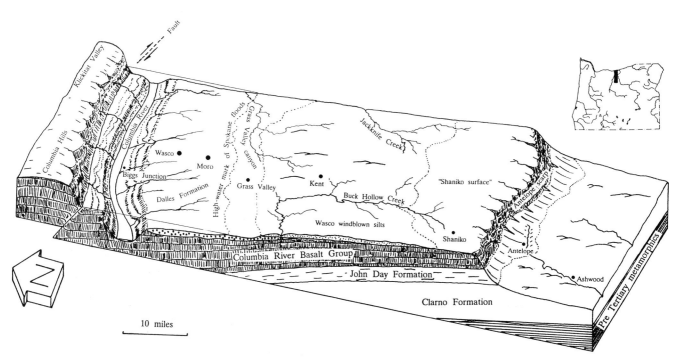

Block diagram across the Columbia River just east of The Dalles

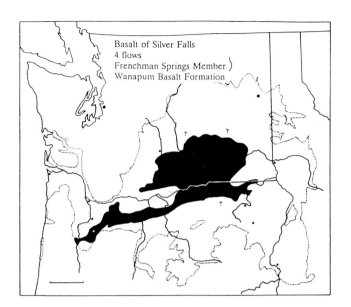

Basalt of Silver Falls
4 flows
Frenchman Springs Member
Wanapum Basalt Formation

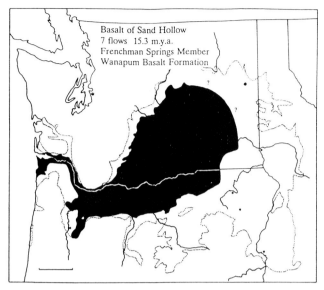

Basalt of Sand Hollow
7 flows 15.3 m.y.a.
Frenchman Springs Member
Wanapum Basalt Formation

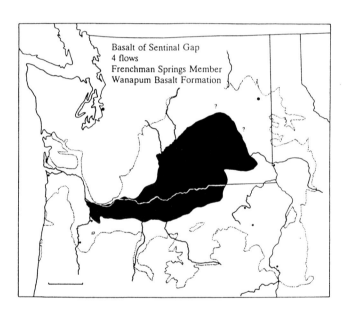

Basalt of Sentinal Gap
4 flows
Frenchman Springs Member
Wanapum Basalt Formation

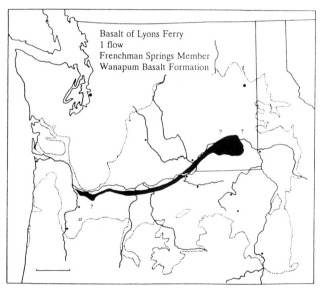

Basalt of Lyons Ferry
1 flow
Frenchman Springs Member
Wanapum Basalt Formation

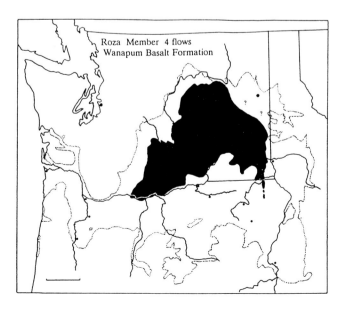

Roza Member 4 flows
Wanapum Basalt Formation

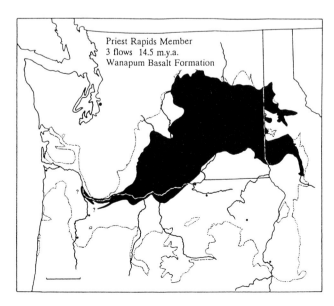

Priest Rapids Member
3 flows 14.5 m.y.a.
Wanapum Basalt Formation

Because of the relative youth and thickness of the Columbia River basalts, rocks immediately below the Miocene lavas are still poorly known. In southeastern Washington small exposures of displaced rocks of the Wallowa terrane imply that this terrane extends far north of its main exposures in the Blue Mountains. Similarly, inliers of the basalt-covered Baker terrane occur well to the west of major exposures in the Blue Mountains. The extent and thickness of Cretaceous rocks as well as Tertiary Clarno and John Day formations, where they disappear under the southern edge of the plateau lavas, may be an indication that they too are much more widespread than presently known. Distinctive rock fragments found imbedded in flows of the Deschutes Formation may have been picked up by the lava as it worked its way to the surface. Although this evidence is tenuous, it suggests that accreted terrane rocks, as old as preCambrian, may be buried beneath the Cascade Range and Deschutes Basin.

Despite the thickness and extent of the lavas, the plateau is not entirely featureless. Over most of its western section, low wrinkles and shallow dimples dot the surface. One of the largest of these is the C-shaped Deschutes basin reaching into the southwest corner of the province. During the early Miocene, the basin received lavas and ash from vents of the Columbia Plateau and adjacent volcanoes of the Western Cascades. Intermittent volcanic activity sent lavas flooding over the vast Deschutes basin, disrupting the flow of streams and rivers. Sediment-choked streams covered 100 square miles of the basin with thin layers of waterlain tuffs, sandstones, and mudstones of the Simtustus Formation. Sedimentation was rapid, and the shallow basin filled with Simtustus deposits up to 250 feet thick. Once designated as part of the Deschutes Formation, volcanic sediments of the older Simtustus Formation are lithologically distinct from the coarse conglomerates of the younger Deschutes. Simtustus rocks are highly significant as they were laid down during emplacement of the Columbia River flows 16 to 12 million years ago.

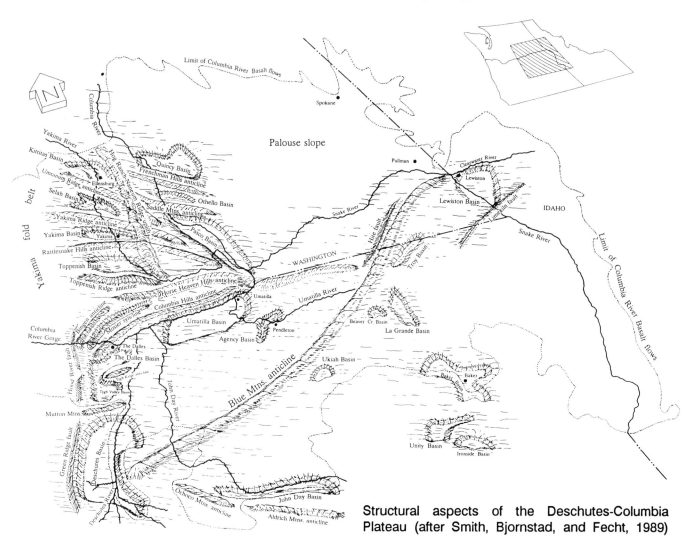

Structural aspects of the Deschutes-Columbia Plateau (after Smith, Bjornstad, and Fecht, 1989)

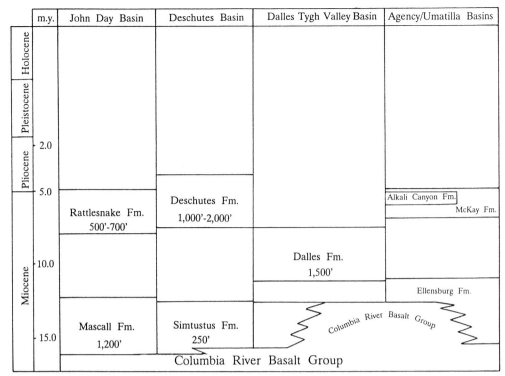

m.y.	John Day Basin	Deschutes Basin	Dalles Tygh Valley Basin	Agency/Umatilla Basins
Holocene				
Pleistocene				
Pliocene — 2.0 / 5.0				
	Rattlesnake Fm. 500'-700'	Deschutes Fm. 1,000'-2,000'		Alkali Canyon Fm. / McKay Fm.
Miocene — 10.0			Dalles Fm. 1,500'	Ellensburg Fm.
— 15.0	Mascall Fm. 1,200'	Simtustus Fm. 250'	Columbia River Basalt Group	
		Columbia River Basalt Group		

Stratigraphy of Tertiary formations of the western Deschutes Columbia Plateau

Following a hiatus of approximately 5 million years, volcanics and sediments again poured into the Deschutes basin 8 million years ago from a chain of early Cascade volcanoes, aligned along the western margin of the basin. Voluminous amounts of basalt, andesitic lava, and hot clouds of ash or ignimbrites accumulated up to 2,000 feet on the west side of the basin thinning to 50 feet adjacent to the Ochoco Mountains. Once the northerly flowing ancestral Deschutes River had been overloaded with volcanic debris, it slowed considerably depositing a mixture of volcanic material along the channel and on the adjacent alluvial plain that extended eastward for 30 miles from the Cascades. Blocked by the rising wall of the Cascade graben along the border of the Deschutes basin, the supply of sediments from the west came to an abrupt end 4 million years ago. Deterred by the high Green Ridge scarp, which is composed primarily of ash, lavas, and ignimbrites, streams began to cut down and deepen their channels.

The rise of the Cascade Mountains in the Miocene and Pliocene created a barrier between eastern and western Oregon preventing the flow of warm moist ocean air and subsequently producing the drier climate of today. Prior to the climate change, the plants would have been familiar, but the mammals would have appeared out of place. Oak was a common tree followed by maple, sweetgum, and cherry in the

higher elevations with willow and cottonwood in the moister valleys. There was a remarkable absence of evergreens in the upper Deschutes region. Fossils of rhinoceroses, camels, elephants, giant beavers, very small horses, and a three foot long turtle lived here before the climate became more arid.

Pliocene

At this point in time, between 4 and 2 million years ago, a number of small volcanic cones in the upper Deschutes and Crooked river watersheds produced over 200 separate lava and ash beds of Pliocene age throughout the region. Shield volcanoes and cinder cones were built up at Squaw Back Ridge, Tetherow Butte, Round Butte, Black Butte, and Little Squaw Back. Lavas from the Tetherow Butte complex of red and black cinder cones near Terrebonne extended northward covering Agency Plains between Madras and Gateway to depths over 150 feet. The most recent eruptions in the basin were lavas from Round Butte 3.9 million years ago along with the fluid andesitic basalts from the broad low volcano at Squaw Back Ridge 2.9 million years ago. After an initial eruptive period, Round Butte continued to build up, eventually exploding to subside with the formation of two small summit cones. With a pleasing symmetrical profile, Round Butte was the site of Indian ceremonies and is still regarded as a sacred site. Little Squaw Back and Black

Four million year old
Pliocene sediments near
Arlington yielded the
carapace of a 3-foot
long land tortoise.

Butte, large shield volcanoes on either side of Green Ridge, probably erupted 3 to 1/2 million years ago. The 6,436 foot andesite cone of Black Butte is the highest in the area.

Pleistocene

The plateau experienced the last phase of volcanism 1.6 million years ago when early Pleistocene eruptions from local vents plugged the Metolius, Deschutes, and Crooked River canyons with basalt. The most extensive of these numerous flows originated near Newberry crater, filling the Crooked River canyon to depths of 800 feet in places, while an intracanyon basalt pouring into the Metolius River valley formed a pronounced bench 600 feet above the present river level. Within the last million years, however, river waters have been able to reestablish their previous channels completely by cutting around the succession of intracanyon flows. The basalts have been reduced to islands midstream or isolated steep cliffs plastered to the older sedimentary and volcanic layers that make up the main canyon walls.

The Ice Ages of northern Oregon were a time of ice build up and repeated catastrophic flooding. Pleistocene continental glaciers advanced as far south as British Columbia and northern Washington infringing on the Columbia River drainage. Glacial ice temporarily plugged the Clark Fork River, a branch of the Columbia River in Montana, resulting in a glacial lake which drained as numerous separate floods. When the dammed up water was suddenly released, it spread through the Idaho panhandle and southeast Washington to reenter the Columbia at Wallula Gateway. After scouring eastern Washington, millions of gallons of flood water continued down river to complete the final sculpting of the Columbia Plateau. Reaching elevations of about 1,000 feet, the floods removed talus and soil

from large areas along the river and covered existing gravel bars with coarse flood-borne material and erratics. On both sides of the gorge from Wallula Gap to The Dalles, flowing water backed up and spread out to create Lake Condon. The lake waters reached the John Day River valley to the south as well as Maupin on the Deschutes where large boulders carried in by icebergs were dropped as the ice melted. Such boulders are common southwest of Arlington, many marking the shoreline of the ancient lake.

Frequent successive flooding left gravel, sand, and silt in the valleys and basins along the Columbia gorge. A thick, structureless, buff-colored Wasco silt, in the vicinity of Moro and Grass Valley, is similar to the Portland Hills Silt. Near Ordinance and Boardman, a large area of unconsolidated sand has been reworked by winds to form dunes. Pleistocene vertebrate fossils are present in the blowouts which reach an elevation of about 600 feet and probably represent fill in the Columbia River valley during high waters. Terraces along the river at The Dalles are composed of gravels that also coincide with flooding stages.

Structure

The tectonics and geologic history of the Deschutes-Columbia Plateau relate closely to the movement of the earth's crustal plates, which brought about structural changes throughout the province. The evolution of the plateau has been divided into three phases that took place 17 to 10 million years ago, a second interval between 10 and 4 million years ago, and occurrences over the past 4 million years. During the first stage, tensional stresses from apparent backarc spreading on the North American plate produced extensive north-south fractures in the southeast corner of the plateau and resultant massive lava flows of the Imnaha Basalt. As the eruption continued, new vents

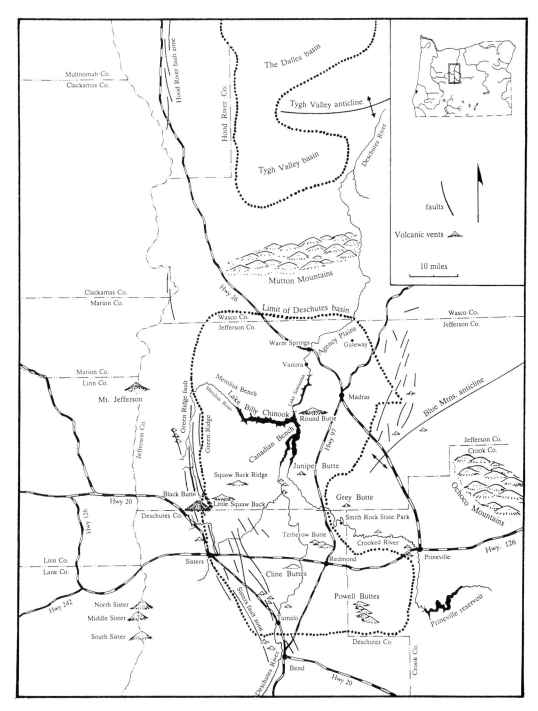

Major geographic and structural features of the Deschutes-Columbia Plateau (after Smith, Bjornstad, and Fecht, 1989)

appeared further north, and the entire plateau was tilted to the northwest by uplift along the Idaho batholith. This tilting allowed the later Grande Ronde flows to extend northward into central Washington and enabled them to reach the Pacific Ocean by way of the Columbia gorge. With as much as 36,000 cubic miles of

Grande Ronde basalts being erupted, the crust at the center of the plateau at Pasco, Washington, began to fail and collapse downward to create a basin over three miles deep.

Concurrently, the deformation of the plateau caused minor folding over the Columbia basin and pro-

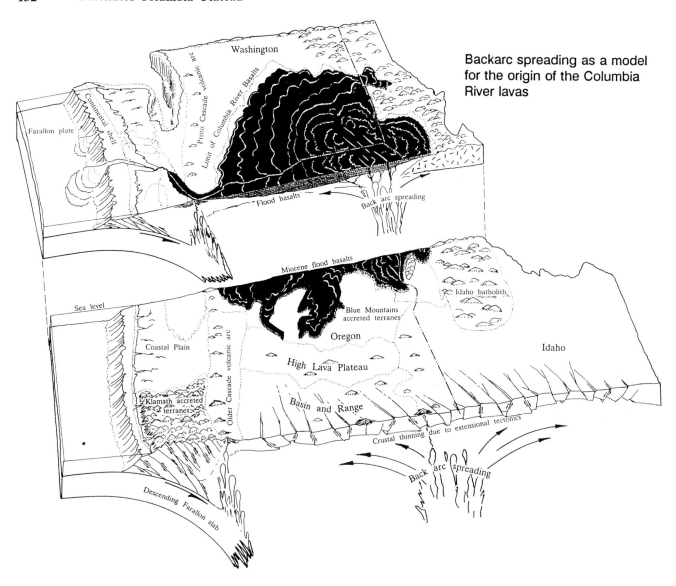

Backarc spreading as a model
for the origin of the Columbia
River lavas

duced the northeast-southwest trending Blue Mountains anticline or arch. Where the Columbia River bisects the central plateau, a number of surface folds run east-west, corresponding to north-south compression. The Columbia River follows one of these wrinkles as it traverses the axis of a major downwarp, the 160-mile long combined The Dalles-Umatilla basins from Wallula Gap to The Dalles. Further east, the Blue Mountains anticline turns north at Meacham, Oregon, to merge and run parallel to the Klamath-Blue Mountains lineament. In Washington, the Columbia Hills and hills of Horse Heaven anticlines follow The Dalles-Umatilla syncline to intersect the Hite fault just south of Milton-Freewater.

Beneath the plateau, major structural features or lineaments converge. These extensive features can be seen in aerial photographs, but they are not obvious at ground level, and what caused them is uncertain. The Klamath-Blue Mountains lineament runs across the state in a southwest by northeast direction, intersecting the Olympic-Wallowa lineament near Wallula Gap on the Columbia River, while the Olympic-Wallowa lineament runs southeast by northwest from Puget Sound across Washington through the Oregon Blue Mountains to the Idaho border. Cutting directly across the Olympic-Wallowa lineament, the Hite fault system trends southwest by northeast.

With phase two between 10 and 4 million years ago, the spreading direction of the Pacific plate rotated 25 degrees clockwise to a more southwest by northeast direction. At this time intense compression produced a series of east-west wrinkles in southcentral Washington and in the vicinity of the Columbia River. Most of the visible folds in plateau lavas along the Columbia, including the Yakima fold belt, Horse Heaven anticline, and Columbia Hills anticline, are related to this

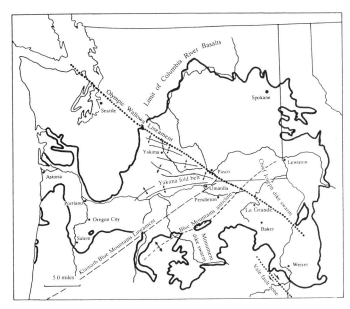

Major structural features of the Deschutes-Columbia Plateau

phase of deformation. Additional uplift of the Blue Mountains anticline, stretching, faulting, and thrusting in the Blue Mountains and John Day region were also part of these tectonic events.

Over the last 4 million years, the plateau was subjected to continual but subdued north-south compression. Cascade volcanics to the west changed in composition, became localized along a narrow archipelago, and greatly intensified. These conditions persisted until faulting began to lower the volcanoes into a graben along the western border of the Deschutes basin. On the eastern flank of the Cascade graben, the Green Ridge scarp is a well-defined fault zone. The Green Ridge belt appears to merge with the Tumalo fault zone to the south which in turn converges with the Walker Rim and Brothers fault zones at Newberry caldera. The Green Ridge and Tumalo fault zones provided avenues for escaping lava and ash between Bend and the Metolius River valley.

Mining and Mineral Industry

Because of its volcanic character, the Deschutes-Columbia River Plateau has not been an area of extensive mining. Along the Deschutes River 6 miles west of Terrebonne, diatomite was mined commercially from a quarry beginning in the late 1950s and continuing until the deposit was exhausted. The diatomite here was up to 67 feet thick, and the layer has been cut through by the Deschutes to expose the strata on both sides of the valley. The soft white material is easily distinguished from the surrounding tuffs, sands, and lavas.

Diatomite, or diatomaceous earth, is composed of the capsule-like skeletons of microscopic, single-celled glassy aquatic plants. Their abundant presence indicates an ancient lake bed once existed here in the late Miocene or early Pliocene. Flourishing in the fresh waters, diatoms build up a layer of skeletons on the bottom of the lake by the millions. Absorbing as much as 300 percent of its own weight, the lightweight opaline diatomite is porous, fireproof, resists chemicals, and is mined for use as cat box filler, to filter drinking water, in swimming pools, and in chemical laboratories.

Geothermal Resources

Geothermal potential for the Deschutes-Columbia River Plateau is moderate to low, similar to that of the Blue Mountains province, although across the river in Washington well waters have recorded high temperatures. At Warm Springs, above The Cove Palisades State Park, the Confederated Tribes of the Warm Springs Indian Reservation operate a resort built around the mineral springs discovered here in 1855. The natural waters flow from Clarno basalts at 140 degrees Fahrenheit and smell somewhat of hydrogen sulfide.

Features of Geologic Interest

The Cove Palisades State Park and the Deschutes Canyon

By persistently cutting downward for millions of years through layers of rock, the Metolius, Deschutes, and Crooked rivers have produced deep gashes in the flat Deschutes-Columbia River plateau to expose 15 million years of geologic history in a 1/2 mile thickness of lavas, volcanic ash, and sand. Once these three rivers merge about 10 miles west of Madras in Jefferson County, the main branch of the Deschutes continues another 80 miles to enter the Columbia River.

Two miles above this juncture, a secluded spot, protected by steep canyon walls, was known as "The Cove" by settlers in the early 1900s. The construction of two dams in this section, the Pelton Dam in the late 1950s to back up river waters into Lake Simtustus and Round Butte Dam in 1964, forming the 3-arm Lake Billy Chinook, changed the configuration of The Cove. The Cove Palisades State Park, originally composed of 3,620 acres when it was purchased as for parkland in 1940 and 1941, now encompasses 7,000 acres.

The geologic history in the park and Deschutes canyon begins in the middle Miocene when the massive lava flows of the Columbia River basalt filled most of the ancestral river canyon to The Cove Palisades park. These lavas originated from fissures in northeast Oregon and southeast Washington and covered vast

"The Ship" at The Cove Palisades State Park is composed of tuffaceous Deschutes beds capped by rimrock basalt. (Photo Oregon State Dept. of Transportation)

areas of eastern Oregon. Layered between the basalts are sediments laid down by the river waters that were impeded by the volcanic debris. Events during this time are recorded in the canyon walls by rocks on the east rim of Lake Simtustus.

After sucessfully cutting through these basalts for 7 million years, the ancient Deschutes river was once again inundated when Cascade volcanoes began to fill the channel with volcanic debris, ash, and cinders, blocking the waterflow and creating lakes in places. Mixed in with this volcanic material are silt, sands, and gravels carried by the Deschutes as it continued to work at eroding these new obstructions. Most of the yellowish-brown sands and black basalts seen in the canyon walls were produced during this interval from 8 million to 4 million years ago. Distinctive ignimbrite layers are common between the river sediments and lavas. Ignimbrites develop when hot air-borne ash falls to cool as a glassy layer. At The Cove Palisades park an ignimbrite comprises the distinctive white prow of "the ship", one of the scenic features, while sedimentary

layers of the Deschutes Formation compose the narrow neck of the isthmus between the two rivers at the south end of The Island. The Peninsula is made up of Pliocene and Pleistocene lavas extruded 2 to 3 million years ago from local vents. Smaller amounts of lavas covered low spots in the valley and diverted the ancient streams and rivers.

Once this eruptive activity had ceased, the river could concentrate its energy renewing and deepening its channel, which followed the same path as it does today. In a final volcanic event beginning 1.6 million years ago, hot liquid lava poured into the Crooked River, entering the canyon upriver near O'Neil. This lava, from vents at the base of Newberry Volcano, flowed for miles downstream from that point, filling the canyon. At least 15 separate flows inundated the canyons as far as Round Butte in the park before coming to a halt. After moving such a distance, the lavas had cooled somewhat to become thick and sluggish. In the narrows here, a dam was built up by successive layers of basalt temporarily blocking the river. Eventually, however, water opened a gap through the lava dam, carving its path into the former channel. In the park, these intracanyon flows can be seen as brownish columnar-jointed basalt along the walls, contrasting to the earlier, lighter-colored beds of ash and sediments. Known to pioneers as the "Plains of Abraham", The Island is a flat ridge which rises as a spectacular 450 foot high cliff in Lake Billy Chinook. This ridge is an isolated remnant of the last lava which flowed down the canyon.

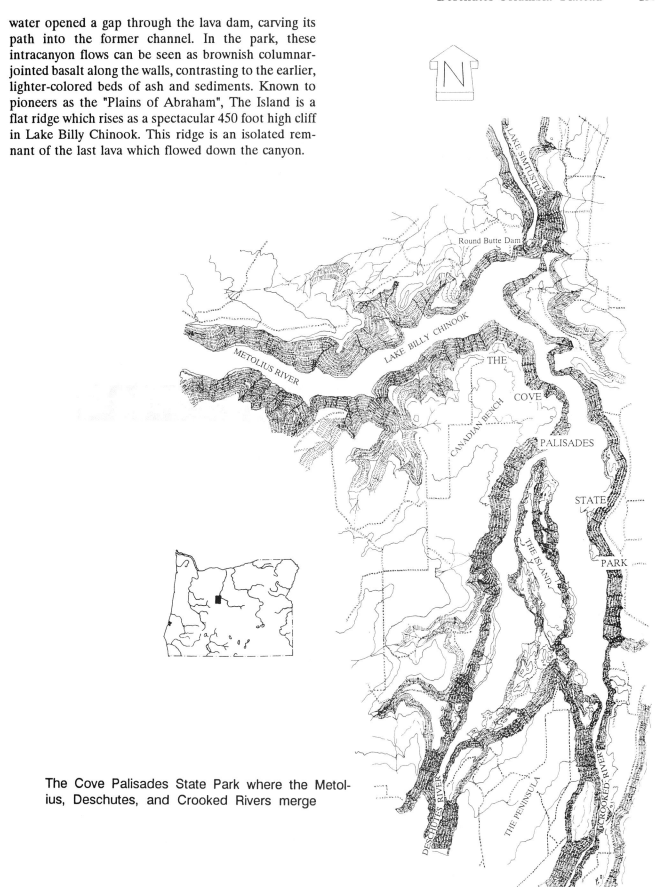

The Cove Palisades State Park where the Metolius, Deschutes, and Crooked Rivers merge

The Island, at the confluence between the Crooked River on the right and the Deschutes River on the left, is a remnant of a Pleistocene intracanyon flow of basalt. (Photo Oregon State Dept. of Transportation)

Intracanyon flow of columnar Deschutes Formation basalts over a valley cut into ash and cinders at the entrance to The Cove Palisades State Park. (Photo Oregon Dept. of Geology and Mineral Indus.)

The filling of the Deschutes Canyon by the waters impounded behind Pelton and Round Butte dams had unusual side effects. As the lakes filled and the local groundwater table rose, many new warm springs emerged. These thermal springs include 14 in Willow Creek canyon alone. Water temperatures of the springs average around 60 degrees Fahrenheit, which is slightly warmer than most other thermal springs in the region. The presence of the springs has also triggered a dramatic change in the modern flora in the immediate vicinity.

Balanced Rocks

John Newberry, on a railroad exploring expedition to Oregon in 1855, examined the Deschutes area in great detail, recording his trip in copious notes. One of his discoveries was a group of precariously balanced rocks on the north-facing slope of the Metolious River just before the river enters Lake Billy Chinook. The balanced rocks of this striking phenomenon, weighing over a ton each, are perched atop tapering pinnacles 20 to 30 feet high. The rocks, resembling the top knots of Easter Island statues, are more massive and greater in diameter than the pillars which support them. Composed of volcanic basalt and interbedded sedimentary siltstones and sandstones of the Deschutes Formation, the columns are the product of differential weathering. Erosion rapidly removed the softer lower layers and left the resistant cap, which armors the striking columns that stand out in relief. Close by, another group of balancing rocks, named Button Head Rocks, are not as high or as massive.

Metolius Springs

One of the swiftest flowing and shortest rivers in the west, the Metolius begins with waters gushing from the springs at the base of Black Butte to form an "instant river". Situated in the southwest corner of Jefferson County about 30 miles northwest of Bend, Metolius Springs are at the junction of the Western Cascade province, the High Lava Plains province, and the Deschutes-Columbia Plateau. The springs issue from two openings at the foot of Black Butte about 200 yards apart, merging after a short distance to flow northward. Total output from the springs measures 45,000 gallons per minute, and the water temperature is a chilly 48 degrees Fahrenheit.

Geologic events associated with the origin of Metolius Springs began with block faulting. Tension along a north-trending fault caused a block to drop down forming the graben valley of the Metolius. The high Green Ridge fault scarp along the eastern side of the valley blocked the flow of surface water and channeled it toward the north and around the end of the ridge creating an ancestral Metolius River. About 500,000 years ago, eruptions from vents along the fault escarpment built up Black Butte cone straddling the Metolius Valley and covering the ancient river drainage. The waters which once flowed above ground now percolate through sands and gravels in the old channel beneath the volcano before they reappear above ground as Metolius Springs.

Basalt slabs are balanced atop pinnacles of Deschutes Formation volcanics on the Metolius River near Lake Billy Chinook (photo courtesy Oregon State Highway Department).

Suggested Readings:

Anderson, James Lee., et al, 1987. Distribution maps of stratigraphic units of the Columbia River basalt group. Wash. Div. of Geology and Earth Res., Bull.77, pp.183-195.

Barrash, Warren, Bond, John, and Venkatakrishnan, R., 1983. Structural evolution of the Columbia River plateau in Washington and Oregon. Amer. Jour. Sci., v.283, pp.897-935.

Beeson, Marvin H., and Tolan, Terry, 1987. Columbia River gorge: the geologic evolution of the Columbia River in northwestern Oregon and southwestern Washington. In: Hill, Mason L., ed., Cordilleran Section of the Geological Society of America, Centennial Field Guide, 1987, pp.321-326.

----Perttu, Rauno, and Perttu, Janice, 1979. The origin of the Miocene basalts of coastal Oregon and Washington: an alternative hypothesis. Oregon Geology, v.41, no.10, pp.159-166.

----Tolan, Terry L., and Anderson, James Lee, 1989. The Columbia River Basalt Group in western Oregon; geologic structures and other factors that controlled flow emplacement patterns. In: Reidel, S.P., and Hooper, P.R., eds., Volcanism and tectonism in the Columbia River flood-basalt province. Geol. Soc. Amer., Special Paper 239, pp.223-246.

Bishop, Ellen Morris, 1990. A field guide to the geology of Cove Palisades State Park and the Deschutes basin in central Oregon. Oregon Geology, v.52, no.1, pp.3-12,16.

Catchings, R.D., and Mooney, W.D., 1988. Crustal structure of the Columbia Plateau: evidence for continental rifting. Jour. Geophys. Res., v.93, B1, pp.459-474.

Fecht, Karl R., Reidel, S.P., and Tallman, Ann M., 1987. Paleodrainage of the Columbia River system on the Columbia Plateau of Washington state - a summary. Wash. Div. of Geol. and Earth Res. Bull.77, pp.238-245.

Hooper, Peter R., and Conrey, R.M., 1989. A model for the tectonic setting of the Columbia River basalt eruptions. Geol. Soc. Amer., Special Paper 239. pp.293-305.

Hooper, Peter R., and Swanson, D.A., 1987. Evolution of the eastern part of the Columbia Plateau. Wash. Div. Geology and Earth Res., Bull.77, pp.197-214.

Peterson, Norman V., and Groh, E.A., 1972. Geology and origin of the Metolius Springs, Jefferson County, Oregon. Ore Bin, v.34, no.3, pp.41-51.

Reidel, Stephen P., and Tolan Terry L., et al., 1989. The Grande Ronde Basalt, Columbia River Basalt Group; stratigraphic descriptions and correlations in Washington, Oregon, and Idaho. In: Reidel, S.P., and Hooper, P.R., eds, Volcanism and tectonism in the Columbia River flood-basalt province. Geol. Soc. Amer., Special paper 239, pp.21-52.

Smith, Gary A., 1987. The influence of explosive volcanism on fluvial sedimentation: the Deschutes Formation (Neogene) in central Oregon. Jour. Sed. Pet., v.57, no.4, pp.613-629.

At Biggs Junction, Oregon, the Columbia River cuts through a thick succession of the Miocene Columbia River lavas [photo courtesy Oregon Department of Transportation].

With Mt. Bailey and Mt. Thielsen on the skyline, this northward view across majestic Crater Lake shows layers of ash and lava exposed in the east rim of the volcano [photo courtesy Oregon Department of Transportation].

Cascade Mountains

Physiography

With its high glaciated volcanic peaks, the Cascade Mountain physiographic province is easily the most dramatic and scenic in the state. Extending south into California and north into Washington and British Columbia, the Cascade range is actually two parallel chains. The older, deeply eroded Western Cascades sharply contrast with the more recent snow-covered High Cascade range that sits on the eastern flank of the older range. The Western Cascades, with heights of 1,700 feet at the western edge to 5,800 feet on the eastern margin, are only half the elevation of the High Cascade peaks which reach altitudes of more than 11,000 feet. This extensive province divides the state into western and eastern parts and touches all of the physiographic regions in Oregon except the Blue Mountains and Coast Range. The Cascades are bounded by the Willamette Valley on the west, by the Deschutes-Columbia Plateau on the northeast, by the High Lava Plains in the central region, with the Basin and Range to the southeast, and the Klamath Mountains to the southwest.

Around 60 to 100 inches of rainfall annually result in deep weathering that is a characteristic of the Cascades where major westward-flowing rivers drain the range. North of the Calapooya Mountains, the Sandy, Clackamas, Santiam, McKenzie, and Middle Fork rivers all wind their way from headwaters in the mountains through an irregular pattern of dissected ridges and valleys before entering the Willamette River. South of the Calapooya divide, streams as the Umpqua and Rogue flow into the ocean. Across the Cascades, streams flowing eastward are small in comparison. Toward the north end of the range, creeks such as Fifteenmile and Threemile as well as the Warm Springs and Metolius rivers all reach the Deschutes River which has its headwaters near Mt. Thielsen. The few creeks that occur south of Mt. Thielsen drain into the Great Basin and ultimately empty into Upper Klamath Lake.

Geologic Overview

The geologic story of the Cascades begins around 40 million years ago with eruptions from a chain of volcanoes just east of the Eocene shoreline.

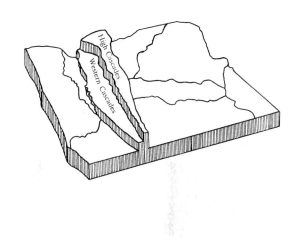

At that time the coast ran northwest through the region of the Willamette Valley. Triggered by movement of enormous crustal plates, intense volcanic activity produced thick accumulations of lava and ash that built up the older Western Cascade volcanoes. Between the eruptions, erosion, assisted by subtropical climates, stripped volcanic material from the slopes to redeposit it along the coastal plain and in the shallow near-shore waters of the ocean.

With uplift of the landmass, the Tertiary sea retreated progressively westward. Volcanic activity shifted to the east in the Miocene and Pliocene when folding and tilting were followed by massive outpourings of lavas in the range as well as throughout Oregon. At this time, large Cascade stratovolcanoes extruded thick lavas. About 5 million years ago the Western Cascades were again tilted creating a sloping ramp on the west side and a steep face on the east. It is this ramp of the Western Cascades and not the dominant and more obvious peaks of the High Cascade range that casts a rainshadow over eastern Oregon, dramatically changing the climate.

As the westward tilting took place, another episode of volcanism from local vents resulted in the more pronounced cones that form the High Cascades today. The High Cascade peaks appear in sharp contrast to the flat lava plains of eastern Oregon and the

Significant volcanic peaks of the Cascade Range

American crustal plate. The early volcanism took place in several stages beginning with lava flows from a volcanic chain lying immediately to the east of the Pacific continental margin. As far back as 40 million years ago, a number of small low volcanoes, irregularly spaced along a northwest-southeast belt in the region of eastern Oregon, deposited a thick accumulation of andesitic tuffs and lava flows that formed the base of the Western Cascade mountains. The broad width of

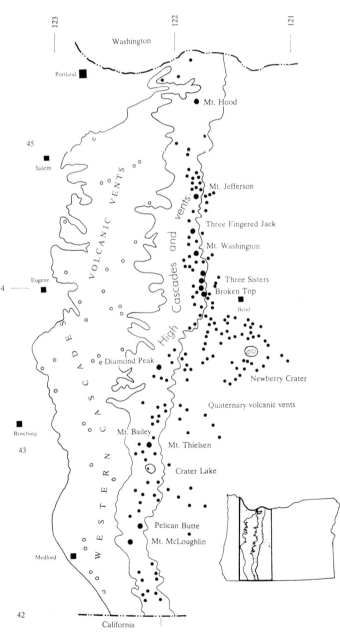

Volcanic vents in the Western and High Cascades (after Peck, et al.,1964; Priest, et al., 1983)

less noticable, eroded Western Cascades. With extensive outpouring of basaltic lavas, the region beneath the volcanoes continued to sink, creating a series of troughs or grabens that extend most of the length of the range. Recent volcanic activity in the province took place as late as 200 years ago when black cindery lava flowed out on the flanks of the younger peaks.

The Ice Ages of the Pleistocene brought glaciers to the range, with ice carving out glacial valleys, daming lakes, and eroding the sharp serrated crests of Mt. Washington, Three Fingered Jack, Mt. Thielsen, and Mt. McLoughlin. Glacial ice is still an active geologic agent at many of the higher elevations.

Geology

Construction of the Oregon Cascade range began with Eocene volcanism that was triggered by subduction of the Farallon plate beneath the North

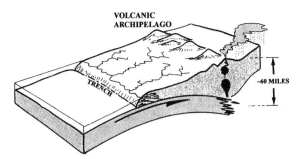

The distance between the subduction trench and volcanic archipelago is believed to be a function of the relative age of the crust being subducted and the rate of plate movement. Young, warm crust moves rapidly, is buoyant, and assumes a low angle of subduction with considerable distance between the arc and trench. Older, cooler crust moves more slowly, is less buoyant, and droops at a pronounced angle upon subduction bringing the arc very close to the trench.

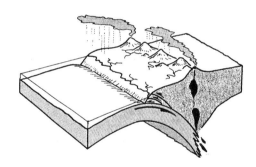

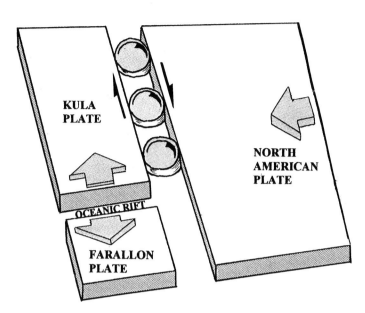

As an alternative to the entire Coast Range block swinging like a clock hand, the "ball-bearing" model accommodates the measured rock magnetic rotations by a simple rifting mechanism [after Wells and Heller, 1988].

this proto-Cascade volcanic belt suggests that the underlying subducting slab was at a shallow angle and that the rate of collision between the North American and Farallon plates was rapid.

Today the origin of the Cascade range is still not well understood. One suggestion for the growth of the range is that a volcanic chain above an actively subducting ocean trench developed behind a slowly rotating Coast Range tectonic block. Breaking loose from the mainland along the Olympic-Wallowa lineament, the block rotated almost 50 degrees clockwise with a pivotal point in the northern Washington Coast Range to its present position. An alternate proposal is that Cascade volcanism was produced by a steepening subducting slab due to a marked decrease in the convergence rate between the North American and Farallon plates.

Cenozoic

During the Eocene and Oligocene epochs the coastline angled northwest by southeast through the area of the Willamette Valley and just to the west of the volcanic vents along the Cascade archipelago. Volcanic ash from the vents was flushed by streams into shallow marine basins along the coast where thick forests of tropical to subtropical plants on the low coastal plain are now preserved as wood, leaves, and pollen. Coal-producing swamps and estuaries formed the alluvial deposits of the Colestin Formation, the oldest rocks in the Cascades. Nonmarine volcanic tuffs, sandstones, breccias, and lava flows of the Colestin reach a thickness of over 4,000 feet along the Cascades from Lane County south to California. From Eugene northward to Molalla, upper continental shelf sands and silts were the final marine sediments to be deposited along the retreating ocean shoreline. Molluscs and a variety of other invertebrates preserved in these rocks of the Eugene, Illahee, and Scotts Mills formations reflect warm subtropical to temperate waters. Broken, worn fossil fragments in some areas testify to stormy coastal conditions, while local accumulations of barnacle shells are so rich in many of these near-shore

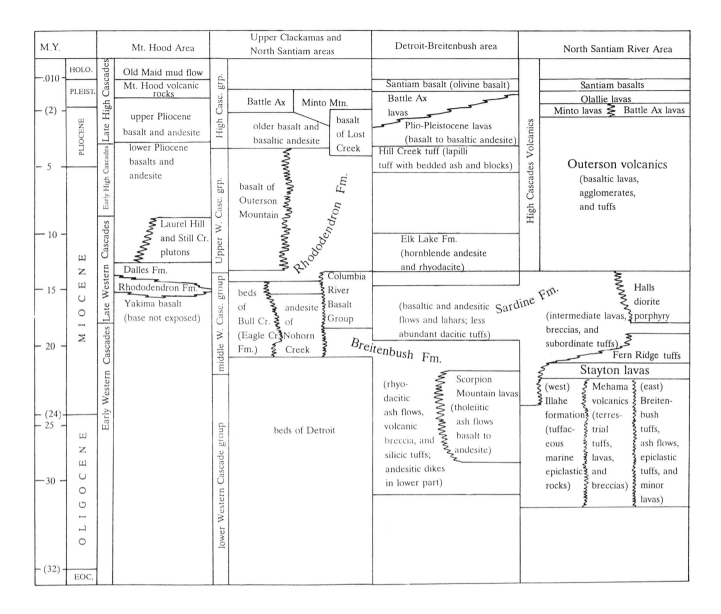

Cascades Stratigraphy (after Priest and Vogt, 1983)

sediments that they built up limestone lenses marking the old shoreline on the eastern margin of the Willamette Valley. By Oligocene time numerous eruptions of andesitic lavas and siliceous tuffs of the Little Butte Formation were interspersed with oceanic sediments along the eastern margin of the valley.

Western Cascades

A period of tilting and folding during the middle Miocene was followed by renewed outpourings of lavas throughout Oregon accompanying the growth of the Western Cascades volcanic arc. Even with a large volume of volcanic debris, the growth of the range was modest as the accumulation sank almost as fast as it piled up. About the same time voluminous

quantities of the Columbia River Grande Ronde basalts began to flow from fissures near La Grande about 17 million years ago. These massive flood basalts blanketed the landscape filling in depressions to produce an even, flat platform of cooled lavas. The eruptions were synchronous in part with volcanic activity taking place in eastern Oregon in the Strawberry Mountains as well as in Steens Mountain. Concurrently in the Western Cascades violent eruptions from composite cones active between 13 and 9 million years ago comprise the Sardine Formation. The combined amount of all this volcanic material has no modern counterpart. By 7 million years ago the belt of active volcanoes had narrowed to an area only slightly wider than the present High Cascade Range.

Columnar jointing in basalts of the Little Butte Formation along Wiley Creek (photo Oregon Dept. Geology and Mineral Industries)

active volcanoes had narrowed to an area only slightly wider than the present High Cascade Range.

The intensity and focus of Cascade volcanism is the by-product of tectonic forces at work deep in the crust. Sitting atop the North American plate, the Western Cascades were rotated clockwise into their present position by the time the earlier eruptions of the Columbia River basalts began approximately 17 million years ago. This was well before the Pliocene episodes of High Cascades volcanism. As the Cascades rotated and the angle of the descending Farallon slab became flatter, the linear centers of volcanic activity simultaneously swept from west to east. The oldest rocks produced in the Western Cascades are approximately 42 million years old, while the youngest at 10 million years are found at the eastern edge of the High Cascades. This change geographically is found within a very narrow range of a single degree of longitude.

Along with a change in volcanic focus, there was a profound difference in amounts of volcanic material erupted across the Cascades, with significantly more lava in the west and less in the eastern portion. In comparing the volcanic activity between the two ranges, there has been six times more volcanism in the

western mountains than in the eastern, younger Cascades. This pronounced decrease coincides nicely with a slowing in the convergence rate of two tectonic plates beneath this area during the same time interval. Initially 35 million years ago the Farallon plate was being subducted at the rate of 3 inches per year, decreasing to about 1/2 inch per year today. In addition, with the clockwise rotation of a portion of the North American plate, the convergence angle between the North American and Farallon plates became progressively more oblique. The more oblique collision resulted in less subduction and consequently less volcanic activity. This trend, which began in the Miocene, has continued into the present.

The Western Cascades were subjected to additional uplift, mild folding, and faulting between 4 and 5 million years ago. Near the center of the range, major north-trending faults and a pronounced McKenzie-Horse Creek fault to the east, mark the boundaries of the uplifted Western Cascade block. What has been called the Coburg Hills fault creates a sharp front range to the Western Cascades just north of Eugene.

High Cascade Development
Two events, the westward tilting of the Western Cascades and the sinking of the Cascade graben or depression signal the beginning of High Cascade volcanism. Numerous, overlapping shield volcanoes and cinder cones built up a platform of High Cascades aligned in a north-south direction on the eastern flank of the older Cascades. Basalts, tuffs, and ash-flows form the base of the range during the early phase of development. With the late High Cascade volcanic episode, from 4 million years ago to the present, a variety of lavas and tuffs again dominated by basalts were expelled from a number of small composite cones constructed on the sides of the earlier, large shield volcanoes. Basalt makes up as much as 85% of the Quaternary High Cascades by volume, although locally silica-rich dacite and rhyodacite lavas have been erupted. Where the Cascades extend south into California and northward into Washington, Quaternary activity decreased, and more andesite was produced. Virtually all of the High Cascade peaks are Quaternary or younger in age, and the eruptions of lavas during this final episode of volcanism are responsible for most of the familiar, present day topography.

Westward tilting of the Cascade block, a process that began in the Miocene, created the Blue River ramp sloping gently upward to the east. The ramp made a rainshadow by trapping the flow of humid oceanic air in western Oregon. As it was elevated over the Western Cascades, the moisture-laden air cooled bringing precipitation of rain and snow. With this

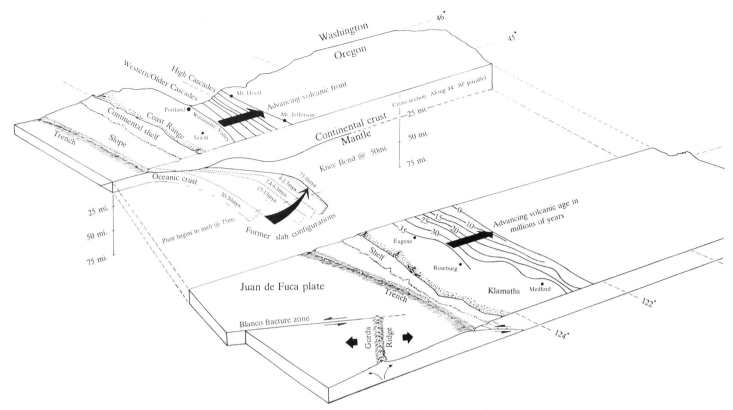

uplift, what was once a moist tropical climate in the eastern part of the state was transformed to the high desert environment of today. The uplift was accompanied by intensive downcutting by streams and erosion of the mountains so that today the older Cascades display deeply cut ridges of resistant lava rather than the rounded volcanoes that once existed.

Concurrent with and resulting from High Cascade volcanic activity was the formation of structural grabens, lying within the eastern margin of the Cascade range. The word "graben" derives from the German "grave" and refers to the sunken trough of earth above a collapsed buried coffin. Structurally a graben is a lowered block sitting between two faults. Beneath much of the younger High Cascades enormous, discontinuous grabens, 10 to 20 miles wide, extend for 300 miles from north to south. In the central Cascades, where volcanoes are densely clustered, both the east and west sides of the graben are well-developed. At the northern and southern borders of the state, where volcanoes are more sparce, the graben is either incomplete or only a "half" graben. Movements of the various faults that make up the east and west sides of the trough are estimated to have dropped down an average of 2,000 feet. The sinking of the grabens, which began between the late Miocene and very early Pliocene, 5 to 10 million years ago, has continued well into the late Pliocene. Collapse of the grabens followed the eruptions of volcanoes pumping

In order to explain the simultaneous eastward advancement and narrowing of the Cascades volcanic front, a popular synthesis calls for a "knee-bend" in the subducting slab that is swinging upward (after Priest, 1990; Duncan and Kulm, 1989; Verplanck and Duncan, 1987).

out magma that left many chambers beneath the area. With the outpouring of basalts and growth of broad shield volcanoes, the heavy volcanic "loading" of the incipient High Cascades may also have contributed to the downward rupture of the grabens.

The Cascade ranges are still too young for erosion to have extensively exposed the batholiths that fed the volcanic eruptions. The siliceous composition of some of the High Cascade volcanoes, however, suggest that these chambers of magma were close to the surface because the highly viscous, molten rock, which forms the massive batholiths, has difficulty travelling upward very far into the crust before cooling. Although these intrusive bodies should theoretically be present at shallow depths beneath the High Cascades, geophysical probes of the range suggest the source magmas are deeper in the lower crust and upper mantle.

Within the Western Cascades, intrusive batholithic rocks are situated in two north-south belts running from Canada across Washington and Oregon. Having been emplaced between 35 and 15 million years ago, the western belt, through Mount St. Helens and ending near Portland, is younger. The eastern belt, which cooled between 50 and 25 million years ago, stretches approximately 45 miles east under Mt. Adams and south through Mt. Hood and into southern Oregon.

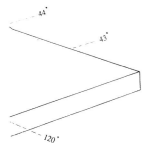

The eruptive history of Cascade volcanoes has followed a loose pattern over the past 20 million years. Since the middle Miocene, major events, each continuing for short periods of only 1 to 2 million years, have occurred at 5 million years intervals. There is evidence that these volcanic episodes in the Pacific Northwest correspond with global volcanism taking place at the same time. World-wide, shorter periods of volcanism were interspersed with longer periods of relative inactivity. The magnitude of volcanism and volume of extrusions has been decreasing steadily since the middle Miocene.

Recent Volcanism

Dramatic eruptions have been absent in Oregon in the 20th century, although at least eleven volcanoes in the entire Cascade range, where it extends north into Washington and south into California, have been active historically. In Oregon Mt. Hood and the South Sister have signalled volcanic activity recently. As in the past, Cascade volcanism today is due to colliding tectonic plates. The Juan de Fuca plate

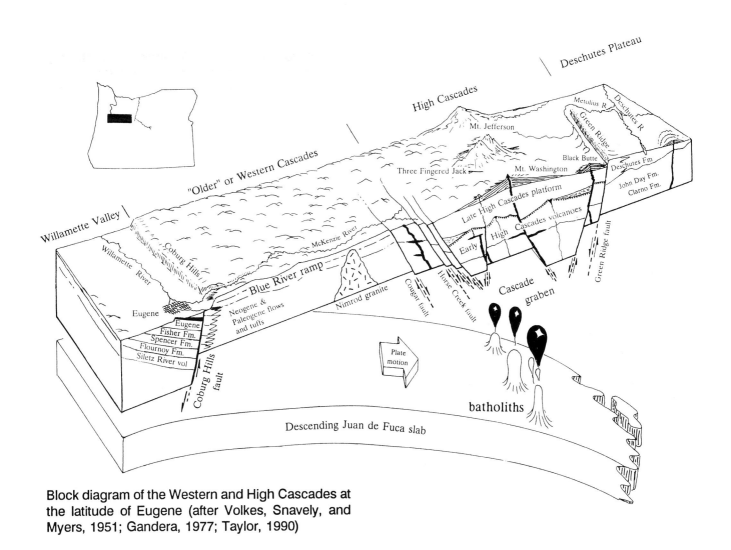

Block diagram of the Western and High Cascades at the latitude of Eugene (after Volkes, Snavely, and Myers, 1951; Gandera, 1977; Taylor, 1990)

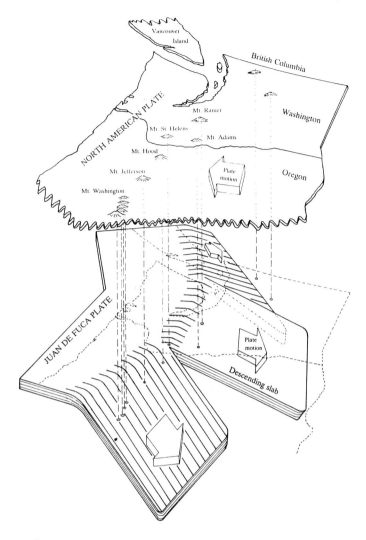

Geometry of the Juan de Fuca slab below Oregon and Washington (after Michaelson and Weaver, 1986; Duncan and Kulm, 1989)

and eroded by ice. Lava as well as ash, steam, and gasses from the stratovolcanoes and small vents mixed with packed ice and snow causing lahars or mud flows down the sides of the mountains. Ice Age glaciation, which brought a massive continental ice sheet south to the northern Washington border, was responsible for extensive valley glacier complexes in the Oregon and Washington Cascades beginning about 2 million years ago. A glacier system on Mt. Hood was situated at the northern end of a nearly continuous ice field that capped the High Cascades and extended south to Mt. McLoughlin as late as 15,000 years ago. Glacial ice stretched from the peak of Mt. Hood down the Sandy River almost to the Columbia. Even the Western Cascades had small ice masses in the valleys where glaciers from the higher peaks reached into the lower canyons, advancing and retreating repeatedly to carve out valleys, steepen walls, and leave behind thick deposits of till.

In the valley floors, moraines of gravel, sand, silt, and clay blocked streams which backed up to create small lakes. Many of the higher Cascade lakes occupy semi-circular basins called cirques, formed by glacial erosion, while some of the lower lakes, such as Suttle Lake near Santiam Pass in Deschutes County, ponded up behind moraines. Virtually all of the naturally occurring lakes in the Cascades owe their origin to glaciation, lava dams, or landslides.

Although the last cold, wet climatic interval ended 10,000 to 12,000 years ago, small scattered glaciers persist today throughout the range. Mt. Hood has nine glaciers with a total volume of 12.2 billion cubic feet of ice and snow. If it were melted, this ice would provide enough water to enable the Columbia River to flow at a normal rate for 18 hours. The volume of ice above Crater Rock alone is 47 million cubic feet. Of the five major glaciers on the volcanic cones of the Three Sisters, Collier Glacier, extending from the north slope of the Middle Sister across to the base of the North Sister, is the largest with a volume of 700,000 million cubic feet. Even though it has been disappearing very rapidly in recent years, 300 feet of ice make this glacier the thickest of those on the Three Sisters. The Middle Sister's steep eastern face is covered by the Hayden and Diller glaciers, while glaciers on the North and South Sisters are very small. Should an eruption occur today in the Cascades, as happened in the case of Mount St. Helens, extensive floods and mudflows would be a by-product of the melting ice and snow.

represents the trailing edge of the older Farallon plate, and the descending slab dips up to 65 degrees beneath the North American plate. As this slab plunges into the upper mantle at about 60 to 75 miles depth, its leading edge begins to melt partially creating the magma chambers that feed Cascade volcanoes. Geophysical evidence suggests that the subducting Juan de Fuca plate splits around 30 miles depth into three eastbound descending tongues. Of these, the Washington portion dips at a shallow angle while the tongues beneath Oregon and British Columbia are much steeper.

Pleistocene

Quaternary volcanic activity in the High Cascades was accompanied by extensive Pleistocene glaciation when the major peaks were being scoured

Structure

The Cascades are scarred by numerous north-northwest fault zones and lineaments, enormous lines

With Black Butte in the background, Suttle Lake, near the summit of the high Cascades, offers fishing, boating, swimming, camping, and other popular pastimes (photo courtesy Oregon State Highway Department).

Forests of pines and firs blanket the lower slopes of Mount Hood in the Cascade Range of Oregon. This view from the southeast side graphically shows the effects of centuries of erosion on the vast lava flows (photo Oregon State Highway Department).

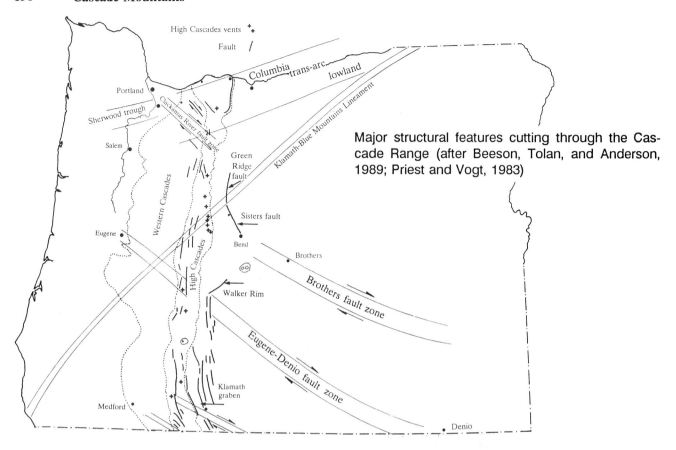

Major structural features cutting through the Cascade Range (after Beeson, Tolan, and Anderson, 1989; Priest and Vogt, 1983)

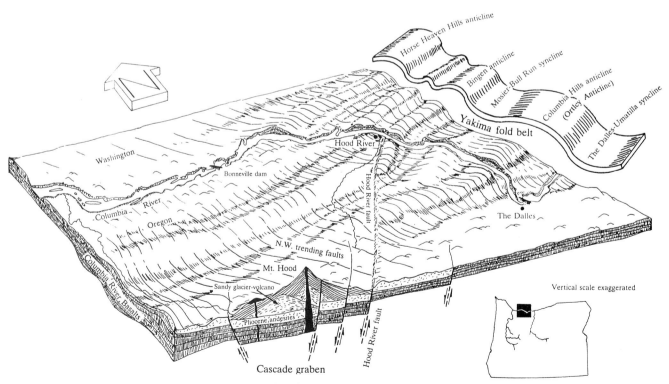

Yakima fold belt and Mt. Hood structure in the Cascades (after Williams, et al., 1982; Tolan and Beeson, 1984)

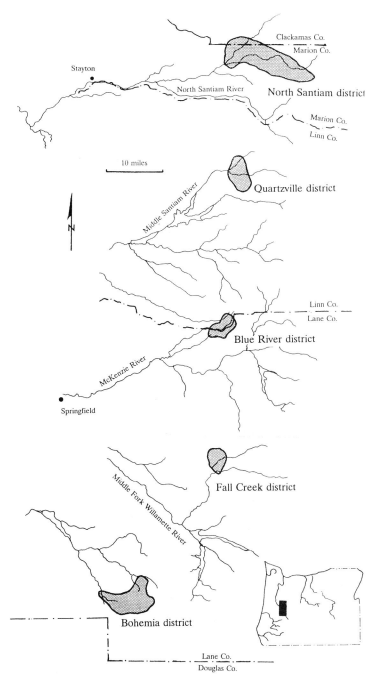

Mining districts situated along the eastern margin of the Western Cascades (after Ferns and Huber, 1984)

Pleistocene less than a million years ago. The Clackamas River fault projects northwesterly to merge with the Portland Hills fault zone. One of the most extensive structures across the Cascades is the Eugene-Denio fault system. Recognized initially as a lineament, the fault zone begins southeast of Eugene on the west side of the Cascades and runs through the range appearing again at Walker Rim east of the mountains. The dominant period of movement along this fault dates back to 22 million years ago.

Running southwesterly into the Cascades from central Washington, the Yakima structural belt is a series of late Miocene east-west by southwest-northeast trending gentle folds in the Columbia River lavas. Along the banks of the Columbia River, the fold belt appears in cross-section where the gorge cuts through the Ortley and Bingen anticlines and Mosier syncline. These folds fit into a broad southwest-northeast trending structure called the Columbia trans-arc lowland which subsided before the Columbia River basalt flows invaded the gorge more than 16 million years ago. This broad trough served as a natural conduit for these lavas as well as for the Columbia River itself. In the Willamette Valley to the west, a farther extension of this lowland is the Sherwood trough.

From the southwest corner of the state, the Klamath-Blue Mountains lineament projects through the Cascades to run parallel to and just north of the Blue Mountains anticline. Whereas many of the lineaments are now recognized as large-scale faults, the Klamath-Blue Mountains structure is of unknown origin.

Jawbone Flats, constructed in 1932 as a mining camp, was part of the North Santiam mining district (photo Oregon Dept. Geology and Mineral Industries).

of unclear origin, running diagonally through the range. From the southeast, the McLoughlin, Eugene-Denio, and Brothers fault zones extend from the Basin and Range and High Lava Plains provinces to the Cascades. The poorly understood McLoughlin belt of faults terminates in the southern High Cascades somewhere in Lake or Jackson counties. Toward the southeast, the Clackamas River belt of faults may be an extension of the Brothers fault zone which was active well into the

The Champion Mill, a 100-ton selective flotation plant, processed ore from the Champion and Musick Mines (photo Oregon Dept. Geology and Mineral Industries).

Mining and Mineral Industry

Gold and silver in the Western Cascades are situated in a 25 to 30-mile wide strip west of and parallel to the volcanic peaks of the younger Cascades. The belt includes five major mining regions, the North Santiam district in Clackamas and Marion counties, the Quartzville district of Linn County, the Blue River district of Linn and Lane counties, and the Fall Creek and Bohemia districts in Lane County. Along with gold and silver, the mining districts yield small amounts of lead, zinc, and copper as pyrite, galena, sphalerite, and chalcopyrite.

Most of the ores occur in veins cutting through older volcanic rocks of the Western Cascades. The veins are often in shattered breccias that have been invaded by mineral-rich hydrothermal fluids. The original volcanic host rocks were faulted and fractured providing an avenue for hydrothermal fluids at temperatures of 250 to 350 degrees Fahrenheit. Today the Western Cascades are deeply eroded so that exposures of the ore- bearing intrusive volcanic rocks within most of the mining districts are at lower elevations along valley floors.

Of all the mining areas in the Cascades, the Bohemia region in Lane County as been the largest and most productive. Mineralization here is confined to nine square miles near Bohemia Mountain 25 miles south of Cottage Grove. The story of the Bohemia district gold discovery began in 1858 when W.W. Oglesby and Frank Bass, miners from California looking for new prospects, discovered placer gold on Sharps Creek. In the spring of 1863 lode gold deposits were located by a man named James Johnson who had

come north to the *"extreemly wild and untenanted wilderness"* to escape punishment for killing an Indian. In the process of dressing out a deer, Johnson found gold quartz, the value and significance of which he recognized immediately. Since Johnson was from Bohemia in eastern Europe, the region was named for its discoverer "Bohemia" Johnson.

Johnson's find attracted a number of miners to this uninhabited terrain. In spite of heavy snows, lack of roadways, and other difficulties, miners organized the Bohemia Gold and Silver Mining District in 1867. Production expanded greatly with the exploitation of the Musick vein in 1891 and the addition of a stamp mill about the same time. The Champion, Helena, Musick, and Noonday mines were the main producers in the Bohemia district, with a total of about 1 million dollars. Along with gold and silver, the Bohemia region has also been mined extensively for copper, zinc, and lead.

The amount of gold and silver taken from the Bohemia district was followed by that of Quartzville and Blue River. At Quartzville in Linn County, gold-bearing veins were located in September, 1864, by Jeremiah Driggs. The complete output of mines and mills here was $181,000, which took into account 8,557 ounces of gold and 2,920 ounces of silver. Less than 200 ounces of gold was removed after 1896, and mines have operated only intermittently since that time. In 1984 the Quartzville Recreation Corridor was reserved for recreational mining as well as other outdoor uses. About 45 miles east of Eugene, the Blue River mining district was largely confined to output from the Lucky Boy Mine located in 1887. Several other mines were opened when a 40-stamp mill was installed, however, practically all operations ceased in 1913 when the Lucky Boy closed. Overall production figures for gold and silver have been estimated at $175,000.

Along with the Bohemia area, the North Santiam district in Marion and Clackamas counties is one of the most versatile, yielding copper, zinc, lead, silver, and gold since it was first opened in the 1860s. Most activity focused on the Ruth Vein found in the early 1900s where zinc and lead have been mined periodically. While amounts of gold and silver from this district remain low in comparision to those of other minerals, 454 ounces of gold and 1,412 ounces of silver were removed. The total dollar figure for all minerals from here is $25,000. Recent exploration involving surveying and drilling by the Shiny Rock Mining Company, which held a large block of claims, has been for copper. The claims were leased in 1980 to Amoco Minerals Company. Situated in eastern Lane County, gold deposits in the Fall Creek district are of

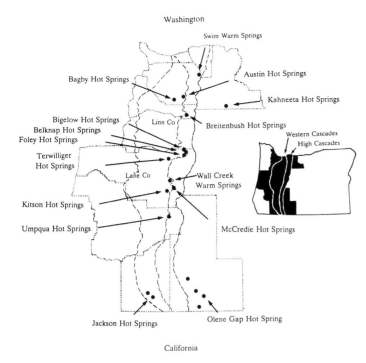

The location of hot springs along the backbone of the Oregon Cascades

comparatively low grade, and mining has been limited.

While production figures for Cascade mining districts can only be estimated because of the lack of records, the total revenues from all minerals extracted since 1858 is $1,500,000, keeping in mind that the price of gold was much lower at that time.

Geothermal Resources

Between the High Cascade and Western Cascade ranges, a belt of hot springs marks a major heat flow change that takes place very abruptly within a distance of less than 12 miles. Low temperature gradients in the Willamette Valley and adjacent Western Cascades increase to high temperatures on the eastern side of the Western Cascades and High Cascades. The reason for the change in heat flow is unknown since the heat source hasn't been located, but higher temperatures may result from a combination of warm groundwater and partially molten material under the Cascade area east of the boundary.

On the surface, the thermal boundary is manifest by springs. Those springs in the central portion of the Cascades have higher water temperatures than those at either end. Many of the thermal springs in the Cascades are located along faults where superheated waters migrate up along the fractures. Although the hot springs that appear on the surface

may be water that is recirculating through a fault system, the specific mechanism producing the springs is unknown. Circulating from east to west, thermal waters reach the surface in rocks of the Western Cascades where erosion has cut deep valleys into the older volcanics and where lavas of the younger Cascades are absent. The springs, situated in this way at the bottoms of deep canyons, serve as a division between the older and younger Cascades. Hot springs are virtually absent directly under the central portion of the higher Cascade range. Heavily fractured younger volcanic rocks here at the crest mix hot springs with cold groundwater long before the waters reach the surface.

Jackson Hot Springs is the most southerly of the Oregon system of springs in Jackson County, while the warm springs at the old town of Swim in Clackamas County are at the northern end of the alignment. Other thermal springs in the chain include Bagby Hot Springs, Breitenbush Hot Springs, Belknap Springs, Foley Hot Springs, McCredie Springs, and Kitson Hot Springs. Temperatures of the spring waters range from around 90 to 190 degrees Fahrenheit. Breitenbush and Belknap Hot Springs have seen extensive commercial development. With the largest discharge of water at 900 gallons per minute from 40 openings, Breitenbush water temperatures reach 180 degrees Fahrenheit as compared to 190 degrees Fahrenheit at Belknap Springs.

Geothermal potential in the Western Cascades is relatively low, while to the east of the thermal boundary higher groundwater temperatures indicate the potential for geothermal exploitation. Of all the regions in the range, Mt. Hood has been thoroughly examined for geothermal resources. The continued tectonic history of the mountain from a number of volcanic centers implies a high heat flow and hydrothermal systems. Although there is little evidence of high temperature waters circulating through near-surface faults here, Mt. Hood remains a target for further exploration.

Features of Geologic Interest

Columbia River Gorge

The Columbia gorge is 75 miles of geologically spectacular scenery along the Columbia River from the narrows at The Dalles west to Portland. Mt. Hood in Oregon and Mount St. Helens and Mt. Adams in Washington overshadow the canyon where it cuts through the Cascades. In its long history, the ancestral channel was repeatedly filled with lava or blocked by landslides and ice dams, but the waters have surmounted these obstacles quickly by frequently changing

From The Dalles to Portland, the Columbia River is entrenched in a gorge 72 miles long across the Cascade Range.

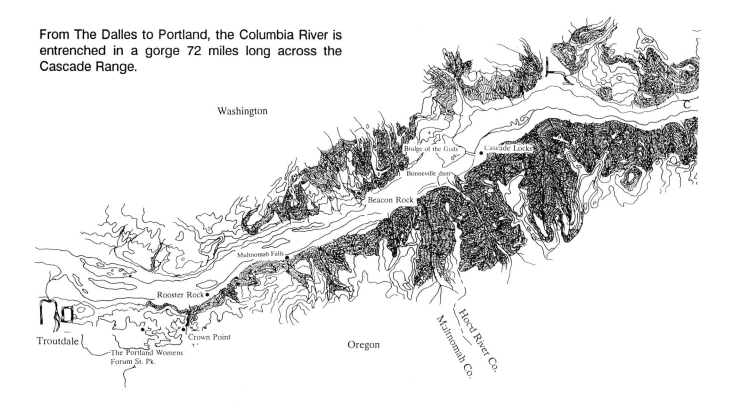

pathways to continue the westward course.

In the middle Miocene, about 16 million years ago, this stretch of the river acted as a conduit for the massive Grande Ronde flows of the Columbia River lavas. The ancient Columbia River canyon, with its gentle slope toward the coast, offered a natural route for lavas making their way to the Pacific Ocean. A succession of flows poured into the gorge to harden and plug the flow of water. With each lava dam blocking its path, the Columbia cut a new channel in its northern bank, so it can be said that the Oregon boundary gradually crept northward leaving a series of lava filled channels to the south. Where they are exposed, the ancient channels have been mapped in the gorge, but most of these have been completely covered over by later flows. Looking upriver at the Portland Women's Forum State Park near Corbett, two previous channels can be seen where they are intersected by the river today. One of these former channels is visible in cross-section at nearby Crown Point. Exposed in the sheer walls of the canyon below the point, a single flow of Priest Rapids basalt, youngest member of the Wanapum Formation, invaded the canyon of the Columbia River. The blocky jointed Priest Rapids basalt is 500 feet thick covering more than 200 feet of glassy, volcanic sediments which had choked the canyon earlier.

Nearby Rooster Rock is a mass of basalt, similar to that at Crown Point, which slid from the canyon wall into soft alluvium where it stands upright

today. The tall columnar basalt pillar rises 200 hundred feet and is a favorite of rock climbers. The rock and surrounding acreage were purchased in 1938 for $10,000 as a state park.

In addition to lavas from the east, smaller flows from local volcanic vents in the Cascades periodically disrupted the progress of the river through the gorge. Large lakes backed up behind these natural dams. Above Cascade Locks the remains of a delta are visible where the ancestral Wind River lavas from the Washington Cascades poured south to create a lake. Before the waters eventually drained away, when the river destroyed the dam, Wind River had constructed a delta 150 feet thick and one mile long in the ponded waters.

Landslides have had significant implications for Oregon's past. It was along this section now covered by the waters of Bonneville Dam that Indians and explorers found that the river dropped *"in continued rapids for three miles, not less than fifty feet,"* as Reverend Gustavus Hines wrote in his diary in 1839. The river cascades were created by a landslide of rocks, soil, and debris into the riverbed less than a thousand years ago. This slide which temporarily dammed the river, produced a land bridge from shore to shore and may be the origin of the Indian legend of the Bridge of the Gods reported at this point.

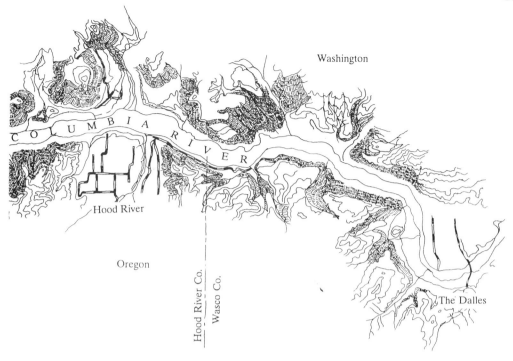

Hood River

Oregon

Washington

The Dalles

Multnomah Falls
The Columbia River gorge is lined with 71 waterfalls in 420 square miles, 11 of which are over 100 feet high. Of these, Multnomah Falls, which is the fourth highest falls in the United States, drops, in what is actually two falls, a distance of 620 feet over a projecting ledge of Grande Ronde Basalt of the Columbia River lavas. Waterfalls tend to be ephemeral features, and most of those in the gorge may have begun during recent glacial floods 13,000 years ago when high waters cut back and carried away softer valley alluvium leaving the streams to fall over the more resistant basalt.

The Columbia River gorge has been intermittently plugged by local lava flows and landslides throughout the Pleistocene and Holocene (after Allen, 1979; Tolan and Beeson, 1984; Anderson and Vogt, 1987).

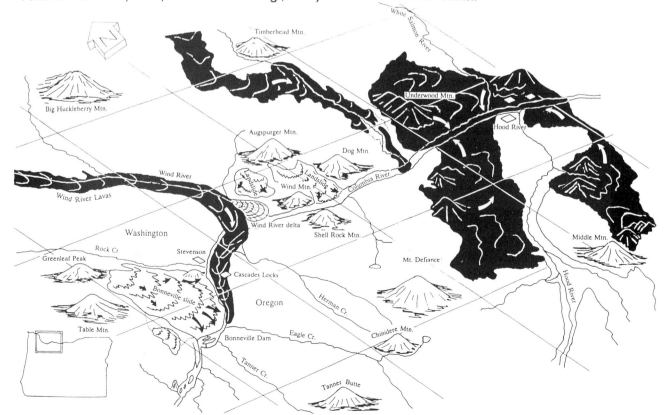

Pathways of separate Columbia River Basalt flows through the Columbia gorge (after Anderson and Vogt, 1987)

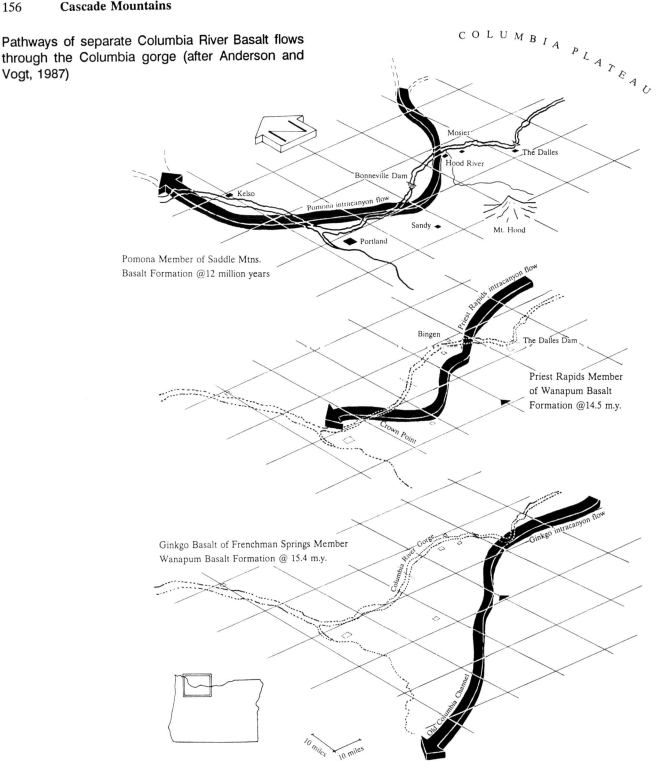

COLUMBIA PLATEAU

Pomona Member of Saddle Mtns. Basalt Formation @12 million years

Priest Rapids Member of Wanapum Basalt Formation @14.5 m.y.

Ginkgo Basalt of Frenchman Springs Member Wanapum Basalt Formation @ 15.4 m.y.

Beacon Rock

On the north bank of the Columbia upriver from Multnomah Falls, Beacon Rock served as a signal to travellers that the last of the river rapids was behind them and the Pacific Ocean lay ahead. The vertical 850 foot high Beacon Rock is part of a north-south extending dike of Cascade andesitic basalt. The resistant dike intruded the surrounding Eagle Creek Formation, which was eventually eroded away leaving the pillar.

High Cascade Mountains

The snow covered Oregon High Cascade mountains offer a variety of spectacular scenery including glaciers, snow fields, thick forested slopes, cold streams and lakes, and waterfalls. The crest of the range averages a little more than 5,000 feet in altitude, but well-known peaks rise considerably higher. The High Cascades, part of a chain of volcanic peaks

The 10,495-foot high Mt. Jefferson, located in the central portion of the Oregon Cascade range, is the second highest peak in the state (photo courtesy Oregon State Highway Department).

extending from California into British Columbia, began to form during the Pliocene when voluminous outpourings of lava from broad shield volcanoes built up along the eastern flanks of the older Western Cascades. Over a period of millions of years these volcanoes were worn down by erosion only to have later eruptions during the Pleistocene construct the volcanic peaks seen today in the same north-south alignment.

Thin, sharp remnants of the older, eroded Quaternary volcanoes, Mt. Jefferson, Three Fingered Jack, Mt. Washington, the North Sister, Broken Top, Mt. Thielsen, Mt. McLoughlin, and several other less conspicuous peaks are easy to distinguish from the later conical-shaped mountains. Of the dramatic younger volcanoes, Mt. Hood, Mt. Bachelor, the South Sister, and Mt. Mazama, now known as Crater Lake, although eroded to a degree, still retain much of the original composite cone-shape. Brown Mountain, 10 miles southeast of Mt. McLoughlin, Tumalo Mountain near the Three Sisters, and numerous low mounds near Mt. Hood erupted as small volcanoes on the flanks of the higher peaks.

Southern High Cascades

The southern-most cone in the Oregon High Cascade range is Mt. McLoughlin at 9,493 feet near the Oregon border. Formerly called Mt. Pitt, this peak was renamed for John McLoughlin, benefactor to early travellers in Oregon and factor of Hudson's Bay Company. Severe glacial erosion, which has carved out a distinct semi-circular basin or cirque in the northeast slope of Mt. McLoughlin, suggests this mountain is probably not one of the younger Cascade peaks. The first eruptions of Mt. McLoughlin may have occurred over 100,000 years ago, and two or three succeeding eruptions built up the cone and surrounding blocky lava flows. In the final stages, andesite lavas oozed from fissures along the base of the main cone and not from the main conduit.

Crater Lake

Oregon's most cataclysmic geologic event took place about 6,900 years ago when Mt. Mazama exploded then collapsed to create the symmetrical caldera of Crater Lake. One of the most serene sights in Oregon today, the crater and surrounding 160,290 acres in Klamath County were declared Crater Lake National Park in 1902. The geologic history of Crater Lake, as recorded in the walls of the caldera, began during the Pleistocene 400,000 years ago with the construction of

Southeast of theRim of the Crater Lake caldera, high slender spires of pumice have been carved out of the volcanic material. Some of the pinnacles are over 200 feet high (photo courtesy of Sam Sargent).

Crater Lake, a caldera with the small volcano of Wizard Island, is one of Oregon's most scenic features.

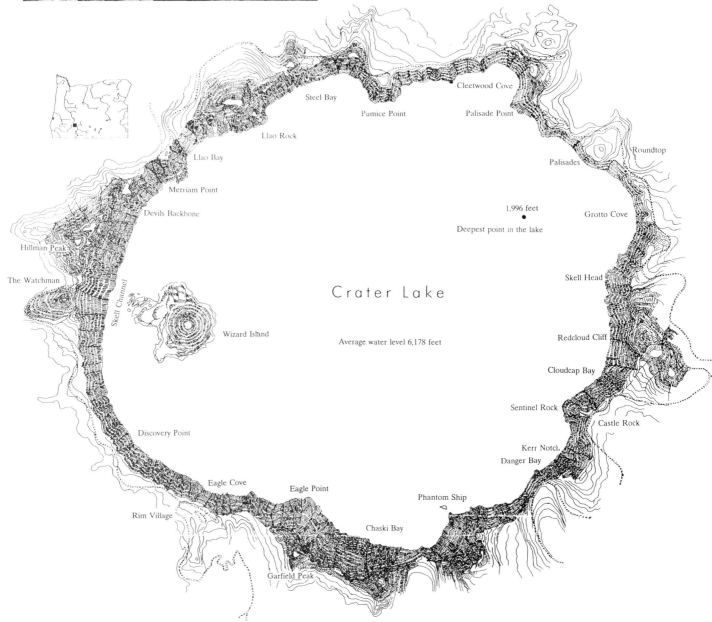

Steel Bay

Cleetwood Cove

Pumice Point

Palisade Point

Llao Rock

Llao Bay

Roundtop

Palisades

Merriam Point

Devils Backbone

1,996 feet

Deepest point in the lake

Grotto Cove

Hillman Peak

The Watchman

Skell Channel

Skell Head

Crater Lake

Wizard Island

Average water level 6,178 feet

Redcloud Cliff

Cloudcap Bay

Sentinel Rock

Castle Rock

Discovery Point

Kerr Notch

Danger Bay

Eagle Cove

Eagle Point

Phantom Ship

Rim Village

Chaski Bay

Garfield Peak

Three Fingered Jack, Mount Washington, and the Three Sisters of the central Cascades looking toward the north (photo courtesy Dave Rohr)

andacite and dacite stratocones and shield volcanoes from numerous flows. These early eruptions spread thin layers of lava 20 to 30 feet thick over the landscape in all directions. Following repeated eruptions and dormant periods, a large magma chamber developed under the Mt. Mazama stratovolcano during the final eruptive stages. At the same time small vents and cones on the flanks of the dome extruded viscous silica-rich rhyodacitic lavas as late as 7,000 years ago.

Shortly after this, a sudden violent explosion from a single vent in the main caldera sent out towering clouds of ash and fiery fragments with such force that layers of debris 6 inches deep covered the area of present day Bend 70 miles away. Following the incandescent ash cloud, hot glowing avalanches of pumice sped down the sides in immense quantities obliterating all topography. Approximately six cubic miles of magma drained from the chamber before the roof collapsed into the caldera. The entire event took place so rapidly that the youngest flow had yet to cool before the caldera had been formed.

Before the eruption and collapse, Mt. Mazama is estimated to have been 10,000 to 12,000 feet high. Today the caldera averages 8,000 feet high, so that approximately 2,000 feet of the cone collapsed into the caldera with the explosion. Several small cones, including Wizard Island, were built up from lava oozing from the floor after the caldera developed. The cold lake waters that occupy the caldera are remarkably clear because of the lack of algal activity. The bright blue

waters reach a depth of 1,996 feet, which is the deepest fresh water body in the United States. A balance between precipitation, seepage, and evaporation has kept the lake level near its present position throughout historic time.

In the late 1980s commercial concerns applied for permission to drill for geothermal power in the vicinity of Crater Lake. They were opposed by environmental groups which feared that drilling, even a distance away from the lake shore, might interfere with yet unknown plumbing systems connected to the springs said to exist on the lake floor. Research conducted with a remote-controlled submersible and later with a manned submersible revealed red and brown pigments in lake floor sediments as well as what appear to be algal mattes, both of which strongly point to the presence of hydrothermal vents. Additional evidence of thermal springs on the bottom of the lake is indicated by the richness of dissolved solids, high amounts of radon and helium gas, and slightly raised temperatures. These discoveries, while not conclusive, have retarded the issuing of permits for hydrothermal drilling.

Of the 40 or more caves of significant size within Crater Lake National Park, 30 are located near the lake level in the caldera. Most of these caves are the result of erosion of the lavas or layers of mudflows between the lavas. Caves within the rim are shallow and may be seen along the water level, whereas those caverns outside the caldera have more depth and topography. All of the caves are hazardous, and climb-

Oregon's rugged Mount Thielsen, called the lightening rod of the Cascades, rises 9,178 feet and forms one link in the chain of mountains which extends from Mexico to British Columbia (photo Oregon State Highway Department).

ing gear is needed to access many of them. Permission to explore the caverns should be requested from the park superintendent.

Central High Cascades

　　To the north of Crater Lake, a line of sharp pinnacles in the central High Cascades belonging an older interval of volcanism have been extensively altered by erosion. Of these, Mt. Thielsen is the most southerly followed by Broken Top, the North Sister, along with Mt. Washington and Three Fingered Jack, which both rise from broad bases to snow-covered spires. The glaciation that shaped these peaks continues even today. The southern-most Cascade glacier is still active on the shady north side of Mt. Thielsen.

　　South of McKenzie Pass, the Three Sisters form a scenic cluster of peaks of rugged beauty. The South Sister, the highest of the cones at 10,354 feet, has a crater with a small lake. The Middle Sister displays remnants of a crater, whereas the deeply eroded North Sister shows no evidence of a former crater. On the south and southwest flanks of the South Sister, an extensive single flow along with the dome at Rock Mesa developed 2,300 years before the present. Clouds of ash and dust expelled from fissures were followed by lava that covered over the vents. A second explosion of pumice then lavas from a series of new vents on the slopes of the South Sister took place

shortly afterward. Historically smoke from the South Sister was noted in 1853 by James Miller, a Presbyterian minister, who "*saw one of the Three Sisters belching forth from its summit dense volumes of smoke.*"

Sand Mountain and Belknap Lava Fields

　　Very recent volcanic activity, in some cases taking place only 400 years ago, has created a vast region of black, cindery lava flows, buttes, and distinct lava fields in the Cascades between North Sister and Three Fingered Jack. Of these, lavas from the Sand Mountain and Belknap flows are the most widespread. The overall eruptive pattern of the region progressed from northwest to southeast with the first eruptions in the Sand Mountain field and the latest from Collier Cone.

　　Slightly to the southwest of Mt. Washington, a linear chain of 22 cinder cones and 41 separate vents make up the Sand Mountain field. From the north, the geographic arrangement of the cones includes Little Nash Crater, the Lost Lake group, Nash Crater, the central craters, the Sand Mountain cones, which are the largest, and a group to the south. The alignment reflects an underlying fracture zone which allowed gasses and lavas to reach the surface. Beginning around 4,000 years ago, volcanic vents discharged close to a cubic mile of lava accompanied by large quantities of ash creating the rugged field of domes and flows. The

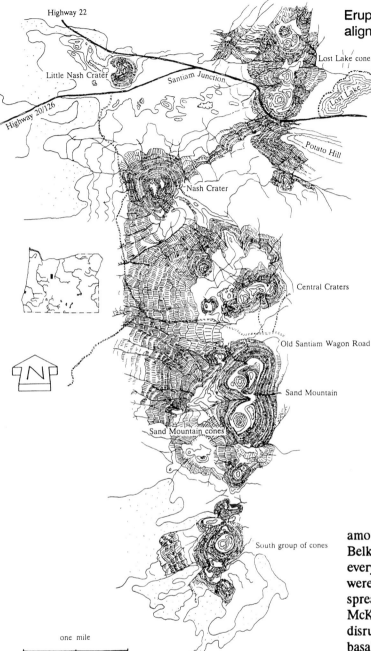

Eruptive cones of the Sand Mountain volcanic field aligned in a north-south direction

one mile

snow-white deposit of microscopic shells of the algae on the bottom.

Three spatter cones in the Sand Mountain field have large pits or conduits from which the volcanic debris ejected. While most of the vents are filled with rubble, one on Nash Crater and two in the Clear Lake flow remain open. The deepest open conduit in the Clear Lake flow is 100 feet to the bottom of the pit, while the Nash Crater pit connects to a 50-foot long lava tube.

Overlapping the Sand Mountain field in the south, the Belknap lava flows were younger and much more extensive. About 1,600 years ago, lava poured from Belknap Crater and the surrounding vents repeatedly innundating 40 square miles. Ash and cinders ejected from the crater were carried in an eastward direction on the prevailing wind. The crest of Belknap volcano is 400 feet above the lava field, and the largest of the two craters in the top is 250 feet deep and 1,000 feet wide.

During the last violent explosions, vast amounts of volcanic material issued from vents around Belknap cone, spreading to the west and engulfing everything along the way. Older Sand Mountain flows were covered and upright trees were immersed by the spreading stream before the lava cascaded into the McKenzie River canyon filling the deeply cut gorge and disrupting the former drainage pattern. The permeable basalts allow river waters, now hidden beneath the surface, to flow through and reappear at Tamolitch Falls. Because of the blockage, Beaver Marsh formed upstream from where the river enters the lava field. Along the margins of the Belknap flow, tree molds from 1 to 5 feet in diameter mark where trees were consumed by the lava before cooling and hardening around the trunk. Trenches up to 35 feet long developed when tree trunks toppled into the fluid lava to be cast as molds. These trees have been dated at 360 A.D.

A number of cones adjacent to both the Sand Mountain and Belknap lava fields are responsible for localized Recent volcanic events. Just to the south of the Belknap field, Collier Cone is probably the most recently active volcano in this region, erupting with lava, cinders, bombs, and other material around 400

oldest cone in this field was probably near Nash Crater, but this feature has since been destroyed by erosion or covered by vegetation.

One of the most extensive flows, emitted from a vent in the south section of the field, sent lava westward to the McKenzie River about 3,000 years ago to dam the river between Sahalie Falls and Clear Lake. The outlet to the basin now occupied by the lake was blocked by a 200-foot high lava dam. Along the river banks, rising waters submerged an entire forest. Today, standing tree trunks, 2 to 3 feet in diameter, can be seen in the unusually clear water. Successions of diatoms living in the waters of the lake have formed a

When central Oregon's lava fields cooled and erosion began to wear away at the tall Cascade peaks, Broken Top crater assumed this craggy pose (photo Oregon State Highway Department).

A view of the recent lava flow at Collier Cone and North Sister looking south with Broken Top in the background (photo courtesy of Delano Photographics)

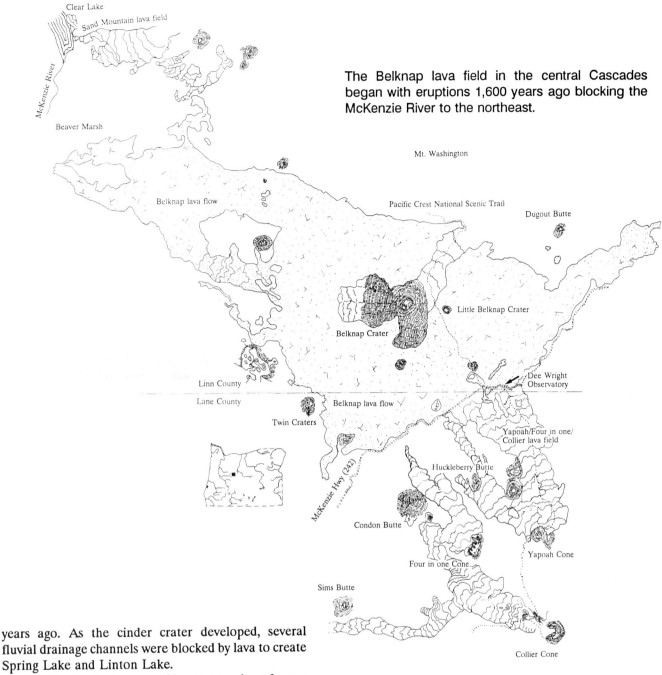

Clear Lake

Sand Mountain lava field

McKenzie River

The Belknap lava field in the central Cascades began with eruptions 1,600 years ago blocking the McKenzie River to the northeast.

Beaver Marsh

Mt. Washington

Belknap lava flow

Pacific Crest National Scenic Trail

Dugout Butte

Little Belknap Crater

Belknap Crater

Dee Wright Observatory

Linn County

Lane County

Belknap lava flow

Twin Craters

Yapoah/Four in one/ Collier lava field

Huckleberry Butte

McKenzie Hwy (242)

Condon Butte

Yapoah Cone

Four in one Cone

Sims Butte

Collier Cone

North Sister

years ago. As the cinder crater developed, several fluvial drainage channels were blocked by lava to create Spring Lake and Linton Lake.

South of the Three Sisters, eruptions from a number of vents in the region of Mt. Bachelor took place about 2,000 to 18,000 years before the present. Coinciding with a time of glacial retreat, a 15-mile long series of cones extruded a total of 7 cubic miles of lava that covered 83 square miles. These volcanic episodes were not continuous but took place sporadically from several fissures in the Mt. Bachelor chain. The oldest flows from cracks alongside Sheridan Mountain were followed shortly by those of Red Crater, Katsuk Butte, and Siah Butte. Once the cones of the Siah group had formed, volcanic activity shifted northward to Mt. Bachelor and Kwolh Butte where the latest flows built a small cone on the north side of the mountain.

Northern High Cascades

Mt. Hood

At the northern end of the Oregon Cascades, Mt. Hood is the highest peak in the state at 11,235 feet. The mountain is named for Samuel Hood, a baron with a successful career in the British Royal Navy who was stationed in American waters during the American Revolution. One of the younger members of the High

Belknap Crater as seen from Dee Wright Observatory (photo Oregon State Highway Department)

Mount Hood, highest of Oregon's Cascade mountains, is visible throughout northwestern Oregon (photo Oregon State Highway Department).

Cascades, Mt. Hood had several very late eruptive periods, an initial phase 15,000 years ago, a middle period 1,000 to 2,000 years ago, and a final stage 200 to 300 years ago which produced the small domes, the main cone, and mudflows.

Feldspar-rich andesite lavas, which built up the bulk of the cone, were the first to erupt sometime before the last glaciation. These lavas cooled and thickened into layers up to 500 feet deep on the south and southwest flanks of the mountain. A second eruption, following a period of dormancy, sent pyroclastic material down the slope melting ice and snow along the way. The hot eruptive material, mud, and melted snow combined to spread nearly eight miles into valleys south and west of the volcano. A very small amount of pumice was scattered lightly on the east and northeast side of the mountain. Between 1760 and 1818 Crater Rock, a dome near the summit of the peak, was emplaced. Today an eroded remnant is all that remains. With the construction of the dome, volcanic material and extensive mud flows filled the Sandy River valley burying large tree stumps upright.

Most recently, escaping gas, pumice, smoke, and flames from the summit of Mt. Hood were recorded between 1800 and 1907 at which time the final historical event took place when steam rose from Crater Rock. Steam vents and fumeroles are present near the top showing that Mt. Hood clearly has the potential to erupt again.

Suggested Readings:

Allen, John E., 1984. The caves of Crater Lake National Park. Oregon Geology, v.46, no.12, pp.143-146.

Bacon, Charles R., 1983. Eruptive history of Mount Mazama and Crater Lake Caldera, Cascade Range, U.S.A. Jour. Volcanology and Geothermal Res., 18, 57-115.

Beeson, Marvin H., and Tolan, Terry L., 1987. The Columbia River gorge: the geologic evolution of the Columbia River in northwestern Oregon and southwestern Washington. In: Hill, Mason L., ed., Cordilleran Section of the Geological Soc. America, Centennial Field Guide, 1987, pp.321-326.

Blackwell, David G., et al., 1978. Heat flow of Oregon. Oregon Dept. Geol. and Mineral Indus., Special Paper 4, 42p.

Bowen, Richard G., Peterson, N.V., and Riccio, J.F., 1978. Low-to intermediate-temperature thermal springs and wells in Oregon. Oregon Dept. Geol. and Mineral Indus., Geol. Map Ser., GMS-10.

Crandall, Dwight R., 1980. Recent eruptive history of Mount Hood, Oregon, and potential hazards from future eruptions. U.S. Geol. Survey, Bull.1492, 81p.

Duncan, Robert A., and Kulm, LaVerne D., 1989. Plate tectonic evolution of the Cascades arc-subduction complex. In: Winterer, E.L., Hussong, D.M., and Decker, R.W., eds., Geol. Soc. Amer., The Geology of North America, v.N: The eastern Pacific Ocean and Hawaii, pp.413-438.

Ferns, Mark L., and Huber, Donald F., 1984. Mineral resources map of Oregon. Oregon Dept. Geol. and MIneral Indus., Geological Map Series, GMS 36.

Hammond, Paul E., 1979. A tectonic model for evolution of the Cascade Range. In: Armentrout, J.A., Cole, M.R., and TerBest, H., eds. Cenozoic Paleogeography of the western United States. Society Econ. Paleont. and Mineralogists, Pacific Sect., Pacific Coast Symp.3, p.219-237.

-----1989. Guide to the geology of the Cascade Range, Portland, Oregon, to Seattle, Washington. Field trip guidebook T306. American Geophysical Union, 215p.

Harris, Stephen L., 1988. Fire mountains of the West: the Cascade and Mono Lake volcanoes. Missoula, Mountain Press, 378p.

McBirney, Alexander, 1978. Volcanic evolution of the Cascade Range. Ann. Rev. Earth Sci., no.6, p.437-456.

Magill, James, and Cox, Allan, 1980. Tectonic rotation of the Oregon Western Cascades. Oregon Dept. Geol. and Mineral Indus., Special Paper 10, 67p.

Peck, D.L., et al., 1964. Geology of the central and northern parts of the Western Cascade Range in Oregon. U.S. Geol. Survey, Prof. Paper 449, 56p.

Priest, George R., 1990. Volcanic and tectonic evolution of the Cascade volcanic arc, central Oregon. Jour. Geophys. Res., v.95, no.B12, pp.19,583-19,599.

-----and Vogt, B.F., 1983. Geology and geothermal resources of the central Oregon Cascade range. Oregon Dept. of Geology and Mineral Res., Special Paper 15, 123p.

Scott, William E., Gardner, Cynthia A., and Johnston, David A., 1990. Field trip guide to the central Oregon High Cascades. Part 1: Mount Bachelor-South Sister area. Oregon Geology, v.52, no.5, pp.99-114.

Sherrod, David R., and Smith, James G., 1990. Quaternary extrusion rates of the Cascade Range, northwestern United States and southern British Columbia. Jour. Geophys. Res., v.95, no.B12, pp.19,465-19,474.

Smith, Gary A., Snee, L.W., and Taylor, E.M., 1987. Stratigraphic, sedimentologic, and petrologic record of late Miocene subsistance of the central Oregon High Cascades. Geology, v.15, p.389-392.

Taylor, E.M., 1965. Recent volcanism between Three Fingered Jack and North Sister, Oregon Cascade Range. Ore Bin, v.27, no.7, p.121-147.

-----1990. Volcanic history and tectonic development of the central High Cascade range, Oregon. Jour. Geophys. Res., v.95, no.B12, pp.19,611-19,622.

Tolan, Terry L., and Beeson Marvin H., 1984. Intra-canyon flows of the Columbia River basalt group in the lower Columbia River gorge and their relationship to the Troutdale Formation. Geol. Soc. Amer., Bull.95, p.463-477.

Verplanck, Emily P., and Duncan, Robert A., 1987. Temporal variations in plate convergence and eruption rates in the Western Cascades, Oregon. Tectonics, v.6, no.2, p.197-209.

A "flight of stairs" formed by raised terraces at Cape Arago in Coos County is a dramatic expression of the rapidly elevating coast here [photo by U.S. Forest Service].

Coast Range

Physiography

A long narrow belt of moderately high mountains and coastal headlands, the Oregon Coast Range physiographic province extends from Washington state and the Columbia River in the north to the Middle Fork of the Coquille River in the south and from the continental shelf and slope to the western edge of the Willamette Valley. Just over 200 miles long and from 30 to 60 miles wide, the province narrows in the center and widens on either end. The crest of the range averages 1,500 feet in altitude with the highest point Marys Peak near Philomath reaching 4,097 feet. Saddle Mountain east of Seaside at 3,283 feet, Trask Mountain at 3,423 feet northwest of McMinnville, and Sugarloaf Mountain at 3,415 feet east of Lincoln City are among the highest of the interior peaks.

Because there is more rainfall, steeper gradients, and more active erosion on the western slope, summits of the passes lie well to the east of the axis of the range. With the close marine influence, the coastal region has the highest average winter temperatures, the coolest summers and the greatest rainfall in Oregon. The western slopes of the mountains receive over 100 inches of rain a year, while east slope precipitation averages only around 30 inches. High rainfall and mild temperatures have contributed to a heavy forest with thick vegetation and mature soils covering most of the range.

Along the western shore of the Coast Range the line of abrupt coastal headlands is fairly evenly interspersed with shallow bays, estuaries, pocket beaches, and sand dunes. Cape Blanco in the southwest corner of the province is a distinctive terraced promontory that extends into the Pacific Ocean as Oregon's most westerly point. Extensive sands are present between Coos Bay and Heceta Head and from Seaside to the Columbia River. Sand has accumulated as bars or spits across the mouths of most rivers hindering the flow of water in and out of the bays. Along the very southwestern edge of the province, wave-cut terraces create a narrow coastal plain with remnants of older marine terraces rising up to 1,600 feet.

Flowing west to the ocean or east to the interior, most of the streams of the Coast Range are small. Only the Columbia, Siuslaw, and Umpqua rivers cut

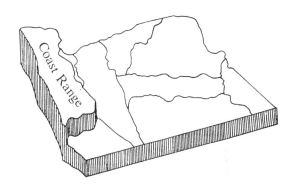

entirely across the province to empty into the ocean. The notable west flowing rivers, the Nehalem, Wilson, Siletz, Yaquina, Alsea, Siuslaw, Umpqua, Coos, and Coquille, end in broad tidal estuaries. Of those streams flowing eastward into the Willamette River, most have less extensive watersheds. Among them, the Long Tom, Marys, Luckiamute, Yamhill, and Tualatin rivers are the largest.

The Coast Range province continues offshore to the continental shelf and slope that descends to the abyssal plain at 9,000 feet below sea level. The surfaces of the shelf and slope are broken by high bedrock scarps, ridges, basins, and canyons. The most significant offshore feature is the subduction zone beneath the base of the continental slope where the Farallon plate is sliding under the North American landmass.

Geologic Overview

At the end of the Cretaceous period, 66 million years ago, Oregon lacked a coastal mountain range, and the ocean shoreline extended diagonally across eastern Washington and Idaho. Southeast of this large embayment the newly formed Klamath Mountains projected well into Idaho. A loose chain of volcanic seamounts was forming in the ocean setting to the west where an active hot spot lay beneath a spreading center

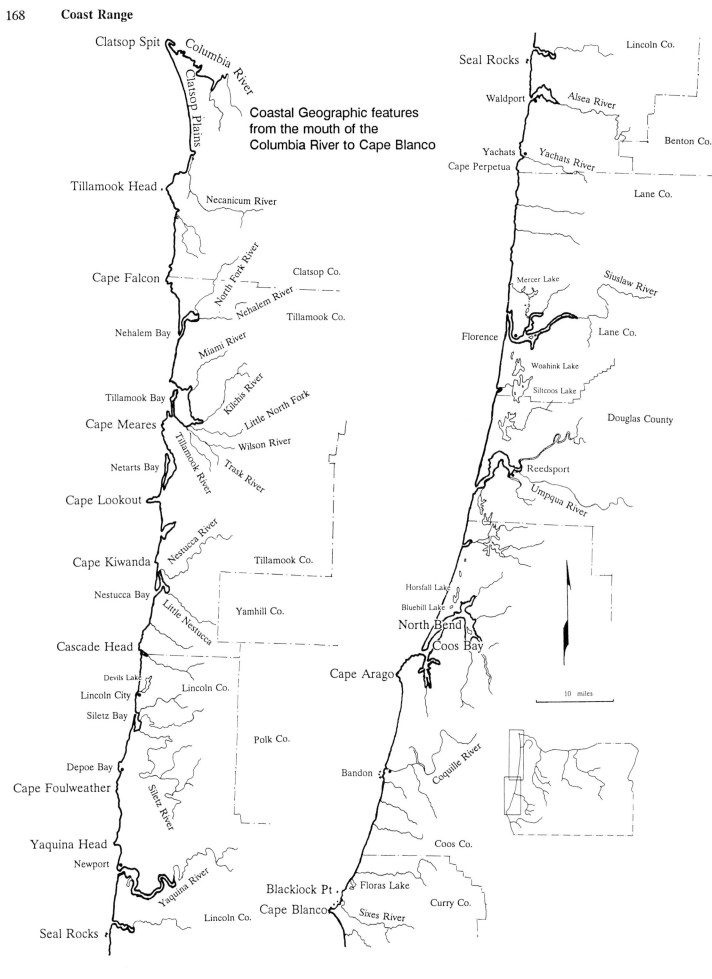

Clatsop Spit
Columbia River
Clatsop Plains

Coastal Geographic features
from the mouth of the
Columbia River to Cape Blanco

Tillamook Head

Necanicum River

Cape Falcon
North Fork River
Clatsop Co.
Nehalem River
Tillamook Co.

Nehalem Bay

Miami River

Tillamook Bay
Kilchis River
Little North Fork
Cape Meares
Wilson River
Tillamook River
Netarts Bay
Trask River

Cape Lookout

Nestucca River

Cape Kiwanda
Tillamook Co.

Nestucca Bay
Little Nestucca
Yamhill Co.

Cascade Head

Devils Lake
Lincoln Co.
Lincoln City
Siletz Bay
Polk Co.

Depoe Bay
Cape Foulweather
Siletz River

Yaquina Head
Newport
Yaquina River

Seal Rocks
Lincoln Co.

Seal Rocks
Lincoln Co.
Waldport
Alsea River
Benton Co.

Yachats
Yachats River
Cape Perpetua
Lane Co.

Mercer Lake
Siuslaw River

Florence
Lane Co.

Woahink Lake
Siltcoos Lake

Douglas County

Reedsport
Umpqua River

Horsfall Lake
Bluebill Lake
North Bend
Coos Bay

Cape Arago

10 miles

Coquille River
Bandon

Coos Co.

Blacklock Pt   Floras Lake
Cape Blanco   Sixes River
Curry Co.

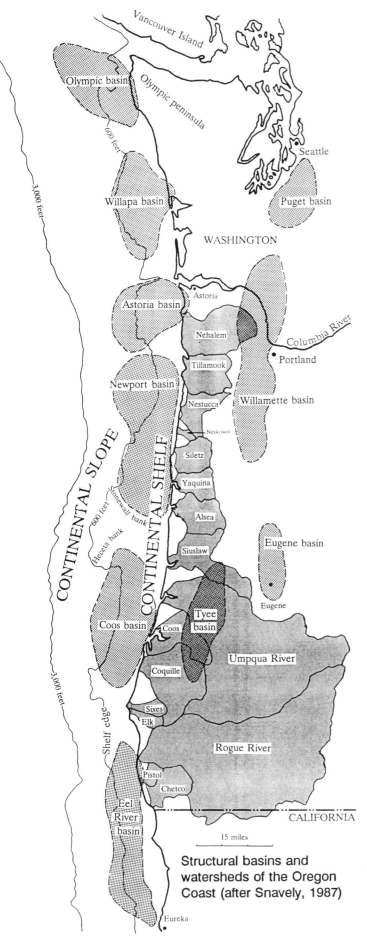

Structural basins and
watersheds of the Oregon
Coast (after Snavely, 1987)

between two moving plates. Around each volcanic
center pillow lavas and breccias built low islands where
the vents continued to eject lava and ash. Riding atop
the moving Kula and Juan de Fuca plates, the island
volcanic chain collided with and was swept up by the
oncoming westbound North American plate.

As the island chain began to subside and a
proto-Cascade volcanic arc was established east of the
block, sediments poured into the newly created forearc
basin. A thick sequence of Eocene through Miocene
marine clastics accumulated in the subsiding basin atop
the platform of older basalts. Eocene sediments in the
basin are derived from several sources. Initially the
Klamath Mountains contributed detritus to the newly
formed basin, but as watersheds extended further east
the primary source of sediments was the Idaho batholith. Finally, volumes of pyroclastics and ash from the
newly formed ancestral Cascade volcanoes covered
earlier sands, silts, and muds in the adjoining basin.

Oligocene seas extended over the northern
Coast Range block, but with uplift of the range during
the Miocene the ocean retreated to the west. As the
western edge of the North American plate was wrin-
kled by pressure from the subducting Juan de Fuca
slab, lava flows from fissures in eastern Oregon reached
the coast where they invaded layers of the softer
sediments. Once the shoreline had withdrawn to the
western edge of the coastal block, the older Cenozoic
formations were shaped by erosion, and river valleys
assumed their present positions. The rivers cut through
many of the later formations laying bare the resistant
Oligocene sills and dikes making up nearly all the
prominent peaks of the central range.

Continued uplift and tilting of the coastal
mountains combined with Pleistocene sea level changes
created raised terraces in the vicinity of Cape Blanco
and Cape Arago. Because rates of uplift vary along the
coast, different sections of a terrace may have elevations that vary by hundreds of feet. Deeply eroded, one
million year old terraces are inland at the highest
levels, while the younger surfaces are near the coast at
lower elevations.

Pleistocene landslide lakes of the inland
mountains and freshwater dunal lakes along the coast
are among the more ephemeral features of the region.
Abundant sand along the shore dams streams creating
lakes of varying sizes and depths. A fragile boundary
between fresh and saltwater within the poorly consolidated dunal sands exists beneath the coastal lakes.
Inland from the coast, massive water-saturated sandstones that are susceptible to landsliding move as
blocks to dam stream valleys and back up waters into
lakes.

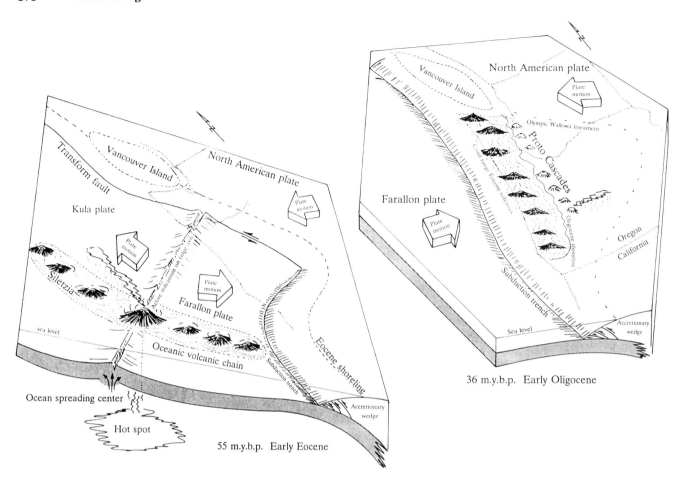

Accretionary island arc model for formation of the Oregon-Washington coast block (after Duncan, 1982)

More than half of the Oregon Coast is bordered by sands resulting from the erosion of sea cliffs or transported and deposited by rivers. While winter beaches are stripped of sand by stormy conditions, the summer months are the period when sands are redeposited. Sand builds up as dunes or spits when it is trapped behind prominent headlands or man-made projections. In spite of the transitory nature of dunes and spits, houses, roads, and even entire towns have been constructed here only to be threatened or destroyed by severe winter erosion.

In the past the coastline has been subject to effects of earthquakes and tidal waves or tsumanis. At many coastal sites in Oregon sediments record land subsidence as buried forests and mud-covered bogs that resulted from powerful earthquakes. Similarly periodic scouring of bays and coast areas by tsumanis is part of the coastal erosion process. A final sculpting of the land took place as the ocean eroded headlands and filled bays creating a variety of spits, bay mouth bars, islands, tunnels, and offshore sea stacks.

Geology

The foundation block of coastal Oregon and Washington mountains began as a volcanic island chain that collided with the North American plate. Some distance to the west of Oregon and far beneath the ocean crust, a hot spot generated deep sea lavas 64 million years ago. This volcanic source was situated beneath an ocean spreading center straddling two active tectonic plates. The Kula plate was being pulled north to collide and slide below North America along Washington and British Columbia. The Farallon plate to the south was being overridden by southern Oregon and California. As these two plates continued spreading or pulling apart, a chain of undersea mountains, strung out to the north and south. Forming a prominent ridge on the sea floor, the line of volcanoes occasionally projected above the water as low volcanic islands. Close to the end of early Eocene time the North American plate, moving westward, collided with the recently formed islands and seamounts. With the capture of the new landmass, North America had grown about 50 miles in width. Volcanic rocks of this island chain today form the backbone of the Oregon and Washington coastal mountains and are the oldest known rocks in the range.

Coastal margin rifting model for formation of the Oregon-Washington coast volcanics (after Snavely, 1987; Babcock, et al., in press)

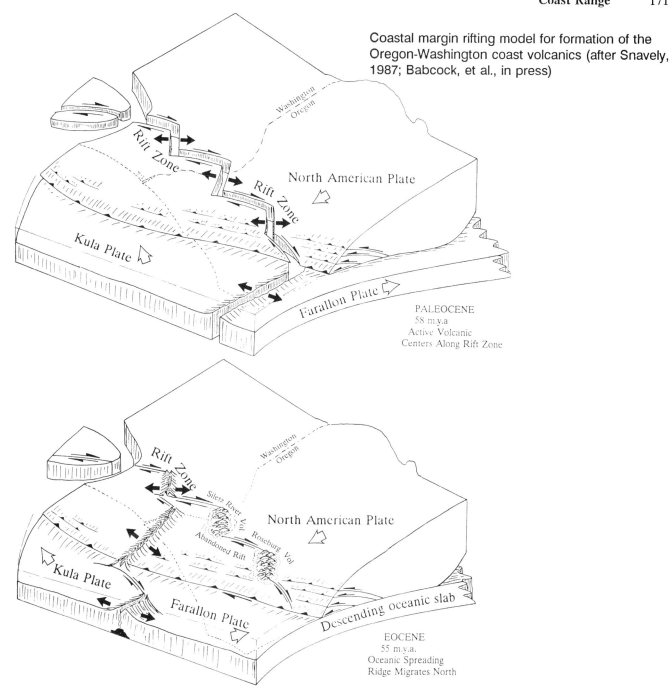

Another possible explanation for the formation of the Coast Range block is that the volcanics developed above a series of rifts or crustal rips along the coastal margin. Differential movement of large-scale blocks would have created a line of overlapping tears releasing lavas that became the base of the Coast Range. Alternately, the rifting could have been caused by a hot spot below the crust. With coastal margin rifting, the Coast block would have been formed in place and not accreted.

Differences in age, texture, and mineralogy of the core Coast Range volcanic rocks have caused them to receive different formation names. From the north to south, they include the Metchosin, Crescent, Black Hills, and Grays Harbor volcanics in Washington, the Tillamook volcanics in the northern Oregon Coast Range, the Siletz River volcanics in the central part of the Coast Range, and the Roseburg volcanics in the south. The oldest volcanic rocks in the range are those of the 64 million year old Roseburg and Metchosin formations at the southern and northern extremes, while the youngest are the 44 million year old Grays River Formation in the central region. Common to all of these volcanics are elliptical bodies called "pillows",

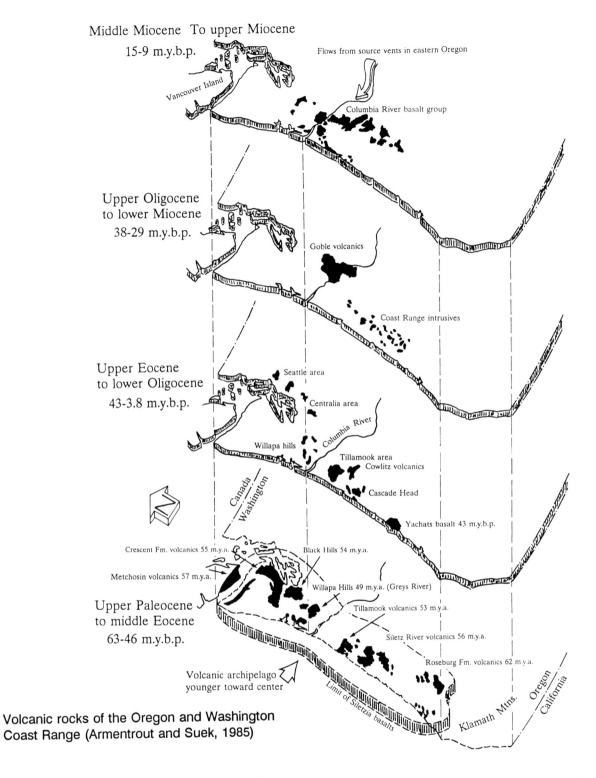

Middle Miocene To upper Miocene
15-9 m.y.b.p.

Vancouver Island

Flows from source vents in eastern Oregon

Columbia River basalt group

Upper Oligocene
to lower Miocene
38-29 m.y.b.p.

Goble volcanics

Coast Range intrusives

Upper Eocene
to lower Oligocene
43-3.8 m.y.b.p.

Seattle area

Centralia area

Willapa hills

Columbia River

Tillamook area
Cowlitz volcanics

Cascade Head

Yachats basalt 43 m.y.b.p.

Canada / Washington

Crescent Fm. volcanics 55 m.y.a.

Black Hills 54 m.y.a.

Metchosin volcanics 57 m.y.a.

Willapa Hills 49 m.y.a. (Greys River)

Upper Paleocene
to middle Eocene
63-46 m.y.b.p.

Tillamook volcanics 53 m.y.a.

Siletz River volcanics 56 m.y.a.

Roseburg Fm. volcanics 62 m.y.a.

Volcanic archipelago
younger toward center

Limit of Siletzia basalts

Klamath Mtns.

Oregon / California

Volcanic rocks of the Oregon and Washington
Coast Range (Armentrout and Suek, 1985)

because of their size and shape, which formed when lavas were extruded underwater onto an ocean floor.

With lava thicknesses up to 2 miles, the island chain was far too large to be simply overridden or subducted, thus it was incorporated into the larger North American landmass following collision. During the 50 to 42 million year interval, the still-forming Coast Range block began to rotate westward in a clockwise direction with a pivotal point in northwest Washington. By Oligocene time backarc spreading had moved the block close to the position of the present day Cascades. Beginning in the Miocene epoch, 20 million years ago, a second rotational phase began. With the process continuing to the present, it is estimated that the Coast Range has rotated 51 degrees since late Eocene, up to 44 degrees since the middle

Isolated pillow basalts are common in the Depoe Bay basalts (photo courtesy of Oregon Dept. of Geology and Mineral Industries).

Eocene tectonics and sedimentary environments of the Oregon Coast range (after Ryberg, 1984; Heller and Ryberg, 1983)

Oligocene, and 32 degrees since the middle Miocene with an average of 1.5 degrees of rotation per 1 million years throughout the entire 50 million year period of movement. This second rotational motion was in response to the extension or extreme stretching of the Basin and Range province which pushed the Coast Range block out to its present location. In accordance with the concept of crustal rifting, small fragments of the Coast Range, caught between moving plates, simply rotated like ball bearings.

Eocene

As the seamount terrane was accreted to North America, a trough or forearc basin, oriented southeast by northwest and partially bounded by the ancestral Cascade volcanoes to the east, developed along its axis. From Eocene through the Miocene the elongate basin was repeatedly choked with sediments before uplift of the Coast Range brought about a shallowing and eventual closure of the trough. Early Eocene sediments were carried into the new marine basin by streams and rivers, blanketing the older submarine volcanics with sands, muds, and silts of the Roseburg, Lookingglass, and Flournoy formations. Rock particles that make up these three formations are lithic in nature. That is, they are composed of pre-existing cherts, metamorphic rocks, and heavy minerals typical of those found in the Klamath Mountains.

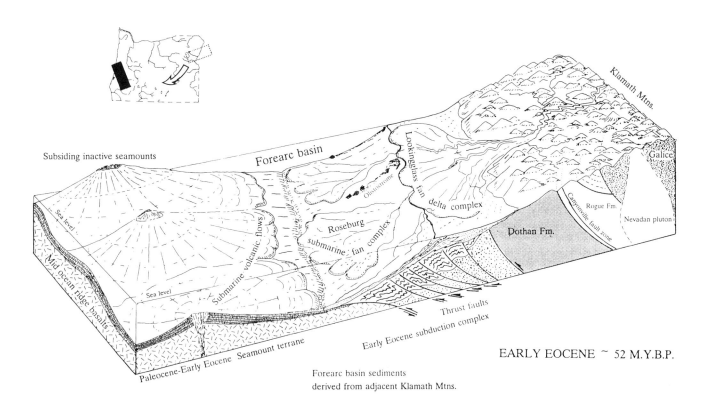

EARLY EOCENE ~ 52 M.Y.B.P.

Forearc basin sediments derived from adjacent Klamath Mtns.

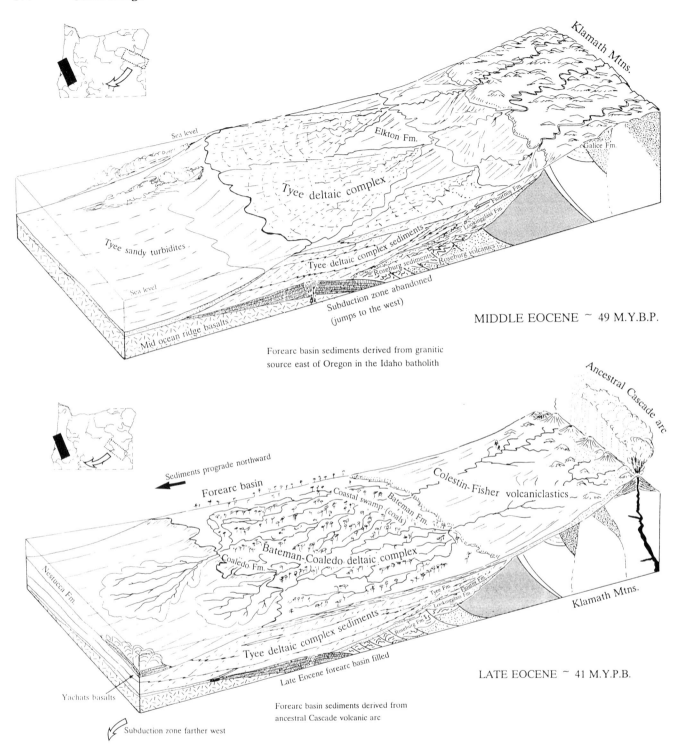

Sea level

Elkton Fm.

Klamath Mtns.

Galice Fm.

Tyee deltaic complex

Tyee sandy turbidites

Tyee deltaic complex sediments

Flournoy Fm.

Lookingglass Fm.

Roseburg sediments

Roseburg volcanics

Sea level

Subduction zone abandoned
(jumps to the west)

Mid ocean ridge basalts

MIDDLE EOCENE ~ 49 M.Y.B.P.

Forearc basin sediments derived from granitic
source east of Oregon in the Idaho batholith

Sediments prograde northward

Ancestral Cascade arc

Forearc basin

Coastal swamp (coals)

Bateman Fm.

Colestin-Fisher volcaniclastics

Bateman-Coaledo deltaic complex

Coaledo Fm.

Nestucca Fm.

Tyee deltaic complex sediments

Tyee Fm.

Flournoy Fm.

Lookingglass Fm.

Roseburg Fm.

Klamath Mtns.

Yachats basalts

Late Eocene forearc basin filled

LATE EOCENE ~ 41 M.Y.P.B.

Forearc basin sediments derived from
ancestral Cascade volcanic arc

Subduction zone farther west

Sand bars were deposited in shallow waters high on the continental shelf and slope in the southern basin. Deep water currents spread sediments in a broad alluvial pattern across the middle of the basin, while fine-grained sediments were carried farthest from the source and into the deepest region of the seaway to the north. In the Roseburg Formation, there is evidence that large cohesive blocks of mud and rocks slumped from a steep upper shelf onto the deep sea fan. In this ancient submarine setting, the huge blocks or "olisto-liths" are found imbedded in chaotic masses of rock indicating they slid down steep submerged oceanic slopes after breaking off an escarpment.

After Flournoy time, the end of Coast Range collision and renewed subsidence of the forearc basin marked a new phase of deposition. White mica, potassium feldspar, and quartz, carried into the basin during the Tyee interval, reflect a major change in the source of sediments as well as the depositional environment in the middle to late Eocene. By late Eocene time rivers feeding the basin had matured extending their watersheds well beyond the Klamaths into the interior of Idaho. White micas and potassium feldspar minerals that make up the Tyee Formation are identical to minerals of the Idaho batholith. In the southern part of the basin, sands transported by rivers created deltas and shallow marine sandbars on the shelf before cascading into the deeper basin where they were carried on turbidity currents to form Tyee submarine fans. The meager fossil record of this interval is due to overwhelming amounts of sand and silt diluting the marine faunas. In localized areas Tyee, Elkton, and Bateman deposits accumulated up to a mile in thickness. Approximately 1,500 feet of Bateman sediments typify a shallow delta in the Tyee seas, while the 2,500 feet of Elkton sands with abundant, shallow water molluscs and microfossils were the last to fill the southern basin.

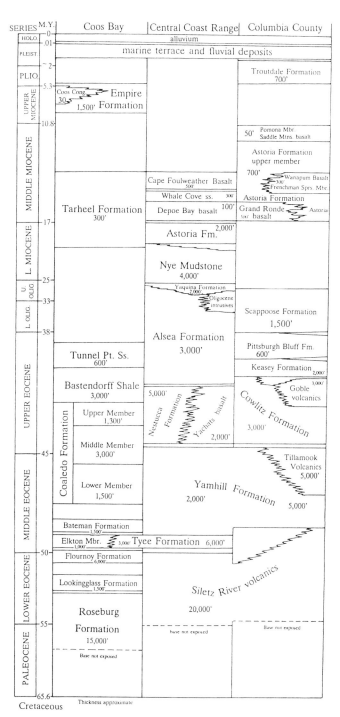

Stratigraphy of the Coast Range (after Correlation of Stratigraphic Units of North America [COSUNA], 1983)

Roseburg Formation south of Sutherlin on Interstate 5 displays deep water turbidite deposits (photo courtesy Leslie Magoon).

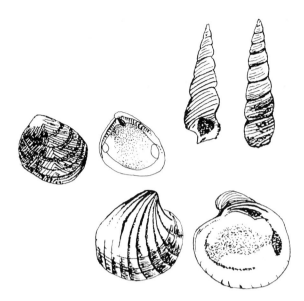

Eocene molluscs, southern Oregon Coast Range
(top to bottom <u>Turritella,</u> <u>Acila,</u> and <u>Venericardia</u>)

As the island arc rotated progressively west-ward, late Eocene swamps, wide coastal plains, and shallow continental shelf conditions mark the third and final phase of sedimentation in the subsiding forearc basin. Since the Klamath Mountains had been previously planed down to a flat surface and the Idaho source rocks were left far behind by the rotating block, the early Western Cascades were the primary source for sediments of the forearc basin. Ash and pyroclastic deposits were in sharp contrast to the feldspar-rich arkosic sands of the older Tyee Formation. In the Coos Bay region, sands accumulated in an immense delta spreading toward the north. Fossil plants and coarse sediments from within the delta are enclosed in the coal-bearing Coaledo Formation. Extensive near-shore swamps characterized by subtropical plants are the source of the coal. Thin-shelled delicate microfossils in the overlying Bastendorff suggest intermediate depths while the Tunnel Point sandstones reflect a shallow, upper shelf environment.

Central and Northern Coast Range

In that portion of the ancient forearc basin which later was to become the central and northern Coast Range, silts, sands, muds, and volcanic debris were deposited in the shallow middle Eocene seas, while aprons of shell banks were built up around the margins of the low volcanic islands and seamounts. Deposited across the continental shelf, Yamhill muds and silts, interspersed with Rickreall and Buell limestone lenses, blanketed the Siletz River volcanics and were in turn covered by about 5,000 feet of Nestucca Formation deep water muds, sands, and silts mixed

with ash and lavas from the nearby Western Cascades. Simultaneously brackish water Spencer Formation sands formed in the shallow, near-shore environment along the eastern margin of the basin. In areas on the coast, the Nestucca Formation is interlayered with lavas of the Yachats basalt. These flows initially erupted under water on a shallow shelf along the coastline where a number of vents were located near Cape Perpetua and Heceta Head. Low profile, shield-like accumulations of Yachats basalt are over 2,000 feet thick.

In the northern Nehalem River basin, conglomerates, sands, shales, and volcanics of the Cowlitz Formation were deposited in shallow, brackish waters over Yamhill muds. Interfingering with the Cowlitz sediments, Goble lava flows west of Portland may have extruded from a late Eocene volcanic arc originally part of the Coast Range block or they may have been remnants of an independent island block. Overlying the shallow water sediments of the Cowlitz, fine volcanic ash forms the deposits of the Keasey Formation that reflect a deep water ocean setting.

Spanning the Eocene-Oligocene boundary in the northern basin, shallow seas with shifting deltas and brackish backwater bays are represented by tuffaceous sands and silts of the Pittsburg Bluff and Scappoose formations. Sands and volcanic debris were derived from the nearby landmass lying to the east. Scappoose

A view of Sunset Bay in the foreground with Gregory Point and Lighthouse Beach in the center and Bastendorff Beach at the top. Simpson Reef, composed of sandstones of the Coaledo Formation, projects offshore (photo courtesy of Ward Robertson).

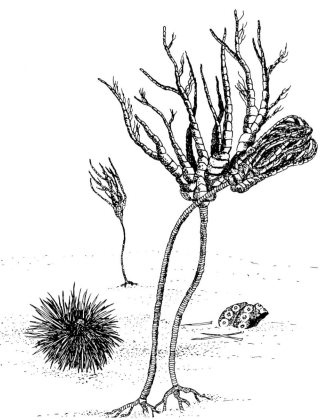

Crinoids (sea-lilys) and echinoids (sea urchins) inhabited the ocean during the Keasey period.

Formation clastics, like those of the Tyee Formation of the southern Coast Range, are "arkosic" or rich in white mica, quartz, and potassium feldspar and probably had a similar source in the Idaho batholith. On the shelf of the central basin, ash mixed with silts and sands of the Alsea and Yaquina formations.

In the interval between 38 to 29 million years ago, a number of intrusive bodies from batholith sources below the forearc basin invaded the softer marine sediments of the Coast Range. Later erosion, aided by coastal uplift, stripped off sediments to expose these Oligocene dikes and sills which form the major peaks of the range today. The most prominent intrusive bodies make up Marys Peak, Saddleback Mountain southwest of Grande Ronde, Laurel Mountain west of Dallas, Fanno Ridge, Sugarloaf and Stott mountains north and northwest of Valsetz, Grass Mountain northwest of Alsea, and Roman Nose Mountain south of Mapleton.

Oligocene

Land areas along the margins of the ocean emerged from the shrinking Oligocene ocean basin. By the late Oligocene and early Miocene, uplift of the Coast Range block was nearly complete with only the central and north end of the coast and northern

Willamette Valley under a shallow seaway. Volcanism had increased considerably in the Western Cascades showering tremendous quantities of ash directly into the sea. Streams draining the mountains washed additional volcanic material into the basin where it was carried out onto shelf and slope by storm waves and turbidity currents.

Miocene

Deposition contined uninterrupted from the Oligocene into the early Miocene. The Miocene is characterized by a rapidly retreating ocean as waters receeded to the present limits with regional uplift and continued rotation of the Coast Range block into the Pliocene. With the shallowing of the forearc region, a number of small basins on the shelf deepened toward the west. Microfossils in the muds and silts of the Nye Mudstone suggest they were deposited in deep cold water of the basin at depths as much as 2,000 feet.

Fossiliferous sandstones and siltstones of the Astoria Formation deposited here are among the best known exposures on the central and northern coast. In addition to a well-preserved fauna of fossil molluscs, corals, barnacles, brachiopods, and crabs, Astoria sediments are famous for remains of marine vertebrates as sharks, turtles, whales, seals, and sea lions. Even occasional ungulates or hoofed mammals were washed

Seal and skull from the Astoria Formation

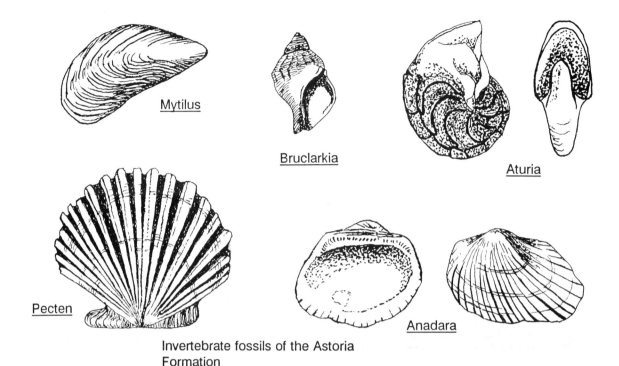

Invertebrate fossils of the Astoria
Formation

into the middle Miocene seaway to become entombed in the Astoria silts. Today the erosion of these sediments along the coast between Astoria and Lincoln City continuously exposes fossils in the sea cliffs.

Middle Miocene time also saw the invasion of Columbia River lavas along the northern coastal area. Fifteen million year old basalts are known from Seal Rocks, Yaquina Head, Depoe Bay, Cape Kiwanda, Cape Lookout, Capes Meares, and Cape Falcon, and Tillamook Head. The basalts found there today are chemically and mineralogically identical to the Columbia River lavas. The striking similarity between the two suites of basalts has always been regarded as an enig-

ma, but it clearly suggests that they were derived from the same magma beneath the Grande Ronde valley of eastern Oregon.

Originally interpreted as having erupted from local volcanoes on the coast, these puzzling basalts now have been designated "invasive" rather than "intrusive". The invasive mode of origin suggests that the coastal Miocene basalts are the distal ends of flows that were extruded elsewhere. Voluminous liquid masses of lavas, flowing down the Columbia River valley and across the coastal plain, temporarily ponded up in bays and estuaries before advancing seaward as nearshore basins filled to capacity. Tremendous hydrostatic pressure by

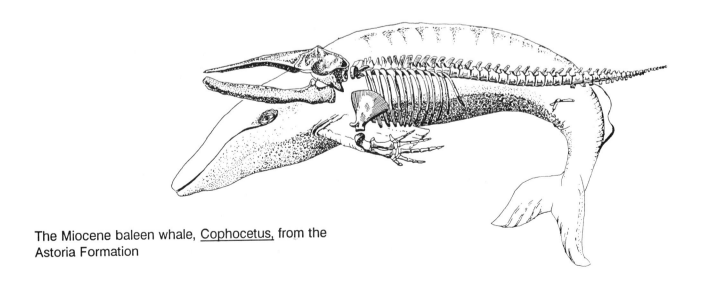

The Miocene baleen whale, Cophocetus, from the
Astoria Formation

Depoe Bay, with the calm inner harbor and the rough Pacific surf, offers the spectacular Spouting Horn and an observation tower above the harbor entrance (photo Oregon State Highway Department).

Yaquina Head north of Newport is composed of Miocene basalt (photo courtesy of Oregon Dept. Geology and Mineral Industries)

the heavy fluid lavas forced them downward into the underlying soft watersaturated sediments. Once the lavas invaded the unconsolidated sediments, they intruded a variety of Eocene and Oligocene rocks in the central and northern Coastal Range as dikes and sills. After cutting across the incipient Coast Range, the lavas terminated as far south as Seal Rocks. Geophysical transects across these coastal basalt exposures reveal that the dikes and sills are "rootless", which further supports an invasive origin rather than a local volcanic source.

At Cape Blanco and Coos Bay sandstones of the Empire Formation are late Miocene. Molluscs, echinoids, and barnacles suggest a shallow water temperate marine environment. Many of the shells show wear and abrasion from action of the waves close to shore.

Pliocene

Few marine sediments record the Pliocene in western Oregon indicating this was a time of erosion when the area was above sea level and the coastline lay approximately in its present position. Pliocene muds, sands, and silts were carried to the shelf and slope where they form vast sheets of poorly consolidated sediments. Up to 1,000 feet of gravels of the Troutdale Formation accumulated along the Columbia River in the St. Helens area and intermittently all the way to the mouth of the river.

Pleistocene

Subduction of the Juan de Fuca plate under North America and continued growth of the wedge of accreted sediments on the leading edge of the North American plate at a rate of about 1 inch per year has pushed the Coast Range upward. Areas offshore closer to the subduction trench have been subject to more uplift than areas farther inland. In a regional sense, the varying uplift produces a tilting effect since the western edge of the coastal mountains are rising, while the eastern margin as well as the Willamette Valley are either subsiding or only rising at a minimal rate. In the early 1980s a Cornell University team conducted very accurate east-west leveling surveys across Coast Range highways in Oregon and Washington repeating measurements which had been taken over 50 years previously. The striking results showed a seesaw effect in every transect, where western coastal areas had risen with respect to the eastern part of the range. Consequently Cape Blanco, only 35 miles from the subduction zone, has the fastest uplift anywhere on the coast at 35 inches vertically every century or 1 inch every 3 years. As Neah Bay along the Washington coast rises at the rate of about 1 inch every 15 years, Seattle and Vancouver on the inland side are sinking at 1 inch every 11 years and 1 inch every 40 years respectively. In Oregon, Astoria rises at the rate of 1 inch in 36 years as Rainier to the east is proportionally depressed.

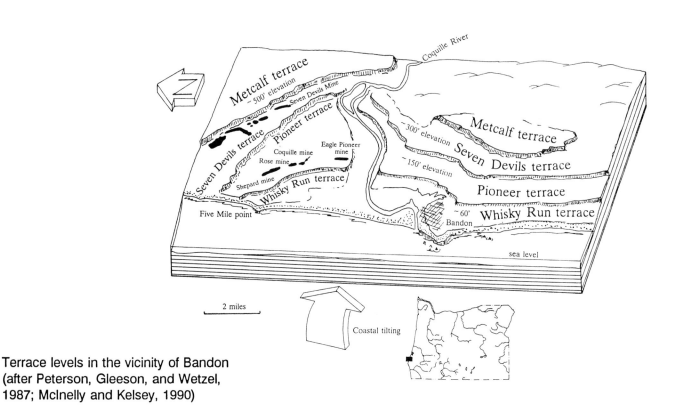

Terrace levels in the vicinity of Bandon (after Peterson, Gleeson, and Wetzel, 1987; McInelly and Kelsey, 1990)

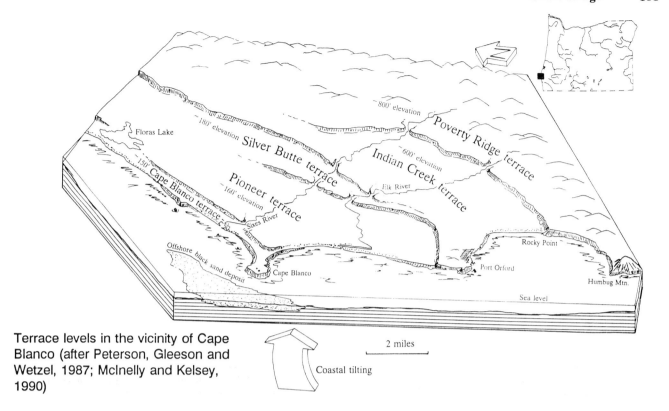

Terrace levels in the vicinity of Cape Blanco (after Peterson, Gleeson and Wetzel, 1987; McInelly and Kelsey, 1990)

Coastal Terraces

Varying from 20 to over 50 feet in thickness, gravel and sand of ancient terraces as high as 1,600 feet have been mapped along the southern Oregon coast between Coos Bay and Port Orford. In the vicinity of Cape Arago and Bandon recognized terraces are Whisky Run, Pioneer, Seven Devils, Metcalf, and the extremely limited, poorly preserved Arago Peak. At Cape Blanco a small terrace covers much of the point, while the extensive Pioneer terrace continues on the landward side toward Port Orford. Silver Butte, Indian Creek and Poverty Ridge terraces rise eastward of Pioneer terrace. The Cape Blanco terrace is dated at 80,000 years, the Whisky Run terrace at 83,000 years, the Pioneer at 103,000 years, the Seven Devils at 124,000 years, and the Metcalf terrace at 230,000 years.

The oldest terraces are inland at the highest levels, whereas the younger recently emergent terraces are on the coast at lower elevations. Terraces cut late in the Pleistocene are ordinarily elevated less than 100 feet. The present elevation of the youngest most prominent terrace ranges from a point below sea level to 225 feet above. It is difficult to correlate or determine the association between the terraces by comparing their elevations because the rates of uplift vary considerably from place to place along the coast, while deformation, faulting, and erosion have obscured the surfaces. A single, continuous terrace may then have separate sections at elevations which differ by hundreds of feet.

More recently these erosional surfaces have been identified by determining the age of invertebrate fossils from within terrace sediments. These fossils are dated by measuring uranium/thorium radioactive decay from the calcite in corals, bryozoa, and sea urchin plates and spines. A second strategy for dating fossils uses the alteration of amino acids preserved in mollusc shells. After an animal dies, organic chemicals in its shell change at a regular rate. Popular molluscs used for dating are the thick-shelled clams, Saxidomus and Mya.

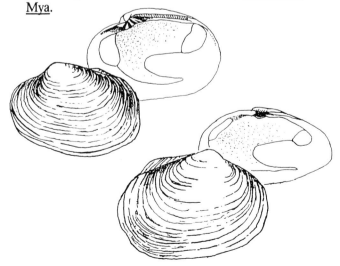

The thick-shelled molluscs Saxidomus (left) and Mya (right) are used for dating terraces.

A Pleistocene fossil whale is being excavated near Port Orford in Curry County (photo courtesy Oregon Dept. Geology and Mineral Industries).

The precise chronology of terrace deposits along the Oregon coast is still incomplete. Terrace development is not a case of a simple steplike uplift of the coast with stationary stages during which the terraces were cut. The terracing reflects both the eastward tilting of the Coast Range block as well as the rise and fall of sea level during Pleistocene time. Glacial stages brought volumetric fluctuations of the ocean that were independent of coastal uplift. Ice, forming on the land as glaciers or ice packs in the polar regions, lowered sea level on the Oregon coast as much as 400 feet. As this ice melted to re-enter the ocean, sea level subsequently rose.

The highest and oldest terraces such as the 1 million year old Poverty Ridge east of Cape Blanco and the slightly younger Indian Creek are thoroughly dissected by erosion making them difficult to trace laterally along the coast. By contrast, lower, younger terraces are in much better condition and form natural flat surfaces ideal for human cultural use. Highway 101, for example, runs for miles in southwest Oregon atop the Pioneer terrace which ranges from only 5 feet in elevation up to 200 feet above sea level.

Coastal Bays and Lakes

A steadily rising sea level as the result of melting glacial ice has produced coves and bays that line the shore. Alluviation by vigorous coastal streams has kept pace with the sea level rise along the Oregon coast south of Bandon so that the tide does not extend far inland. From Bandon northward, however, drowning has produced bays and marshes, and the saltwater tide reaches as much as 20 to 30 miles upstream. Many of the broad fertile plains and natural levees along Oregon's coastal streams may be attributed directly to alluviation during postglacial drowning. Were it not for this, the coast might have narrow valleys less hospitable to habitation.

Where the mouths of streams have not been filled in with sediments, bays have formed. Varying in size from larger inlets at Coos Bay and Tillamook to smaller scenic coves at Depoe and Boiler bays, few are very deep, and at low tide most display great areas of mud flats interlaced by channels. Because of runoff and sedimentation, a number of the bays must be constantly dredged to allow passage of ships, and only the major bays are used as sea ports.

One of the youngest and perhaps more ephemeral features of the Oregon Coast is the chain of freshwater lakes within the sand dunes that occur in a 50 mile strip from Floras Lake at Cape Blanco to Lily Lake near Heceta Head. Blocked by sand dunes occupying low spots or former river valleys, the lakes are fed by streams draining the western Coast Range. At 3,164 acres, Siltcoos Lake south of Florence in Lane County is similar in size to nearby Woahink, Tahkenitch, and Tenmile lakes, making them among Oregon's largest dunal lakes. The smaller string of lakes north of Florence, Nott, Mussel, Alder, Dune, and Buck, each less than 10 acres, are all comparatively deep at 25 to 35 feet. Clear Lake south of Reedsport in Douglas County and the similarly named Clear Lake north of Florence are the two deepest lakes on the coast at 119 feet and 80 feet. Dunal lakes vary in elevation as well. The surface of Siltcoos Lake is 5 feet above sea level, while that of Woahink is 38 feet.

Despite their diversity in size, the dunal lakes all share an exceedingly fragile environment because of the freshwater saturated sands beneath the lakes and the nearby mixing of saltwater and freshwater to the west. Around the lakes, the poor consolidation of sands, which project 60 to 70 feet above the water surface, makes them particularly susceptible to pollution, silting, and mudflows. All terrain vehicles, logging clearcuts, and rampant development for lakeside homes have already deleteriously affected most of these lakes.

The steep walls and narrow valleys of the Coast Range in conjunction with the high rate of

Ever-changing sand dunes, sheltered lakes, and dense forests come together along the southern Oregon coast. Lake Cleawox is in the foreground, Woahink is in the upper center, and a portion of Siltcoos Lake is at the upper right (photo Oregon State Highway Department).

precipitation have combined to form landslide lakes. Where the strata involved is soft, numerous landslides occur as the hillside moves more or less continuously in waves of viscous debris. Saturation during rainy periods increases the weight of the material. Landslides are easily recognized by their hummocky surface, tilted trees, small ponds, and swampy depressions.

During the Pleistocene, some of the massive sandstone formations, in particular the Tyee and Flournoy which are susceptible to landslides, moved as large blocks to dam creeks and form lakes. Such was the case in the formation of Triangle Lake near Blachly, Loon Lake near Scottsburg, and perhaps for the filled lake basin at Sitkum. Triangle Lake in Lane County was damed by a tilted mass of Flournoy sandstone that slid from high on the north slope of the valley to block Lake Creek. The creek eventually found an outlet against the south wall where it is incised in the bedrock rather than in the slide material. This condition will insure the long life of the alluvial valley but not necessarily of Triangle Lake which is only the remnant of a former larger body of water that extended upstream beyond Horton. Today most of the old lake has been filled in by alluvium. Triangle Lake is reportedly 97 feet at its deepest, but this is much less than it

was originally, and only a small volume of alluvium would be necessary to fill it.

Loon Lake in southwestern Douglas County has a similar history. Great blocks of Tyee sandstone that slid from the west wall of the valley lie as a jumbled mass to dam a creek. The creek, also named Lake Creek, is working on the fill, not on the adjacent bedrock, but such massive blocks are resistant to erosion. Loon Lake also was originally a much longer lake extending several miles upstream from the town of Ash. Tree trunks standing on the floor of Loon Lake were drowned at the time of damming approximately 1400 years ago. Ancient Lake Sitkum is a name applied to the lake that once occupied Brewster Valley at Sitkum in eastern Coos County. The history of this alluvial valley is almost identical to that of Loon Lake. Sandstone slipped northward in the valley to dam the river. A beautiful lake undoubtedly occupied this steep-walled valley at one time, but sediment has filled in the depression.

These lakes are relatively large. Numerous smaller lakes of similar origin have had an ephemeral existence in many of the stream valleys where the landslide material was either less extensive or the stream was large enough to cut through, draining the

184   Coast Range

Beach Processes

lake. Even with much of the damming material removed, larger blocks of rock form rapids in these streams as evidence of the existence of former landslide dams.

With at least 140 miles or almost one-half of the 310 mile long Oregon coast bordered by sand dunes, the state boasts much more of this fragile resource than either Washington or California. The main source of beach sands is erosion of sea cliffs supplemented by material transported and deposited by rivers in coastal estuaries. Except for a small percent of sand carried to the Clatsop Plains, Columbia River sediments move north along the Washington coast.

Oregon beaches go through a cycle that is dependent on longshore currents and shifting, wind-blown beach sands. During the winter, storm waves strip sand from many beaches, depositing it offshore in submerged bars. During the summer months the sand is redeposited on the beaches. Dunes build up when moving sands, carried northward with the winter storms and south during the summer, are trapped behind headlands. The amount of sand moved north is about the same as that carried south, so the supply remains approximately the same. This reuse of sand back and forth, year after year, has important implications for beach mining. As the sand is recycled, most minerals, except quartz, are removed to deeper water so that the process argues against economic sand mining ventures.

Today the Oregon coast is retreating by as much as 2 feet per year. The construction of riprap, piers, and breakwaters often accelerates this coastal erosion. With poorly designed jettys, sand piles up unevenly against man-made obstructions or is flushed out to sea. The jetty on the north side of Tillamook Bay has drastically retarded the flow of sand transported along the coast here so that the deficiency of sand south of the jetty has caused an increase in erosion of Tillamook Spit, which retreated by 50 feet from 1936 to 1960. At Cape Meares directly to the south the coastline was cut back 320 feet during the same period, and the winter storms of 1960 to 1961 moved the cape back by 75 feet. Along the southern coast, the south jetty at Coos Head has in turn altered Bastendorff Beach. Prior to construction, beach sands were limited to the area between Yoakam Point and Tunnel Point. With the jetty in place, sand filled in behind the new

Dominated by high cliffs, the small cove at Cape Meares is viewed with the wide Tillamook Bay in the background (photo courtesy Oregon State Highway Department).

structure, extending Bastendorff Beach seaward to obscure the original configuration. Dune sands here have been further stabilized by grass plantings.

Among the erosional processes on the coastline, landsliding is one of the most significant. Once a section of land has slumped and come into contact with the ocean, wave activity cuts away the supporting material at the toe. A steep cliff which is even more vulnerable to repeated sliding and erosion is produced. Landsliding is readily apparent along 70 of the 150 miles of the northern Oregon coast as far south as Florence. In February, 1961, a large mass of earth began to move at Ecola State Park in Clatsop County dumping much of the debris into the sea. Over a period of two weeks approximately 125 acres was carried downslope at a rate of 3 feet per day. A classic case of coastal sliding took place near Newport when a large block slid seaward from 1942 to 1943 taking 15 houses with it. The opening widened, breaking up the pavement, roads, and water mains that formerly connected to houses on the slumping ground. Today the sea is attacking the seaward side of the mass, and eventually the bulk of the debris will disappear.

Sand Spits

Sand spits partly block the entrance to a number of rivers, and frequently the sands have narrowed the opening to restrict tidal movement in and out of the bay. Receiving sands carried on currents, sand spits develop where a projecting headland diverts

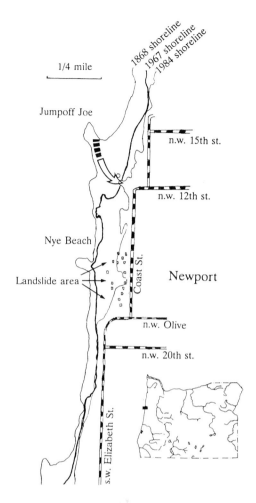

Landslides near Newport began in 1922, but major movement occurred from 1942 to 1943 (photo courtesy E. M. Baldwin).

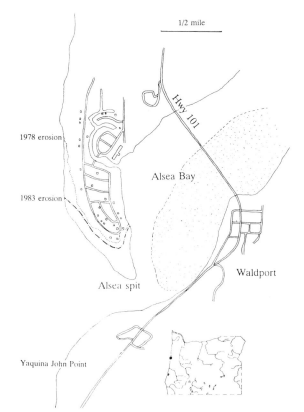

Erosion of the Alsea Spit in Lincoln County

until 1948 when storms cut three gaps in the peninsula. Only a few concrete foundations remain today.

Records kept since the early 1800s indicate Siletz Spit in Lincoln County had been relatively stable. With the early 1960s, a road the length of the spit, houses, and artificial lagoons were added. Storms in the winter of 1972 to 1973 destroyed one partially constructed house and threatened others on the spit. For a time it was feared the entire spit might be cut in two, but riprap was successfully placed on the ocean side to protect the development. The tip of Alsea Spit was lost as a result of winter storms and high wave action

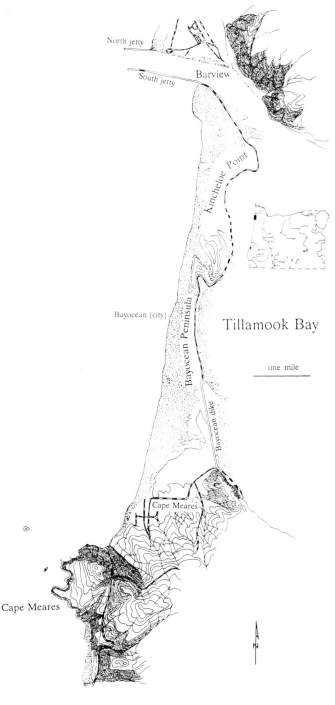

the current as at Cape Lookout and Cape Arago. The spit from Cape Kiwanda across Nestucca Bay deflects the Nestucca River southward, and at the mouths of the Coos and Coquille rivers long sand spits have grown down from the north pushing the rivers up against the rocks on their south banks.

Even with their inherent instability, sand spits attract commercial development. Unfortunately storms and the lack of sand on winter beaches cause severe erosion, attacking buildings close to the strand. Jutting northward from Cape Meares for 4 miles in Tillamook County, Bayocean Spit across Tillamook Bay is the eroded remains of an elongate dune probably originally connected to the mainland at the north while the southern end was open for drainage. Bayocean, or Tillamook Spit, has had an extensive history of erosion, and in 1952 a mile-long break appeared. A dike was built along the spit to seal the gap. In 1907 this spit was the focus of real estate developers who advertised the city of Bayocean as "The Queen of Oregon Resorts." Nearly 2,000 lots were sold, houses built, and streets paved. A natatorium and the beginnings of a hotel with a downtown grocery, bowling alley, and rock shop appeared on the spit before legal problems halted construction by 1914. Winter storms in 1939 inflicted considerable damage on the road and natatorium, even though a few permanent residents held on

during 1983, and within 5 weeks the opening to Alsea Bay had widened to over 1,900 feet by shifting erosional patterns. Houses on the the spit as well as the city of Waldport were threatened. At Netarts Bay, the long narrow sand bar jutting northward from Cape Lookout is forested in the central section, but the north end is bare and subject to erosion.

Offshore Continental Shelf and Slope

The continental shelf and slope are of increasing importance as available technology makes exploration and exploitation easier, and the search for food, fossil fuels, and metals expands into the ocean basins. The edge of the continent does not stop at the beach but extends to the shelf that dips gently seaward to a depth of 600 feet. The steeper continental slope descends to abyssal seafloor depths of 9,000 feet below sea level. The combined width of the shelf and slope is approximately 70 miles off Astoria and 40 miles off Cape Blanco. Shelf and slope geology adjacent to continents are linked directly to onshore geologic

systems, and the Oregon coastal margin is no exception. Sedimentary formations ranging from Eocene through Pleistocene exposed in the Coast Range can be traced offshore where the oldest sediments on the continental shelf grade into younger deposits on the slope.

The continental margin of the offshore shelf, slope, and abyssal plain are broken by features as banks, ridges and basins, canyons, and channels. On the outer edge of the shelf, prominent submarine fault-bounded escarpments as Nehalem, Stonewall, Perpetua, Heceta, and Coquille banks are exposed Miocene through Pliocene mudstones, sandstones, and clays that project as much as 200 feet above the surrounding seafloor. The steep escarpment at Stonewall Bank is situated approximately 19 miles off Yaquina and Alsea bays in water less than 120 feet deep. Another submarine projection, Heceta Bank lies about 35 miles west of the mouth of the Siuslaw River and rises within 360 feet of the surface, while Coquille bank, a shoal approximately 3 miles wide and 8 miles long between Coos Bay and Cape Blanco, exhibits 198 feet of relief.

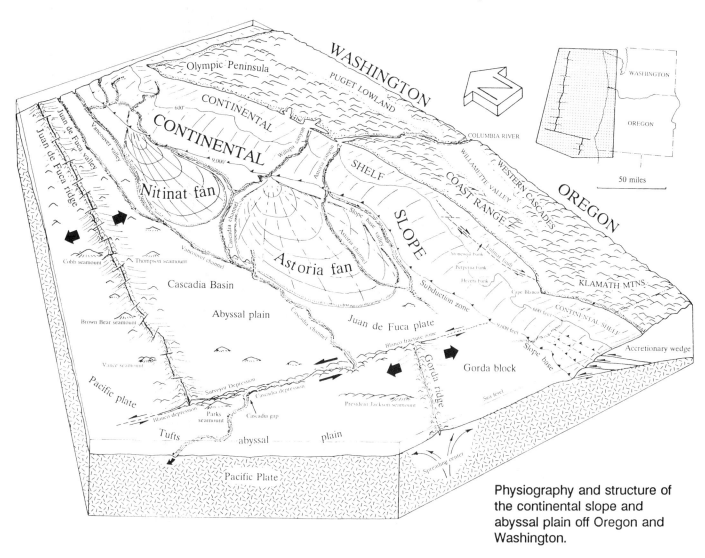

Physiography and structure of the continental slope and abyssal plain off Oregon and Washington.

In contrast to the mud seafloor below, the banks often have a rocky substrate where dense populations of invertebrates and fish dwell. Seaward from these banks, the continental slope is interrupted by prominent north-northwest ridges and intervening troughs. The longest ridge between Cascade Head and Tillamook Bay extends up to 2,000 feet while a similar ridge south of Cape Sebastian is a northward continuation of the Klamath plateau.

Cutting completely across the continental shelf and slope, submarine canyons and channels are important avenues for dispersing sediments by turbidity currents that originate in shallow water. Of these, Astoria Canyon, a major submarine feature off the Oregon coast, heads 10 miles west of the mouth of the Columbia River and extends across the Astoria submarine fan. The canyon eventually reaches a depth of over 9,000 feet where its identity is lost in the abyssal plain. Another noteworthy feature, the Rogue Canyon, with its head near the edge of the southern Oregon shelf, cuts across the continental slope to connect with the abyssal plain off the Rogue River. The longest deep-sea channel known in the Pacific Basin is the Cascadia channel. About 1,200 miles in length, the channel connects to the edge of the continental shelf northwest of the mouth of the Columbia River before angling in a southerly direction to breach the Cape Blanco fracture zone. The great volume of deep sea sediments carried from the Washington continental slope have pushed this channel farther and farther out onto the abyssal plain since the Pleistocene.

The focal point of Oregon offshore geology is the subduction zone at the base of the continental slope. Throughout the collision and accretion of the Coast Range island archipelago, the two main slabs, the Farallon and North American plates, continued to grind slowly toward each other. With the collision event, the subduction zone separating the two was abandoned and a new zone established west of the island chain. Today the subduction zone extends north and south along the foot of the Oregon continental slope, but the trench itself is largely obscured because of a rapid sedimentation rate. For this reason Oregon does not display the pronounced offshore subduction

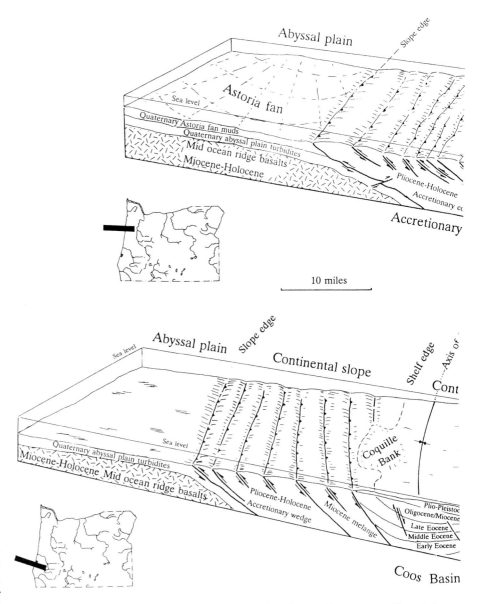

Block diagram of the Coast Range, continental shelf, and slope at Cape Arago.

trench or depression typical of much of the Pacific basin. Off southern Oregon, upper slope deposits are accumulating at a rate of 4 to 6 inches per 1,000 years, whereas lower slope deposits both to the north and south are accumulating at a phenomenal 8 to 26 inches every 1,000 years. The distinctive Crater Lake ash and pumice layer, dating back almost 7,000 years, is easily recognized in the deep marine environment off Oregon and provides a useful milepost for estimating rates of sedimentation.

The great influx of sediment from the Columbia River piles up at the base of the slope in huge submarine deposits called fans. The Astoria fan extends from 6,000 feet along the continental slope to the abyssal plain below 9,000 feet and covers more than 3,500 square miles. Only gently sloping, the Astoria fan

Block diagram across the Coast Range, continental shelf, and slope at Siletz Bay

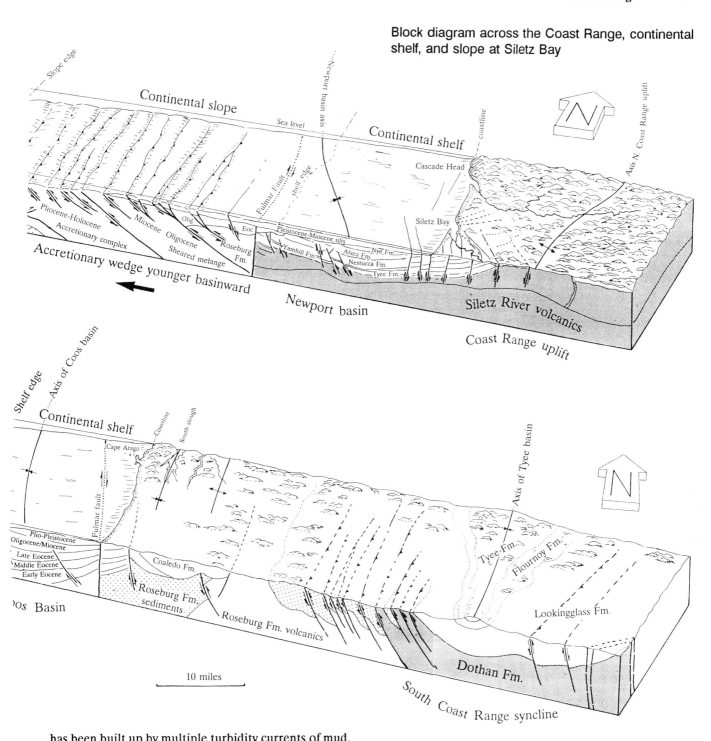

has been built up by multiple turbidity currents of mud, clay, and sand from the Astoria and Willapa canyons.

Pressure along the subduction zone produces a series of oblique faults and broad folds which continue from the lower to upper continental slope. Miocene and younger oceanic basalts as well as the veneer of upper Miocene, Pleistocene, and Holocene clays, silts, and sands here are deformed into a wedge or an accretionary melange. Sediments of the descending Juan de Fuca plate have been scraped off to pile up on the leading edge of the North American plate.

Across the abyssal plain to the west, the major boundaries between the subducting plates or tectonic blocks are the Juan de Fuca and Gorda spreading ridges and the Blanco and Mendocino fracture zones. In harmony with the Coast Range block, this entire network of spreading ridges and fracture zones shows evidence of having been rotated as much as 20 degrees clockwise in the past 10 million years, since late

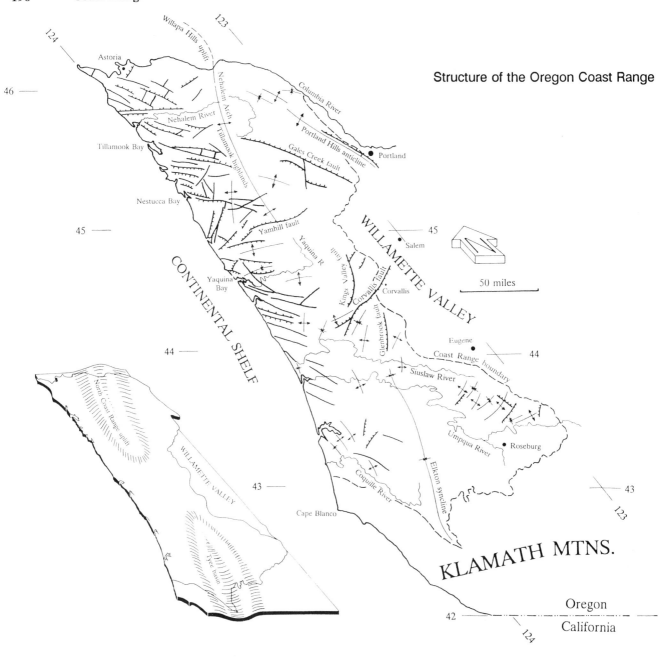

Structure of the Oregon Coast Range

Miocene time. The Blanco and Mendocino fractures are the source of an ongoing series of magnitude 5 and 6 earthquakes in southwestern Oregon and northern California. Although the quakes have occasionally moved houses off foundations, to date no catastrophic damage has occurred. Some commercial interest has been shown in the offshore spreading centers for their mineral potential, but technology, economics, and environmental considerations may temporarily preclude the exploitation of these massive sulfide deposits.

Structure

Evidence throughout the Coast Range reflects the ongoing uplift of the entire province. Rock fractures, faulting and folding, terraces on the western

slope, and tilting phenomena all suggest a close relationship between uplift of the Coast Range block and the subducting slab of the Juan de Fuca tectonic plate beneath the North American plate.

Structurally the Coast Range is a large crustal wrinkle with a north-south axis. The terms anticlinorium and synclinorium are used to describe extremely large folds of this type with many smaller crenulations across the flanks or limbs. At the north end of the coastal mountains the Willapa Hills of southwestern Washington is an upfold or arch that crosses the Columbia River into northeast Columbia County. Just south of the river at Mist, rocks of the coast are domed into a low anticline at the Nehalem arch. To the southwest, this fold becomes the Tillamook highlands.

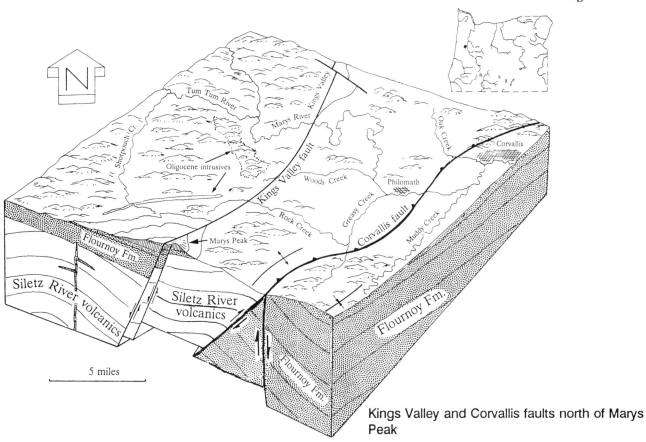

Kings Valley and Corvallis faults north of Marys Peak

South of the 45th parallel at the latitude of Salem the fold gradually inverts to a broad basin or syncline. The southern part of the Coast Range synclinorium extends all the way to the limit of the physiographic province where the axis swings southwest near its extremity in the Klamaths.

Because the Coast Range has been severely flexed and bent by folding, rocks in the province are cracked and broken by a network of criss-crossing faults and joints. When viewed from the air, the major faults follow three patterns. One group runs in an east-west direction, one in a northeast by southwest direction, and one in a northwest by southeast direction. Across the continental slope and into the shelf most of the faults are steeply angled thrust faults where the overhead or hanging wall has gone up and over the bottom or footwall. This kind of fault, typical of crustal shortening or telescoping, is produced as the oceanic Juan de Fuca plate is thrust eastward beneath the North American continental plate. Thrust faults along the coast, continental shelf, and slope reflect the pressure on rocks caught between these two large-scale plates. Within the Coast Range itself, most faults, known as normal faults, are nearly vertical or dip steeply to the east or west. These faults ordinarily have their overlying part or hanging wall dropping with respect to the underlying footwall. This kind of rock

failure represents crustal extension or stretching and reflects the tension placed on rocks by folding and flexing. Examples of such enormous structures are the Kings Valley, Corvallis, Gales Creek, and Portland Hills faults. Displacement on the Corvallis fault is as much as several miles. A third type of fault is exemplified by the north-south Fulmar fault midway across the Oregon continental shelf. This fault is a shear in the crust where the east side moves south and the west side moves north much like the San Andreas fault in California. Called a strike-slip fault, this large-scale feature may have had as much as 120 miles of movement in middle and late Eocene time.

Earthquakes

Oregon's recorded history of earthquakes is minimal and within the written records, which go back only 175 years, no significant quakes have been noted. However, evidence of catastrophic Holocene seismic activity has come to light as recently as 1987 with studies of buried coastal lowland swamps and bogs as well as offshore turbidity flows in Oregon and Washington. Along the coast the land dropped abruptly from 2 to 8 feet a number of times over the past 7,000 years. The subsidence of sections of the land below intertidal muds must have occurred very quickly as many of the trees in the swamp deposits are still preserved in an upright position. Subsidance of water-saturated coastal sediments causes the mud and silt to liquify and flow. In some localities these liquified sediments burst out as a "sand volcano" or sand boil. In Oregon, 15 locations between the Coquille River and Seaside record the distinctive pattern of land subsidence, buried forests, and bogs with a mud covering. At South Slough in Coos Bay, marshes with at least 8 burial events over the last 5,000 years demonstrate clear evidence for large-scale coastal earthquakes.

An analogy can be made between this rapid subsidence along the coastal region and a deck of cards. As the Juan de Fuca plate is subducted beneath the North American plate, the lower descending plate may bind and hang up. As plate movement proceeds, the plate begins to arch up like a bent playing card, causing the coastal area to rise slowly. Once the slab slips, the uplifted section snaps down bringing about wide-spread and rapid dropping of the land accompanying catastrophic earthquakes. These large-scale earthquakes are also recorded as displaced submarine sediments deep on the shelf and slope. With the vigorous shaking implicit in a strong quake, piles of unconsolidated sediment perched at the edge of the continental shelf are loosened and sent further into deep water as a roiling, muddy density current or turbidity flow to spread out over the slope and abyssal plain. Over 14 of these separate turbidity flows that correlate with depressed coastal swamps have been identified. Each turbidity flow shows up as an unmistakable sequence of coarse sand at the base grading toward the top into fine-grained, deep sea clays.

Tsunamis

Earthquakes from fault activity on the ocean floor can trigger enormous seismic waves called tsunamis that cause considerable damage once they reach the coast. Travelling up to 600 miles per hour, tsunamis appear as low waves in the open ocean but pile up as high as 100 feet when the wave energy is concentrated in shallower water near the coast. Evidence for prehistoric tsunamis is preserved in estuaries and coastal marshes where careful excavation shows traces of wave damage far beyond that caused by storms. Sediments record periodic scouring of bays and coastal areas that are covered by a distinctive deposit identifying the tsunami event.

Oceanic seismic events typically occur as a series of destructive waves over the span of several hours or longer. Since its installation in 1967, the instrument to record tides at the Oregon State University Marine Science Center at Newport has measured only three small tsunamis, even though one group of

waves lasted up to 2 days. One of the highest waves to reach Oregon in the early 1960s was measured at 95 feet high by an offshore wave-monitoring program concerned primarily with oil exploration. On March 27 and 28, 1964, the large seismic waves that struck the Oregon coast were generated by an earthquake in the Prince William Sound of Alaska. When the enormous waves reached Oregon, they tossed logs and debris onto beaches, across highways, and into nearby buildings. One wave caused the water level to drop exposing vast areas of the upper shelf before seawater cascaded over it. Although property damage in Oregon was light, tragically, 4 children sleeping at Beverly Beach State Park were drowned. A seismograph in operation at Newport since 1971 measured waves as high as 21 feet in the winter of 1972 to 1973.

Mining and Mineral Resources

Black Sands

Along the southern Oregon coast, heavy black sands in marine placers contain an assortment of economic minerals. During the Pleistocene minerals from the interior were transported by streams draining the Coast Range to the ocean where they were sorted by winnowing action of the waves. Heavier mineral grains such as gold, platinum, chromite, magnetite, garnet, and zircon were left along the beach, whereas the the lighter fragments such as quartz, feldspar, and mica were carried out onto the continental shelf. Many of the heavy minerals are black, and the deposits are referred to as "black sands". The original source for this placer gold is the Jurassic Galice Formation drained by the South and Middle forks of the Sixes River. Shales, sandstones, and greenstones of the Galice have been intruded by small grantitic dikes that precipitated the minerals into the host rock.

Heavy mineral concentrations are found from Cape Blanco north to Cape Arago with the major economic deposits located at the mouths of rivers and on elevated terraces. Early day prospectors were mainly interested in gold, which was discovered by Indians in 1852 near the mouth of Whisky Run north of the Coquille River. Shortly after news of discovery had spread, miners erected a network of long sluice boxes in the small creeks and along the beach. Sand, shovelled into the sluices, was washed down the trough with a stream of water so that the heavier gold particles were caught in the riffles. On higher terraces miners tunnelled deep into the basal black sand lenses. The 50 to 60 foot covering of loose, barren sands above the placer layer and the extremely fine-grain of the minerals made recovery difficult. Near Whisky Run, Randolph, a boom town of tents, stores, saloons, and houses,

A bucket line dredge mounted on a wagon with 6-foot steel wheels was designed to work beach placers in and out of the tide (photo courtesy Oregon Dept. Geology and Mineral Industries).

had sprung up only to be abandoned after a disasterous storm destroyed most of the mining operations on the beach in 1854. Shortly after that, prospecting here was only carried out sporatically. Although gold remains in the thin layers of black sands today, it is doubtful if the amounts are economic.

North of the Coquille River the two largest placers occur on the uplifted Seven Devils and Pioneer terraces. Here north-flowing longshore currents carried the minerals to be deposited near old seacliffs. At the Pioneer Mine, the 3-foot thick placers atop the terrace are covered by recently deposited sand dunes. Cutting three long tunnels, the longest of which was 1,340 feet, miners recovered gold and platinum from the mineral-rich layers. The Seven Devils or Last Change Mine south of Cape Arago was originally opened for gold and platinum, but from 1942 to 1943 chromite was extracted before the operation closed after World War II.

Production amounts from Coos and Curry county beach placers were never systematically reported, but $60,000 in gold and platinum is estimated to have been extracted between 1903 and 1929. In 1988 a joint federal and state Placer Task Force examined the Oregon shelf for economic deposits of metalliferous black sands, but few mineral resources were discovered. The search for economic minerals in the black sands was pursued as recently as 1992 when a mineral resources company leased the rights on 4,000 acres south of Coos Bay to investigate the possibility of extracting chromite, garnet, zircon, and titanium.

Coal

Although coal deposits are scattered across Oregon, there are only two significant coal producing regions in the state, the Coos Bay and Eden Ridge fields, both in Coos County. Of these, only the Coos Bay field has been extensively mined. Eocene coal beds formed as a result of a chronically swampy environment that existed along the edge of the ocean covering western Oregon. As plants growing in the lowland swamps accumulated in the brackish water, layers of peat developed. With the land sinking locally, sediments covered the bogs compressing and compacting the organic material into coal. Where there was not enough compression and the coal contains too much water, it is ranked as sub-bituminous and of little economic value. Most of Oregon's coal is of inferior quality, even though deeper parts of the basin have bituminous zones where pressure and heat have reacted with the sub-bituminous material to make a higher rank of coal capable of generating in excess of 9,000 B.T.U.'s. In spite of its low rank, coal here spontaneously combusts. Coal loaded in a bunker or gondola

Historic Libby Coal Bunker on Coos Bay (photo courtesy of J. L. Slattery)

The Beaver Hill Coal Mines in the Coaledo Formation reflect an industry from Oregon's past (photo courtesy of Oregon Dept. Geology and Mineral Industries)

car more than 3 to 4 feet deep will often begin to smolder and burn.

The Coos Bay coal field in the Beaver Hill bed of the Coaledo Formation, 30 miles long and 12 miles wide, is situated with the long axis in a north-south direction. After its discovery in 1854, the Coos field was mined sporatically through the 1940s at which time modern machinery was installed. Between the middle 1800s and 1920, the recorded production from 37 mines was 3 million tons. The potentially dangerous shafts and portals of the mines were sealed and covered in 1985 as a safety measure completed by the U.S. Office of Surface Mining, temporarily ending the Oregon coal mining era.

Coal in the Eden Ridge field in southern Coos County is part of the Tyee basin sediments. Here significant amounts of sub-bituminous coal were located by extensive drilling during the 1950s, but commercial exploitation has never been feasible.

Oil and Gas

Exploration for oil and gas in the Pacific Northwest began about 1881 with reports of "smell muds", oil, and gas seeps along the Washington coast. Almost all of the 500 exploratory wells were drilled prior to World War II, but only about 80 penetrated below 5,000 feet. Drilling activity in Oregon has been sporatic since the early 1900s. Wells drilled near Coos Bay after 1930 record minor amounts of both oil and

gas. In the Willamette Valley a number of drill sites showed oil and gas, and in 1981 a well near Lebanon in Linn County tested 170,000 cubic feet of gas per day before it was plugged and abandoned. Oil and gas have also been reported in the Tyee basin of the southern Coast Range.

Drilling of the Mist region in Columbia County began in 1945, but it wasn't until 1979 that the gas field here was discovered by the combined efforts of Reichold Energy Corporation, Diamond Shamrock Corporation, and Northwest Natural Gas Company. Today the field consists of 18 wells, producing 38.4 billion cubic feet of gas to date, with no significant oil. Most of the reservoirs in the field are fault traps within a larger anticline structure. Hydrocarbons found in sandstone reservoirs of the upper Eocene Cowlitz Formation initially accumulated in organic-rich marine deltaic sediments before being transformed by heat into gas.

Throughout the Pacific Northwest, several fine-grained sedimentary rocks are regarded as potential sources for petroleum. However the lack of porous reservoir rocks is largely due to clays and zeolite minerals plugging the sandstones. The most likely reservoir sands are those deposited in high energy upper continental shelf environments where volcanic debris is limited. The Cowlitz, Yamhill, Coaledo, Spencer, Eugene, Yaquina, and Astoria formations meet these conditions in part. Similar requirements for reservoir rocks need to exist for the successful exploration for offshore oil. Even with the limited success of Oregon petroleum exploration, it would be premature to say all available resources in the state have been expended.

Mercury

In western Oregon cinnabar or mercury ore occurs scattered within a belt 20 miles wide that extends from Lane, Douglas, and Jackson counties in the southern Coast Range to the California border. In Lane County, the Black Butte and Bonanza mines are responsible for about one-half of Oregon's quicksilver production. Mining here exploits Eocene marine sediments as well as volcanics of the Fisher Formation. Discovered in 1890 by S.P. Garoutte, the Black Butte Mine yielded a reddish cinnabar ore. Sulfur, combined with the mercury in the ore, was burned off in a 40-ton-a-day furnace. The mine was operated off and on until the early 1970s when the land containing the mine was sold for its timber assets.

Slightly to the south in Douglas County, cinnabar at Bonanza Quicksilver Mine was found in the 1860s. The mine was operated under a succession of owners during World War II when wartime demands

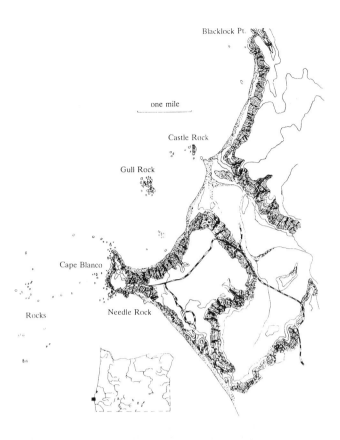

one mile

As James Stovall frequently remarked to his geology classes at the University of Oregon, the Devil nominally owns much of the coastal real estate. Devils Elbow, Devils Punchbowl, and Devils Lake are among some of the more scenic spots. While many of the names were given for imaginative reasons only, other places have been so designated because of specific beliefs or events. Unusual rock formations, lakes, or groves of trees attributed to mythical influences or which display dramatic physical features were feared by coastal Indians who created legends about evil spirits or skookums which may have haunted them. Devils Lake is so named because of an Indian legend concerning a giant fish or marine monster which periodically rises to the surface to attack passing California tourists.

Headlands, Sea Stacks, and Islands

In the cycle of coastal erosion, softer rocks and sediments are rapidly removed, and harder volcanic rocks emerge as headlands projecting from the mainland. Continued erosion of the durable headlands may isolate these resistant bodies to produce sea stacks, shoals, islands, and knobs. A distinct promontory noted on early maps, Cape Blanco is Oregon's most westerly

forced prices upward. Peak years were from 1940 to 1943 after which output dwindled until the Bonanza closed in 1960. The Bonanza Quicksilver Mine was Oregon's largest producer with 39,488 flasks.

Features of Geologic Interest

Long famous for its unique and beautiful scenery, the Oregon coast provides a marvellous combination of pleasing beaches and dramatic rocky headlands with steep cliffs interrupted by quiet coves. Travelling along the coast, impressive vistas that appear with each curve rival those in the Cascades as Oregon's most impressive visual treasure. High flat marine terraces dominate the coastline from Cape Blanco to Coos Bay. North and south of the Cape, these terraces occur in a discontinuous belt recording episodes of uplift and wave-cut erosion during the Ice Ages. Towering dunes at Florence make up a long section of beach sands, stretching 50 miles from Coos Bay to Heceta Head. North of Florence, volcanic rocks have been carved by wave action into cliffs, shoals, islands, sea stacks, and caves which interrupt the stretch of beaches. Where sedimentary rocks and sand dunes are bordered by low cliffs near Yachats and at Newport, a narrow coast with rolling topography is created. Northward to Tillamook Bay, once again a rough, rocky coastline is interspersed with long arcuate sandy beaches, spits, and bays, ending at the mouth of the Columbia River with the 17 mile long Clatsop Plains.

The raised terrace is clearly visible above Cape Blanco (photo courtesy U.S. Forest Service).

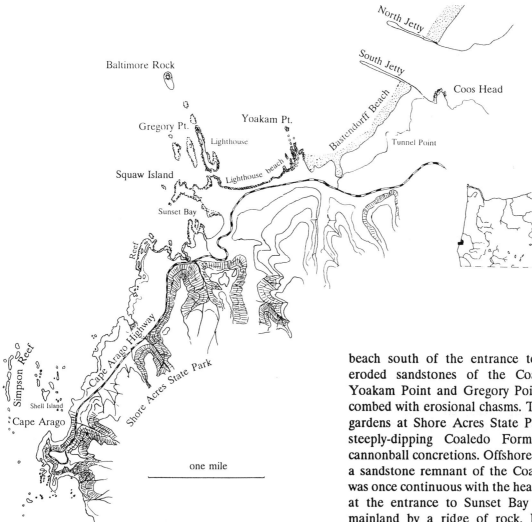

point as well as being the most southern projection of the Coast Range province. The cape was named "White Cape" by 16th century Spanish explorers who noted the white fossil mollusc shells that wash out of buried Pleistocene terraces high on the bluffs. Rocks of the promontory are a complex of Jurassic Otter Point conglomerates covered in turn by Eocene mudstones capped by the the highly fossiliferous late Miocene Empire Formation. North of Cape Blanco a number of stacks and large islands are found from Bandon State Park to Cape Arago. A field of durable Jurassic rocks of the Otter Point Formation are common here creating such landmarks as Face Rock and Table Rock. Blacklock Point, immediately north of the cape, is also formed of resistant Jurassic rocks.

The most dominant promontory north of Cape Blanco is found at Coos Bay, where a complex of headlands at Cape Arago, Gregory Point, Yoakam Point, and Coos Head are braced by massive sandstones. Coos Head is supported by the Empire Formation, while fine-grained sandstones of the Tunnel Point Formation compose the point that projects onto the

beach south of the entrance to Coos Bay. Deeply eroded sandstones of the Coaledo Formation at Yoakam Point and Gregory Point have been honeycombed with erosional chasms. The impressive formal gardens at Shore Acres State Park are underlain by steeply-dipping Coaledo Formation studded with cannonball concretions. Offshore here, Simpson reef is a sandstone remnant of the Coaledo Formation that was once continuous with the headland. Squaw "Island" at the entrance to Sunset Bay is connected to the mainland by a ridge of rock. During low tide the exposed bedrock creates a causeway to the island. Sand carried by longshore currents is accumulating along the bedrock exposures to build up the causeway. North of Florence, Eocene volcanic rocks of Yachats Basalt form resistant promontories at Sea Lion Point, Heceta Head, and Cape Perpetua. Narrow ledges are carved along the base where the fractured basalts are being cut away. At Heceta Head and Devils Elbow State Park several large rock masses appear offshore as isolated stacks that were once part of the mainland. A series of flows in the basalt and the nearly horizontal layers suggest the lavas issued from low profile shield volcanoes.

Basalt eroded into a variety of rocks shapes and sizes makes Seal Rocks among the most picturesque along the coast. Here the southern coastal extent of Columbia River basalt flows, originating in eastern Oregon, have been eroded to expose dikes and sills. Of these, Elephant Rock is the most conspicuous. Now part of the mainland, this rounded knob was an isolated sea stack during Pleistocene interglacials when sea level was higher. A distance offshore, steep-sided Otter Rock is an isolated block of the same invasive basalt.

Northward from Seal Rocks, headlands as

Elephant Rock, an elongate knob of basalt at Seal Rocks State Park (photo courtesy Oregon Dept. Geology and Mineral Industries)

Yaquina Point, Cape Foulweather, Cape Lookout, Cape Meares, Cape Falcon, and Tillamook Head are all of resistant exposures of Miocene Columbia River Basalt. Cape Kiwanda, which is bracketed by softer sediments, and Cascade Head, composed of local basalts, are the exceptions. Covering about 7 square miles between Neskowin Creek and Salmon River, Eocene basalt from a volcanic vent forms Cascade Head, rising 760 feet above the sea surface. Proposal Rock, now a tree covered island, was once part of Cascade Head before being cut off from the coast. Cape Kiwanda is unusual in that it is made of resistant sedimentary rocks of the Astoria Formation. Offshore Haystack Rock of Miocene basalt protected the headland from erosion until it was separated from the cape by wave action. Since that time the cape has receded to its present position.

Jutting dramatically for three-fourths of a mile from the shore north of Cape Kiwanda, the narrow Cape Lookout is composed of Miocene basalts. Because the layers that make up the cape are tilted toward the north, the sheer cliffs to the south are 800 high, while those to the north rise only 400 feet. Maxwell Point and Cape Meares are also part of the same invasive Columbia River basalt flows. A pocket beach is located between the smaller southern lobe at Maxwell Point and the more northerly projection at Cape Meares. Sands in these pocket beaches are protected from erosion by dominant headlands. Here three points of rock, separated by deep coves, rise vertically to elevations of 400 feet. The sea stacks, knobs, and arches of Pillar Rock, Pyramid Rock, and Three Arch Rocks offshore from Cape Meares and Maxwell Point are remnants of Columbia River basalt flows once connected to the mainland. Of these, Three

Arch Rocks were most likely tunnels at one time. Today these offshore rocks are inhabited by countless marine birds and sea lions.

To minimize hazards to shipping from the projecting rocks, a lighthouse was constructed in 1890 on the longest headland at Cape Meares. Sheet iron and bricks that were formed and baked on site at the cape were used in the building. The unusual lens in the lighthouse has eight sides, four with deep red panes and four with white ones, capable of projecting a beam 21 miles to sea. The only other light of this type operates in the Hawaiian Islands. The giant lens, weighing more than a ton, was ground in France, shipped around Cape Horn, and hoisted up a 200-foot cliff using a crane made from local spruce trees. An older oil-vapor lamp replaced the original 5 wick kerosene lantern, and in 1934 electricity reached the cape. The lighthouse was taken out of service in 1963.

Wave action at Tillamook Head separated the 100 foot high Tillamook Rock, an offshore basalt sea stack, from the mainland. One of Oregon's largest coastal monoliths, Tillamook Head is a complex of Miocene basalts and Astoria Formation sediments eroded into headlands, coves, stacks, and arches. Two promontories, connected by a crescent-shaped indentation, make up the 1,136 foot high Tillamook Head. The famous lighthouse atop Tillamook Rock was built in 1878 with material shipped in and landed by derricks. Improved navigation brought about the closure of the lighthouse in 1957. Purchased by Las Vegas businessmen in 1959, it remained vacant until 1980 when a Portland real estate consortium bought the island to remodel the lighthouse structure for use as a columbarium.

Caves and Trenches

Rocks, with fractures and faults running through them, have been eroded as caves, trenches, and tunnels seen in coastal areas. As waves flush away softer zones of rock along the faults, long straight trenches or caves result. When the roof collapses, as happens frequently, a churn, punchbowl, or water spout may develop. Between Sunset Bay and Cape Arago, narrow trenches and chasms are common. At Shore Acres State Park, the roof of a cave has collapsed leaving a natural bridge. Tunnel Point, the sharp headland immediately west of Coos Head, once had a natural tunnel and cave created by erosion of the sea cliff. Devils Elbow near Heceta Head and Devils Churn south of Neptune State Park are sea caves where the roof has collapsed along a trench that terminates in a cave beneath the headland. The tunnel at Devils Elbow extends 600 feet through Eocene Yachats Basalt. Near Neptune State Park, what was once a sea cave or

The dramatic promontory of Cape Lookout with Netarts Bay to the left (photo courtesy of the Oregon State Highway Department).

tunnel through rock, has been eroded to the degree that it now resembles an arch.

The largest of the caverns just south of Heceta Head are Sea Lion Caves, home to a large population of these marine mammals. The caves have been cut in bedrock at the intersection of a system of east-west and north-south fractures and faults through Eocene Yachats basalts. The largest west facing opening is used by the sea lions to move in and out of the cave. The southern part of the tunnel is below sea level, while the northern section is the location of the public viewpoint into the cave. Fractures along which the cave developed can be seen in the ceiling.

Among the best-known of the coastal erosional features are Devils Punchbowl at Otter Rock and the spout at Depoe Bay. At high tide and on windy days, waves, forcing their way through a tube cut into fractured Columbia River lavas at Depoe Bay, erupt upward in an opening along Highway 101 sending a column of water high into the air. These same invasive Miocene basalts form the wall that separates the picturesque inner and outer harbors here. At Devils Punchbowl, the rock where two sea caves met collapsed to form a pit. During high tides or stormy conditions, sea water foams and froths as in a boiling pot.

Sand Dunes

Of the several areas of extensive sands along the Oregon shoreline, the belt of dunes between Coos Bay and Sea Lion Point north of Florence is the longest, extending for a distance of 55 miles. Averaging 2 miles in width, the strip is 3 miles at the widest point near Florence, where it was declared the Oregon Dunes National Recreation Area in 1972. From Tillamook Head northward, another strip of sand dunes that makes up Clatsop Plains is a mile wide at the southern end and about 3 miles wide in the north where it extends across the mouth of the Columbia River to Cape Disappointment in Washington. Rising from 10 to 100 feet above sea level, long ridges of these dunes run parallel to the shore for miles.

The Clatsop Plains developed during the late Quaternary when sand began filling in the ancient shoreline that lay east of where it is now. Spits between Cullaby Lake and the mouth of the Columbia increased the deposition of sand between 1400 to 400 years ago when major accumulations took place as the coast migrated westward. At the northern end of Clatsop Plains, Clatsop Spit is the result of recent in-filling behind the jetty built on the south side of the Columbia River in 1885. The spit has been stabilized by

Jumpoff Joe along the coast near Newport showing erosion in 1900, 1913, and 1926. Only a small knob of the original rock remains today (photo courtesy Oregon Dept. Geology and Mineral Industries).

marram grass growing on the surface.

The unusual number of lakes found on the Clatsop Plains are elongate depressions between the ridges of the dunes. The 1 to 2 mile long basins are filled by groundwater percolating through the sand, and most of the lakes, as Cullaby, Smith, and Coffenbury, are independent of the local stream and river system with no surface water running in or out.

Coastal Mountains

Dominating the skyline for miles in either direction, Saddle Mountain, one of the highest peaks in the Coast Range, projects 3,283 feet above the surrounding countryside. Located in Clatsop County just east of Seaside, Saddle Mountain was first noted and named by John Wilkes of the U.S. Exploring Expedition in 1841. Now part of Saddle Mountain State Park, the towering cliffs and surrounding forests make it one of Oregon's most remarkable spots. A climb to the top provides vistas of the Cascades, Mount St. Helens, the Columbia River, and the city of Astoria on the Pacific Ocean.

The 2,000 feet of dark basalts that make up Saddle Mountain are part of the Miocene Columbia River series extruded from fissures in northeastern Oregon and southeastern Washington. Moving westward down the ancestral Columbia River gorge, the lavas eventually reached the Oregon coast. Where the flowing lava encountered water, as is the case with the basalts making up Saddle Mountain, breccias or broken fragments of basalt and characteristic oval-shaped pillow lavas resulted. Once the lavas had filled the local stream valley and cooled, sedimentary rocks enclosing the basalt were dissected by streams to expose the mass of breccias at the crest of Saddle Mountain. Also evident on the mountain are extensive vertical dikes of basalt running through the breccia for hundreds of feet. The dikes formed when the lava penetrated soft sediments. Miocene basalts making up Mt. Hebo and Neahkahnie Mountain are also part of the same coarse-grained Columbia River series. North of Nehalem Bay, the thickness of the layers making up Neahkahnie Mountain suggest that lavas must have ponded up in a bay or similar low spot prior to cooling.

In contrast Marys Peak, the highest elevation in the Coast Range, is capped by a resistant 1,000-foot thick, flat-lying sill of medium-grained gabbro that intruded Eocene sandstones of the Flournoy Forma-tion. Oligocene intrusions of this type are common in the central part of the range, and their emplacement took place during the same time as coastal rotation about 30 million years ago. An easy drive to the crest of Marys Peak yields scenes of grassy alpine meadows and a magnificent view. To the east the Willamette Valley and Cascade volcanoes form the skyline while to the south prominent peaks are Flat Mountain and Green Peak, both intruded sills. Westward the Coast Range and Pacific Ocean stretch to the horizon.

Marys Peak looking south. The Kings Valley fault trending north-south cuts directly across this promontory (photo courtesy U.S. Forest Service).

Suggested Readings

Adams, John, 1984. Active deformation of the Pacific northwest continental margin. Tectonics, v.3, no.4, pp.449-472.

-----1990. Paleoseismicity of the Cascadia subduction zone: evidence from turbidites off the Oregon-Washington margin. Tectonics, v.9, no.4, pp.569-583.

Addicott, Warren O., 1983. Biostratigraphy of the marine Neogene sequence at Cape Blanco, southwestern Oregon. U.S. Geol. Survey, Prof. Paper 774-G, pp.G1-G20.

Armentrout, John M., 1987. Cenozoic stratigraphy, unconformity-bounded sequences, and tectonic history of southwestern Washington. Wash., Div. of Geology and Earth Resources, Bull.77, pp.291-317.

Baldwin, Ewart M., 1964. Geology of the Dallas and Valsetz quadrangles, Oregon. Oregon Dept. Geol. and Mineral Indus., Bull.35, 52.

-----1974. Eocene stratigraphy of southwestern Oregon. Oregon Dept. Geol. and Mineral Indus., Bull.83, 40p.

-----1984. The origin of olistostromes in the Roseburg Formation in southwestern Oregon. Oregon Geology, v.46, no.7, pp.75-76, 82.

Beeson, Marvin, Perttu, R., and Perttu, J., 1979. The origin of the Miocene basalts of coastal Oregon and Washington: an alternative hypothesis. Oregon Geology, v.41, no.10, pp.159-166.

Brooks, Howard C., 1963. Quicksilver in Oregon. Oregon Dept. Geol. and Mineral Indus., Bull.55, 223p.

Brownfield, Michael E., 1981. Oregon's coal and its economic future. Oregon Geology, v.4, no.5, pp.59-67.

Chan, M.A., and Dott, R.H., 1983. Shelf and deep-sea sedimenation in Eocene forearc basin, Western Oregon - fan or non-fan? Amer. Assoc. Petroleum Geol., Bull.67, pp.2100-2116.

Cooper, William S., 1958. Coastal sand dunes of Oregon and Washington. Geol. Soc. Amer., Mem.72, 163p.

Couch, Richard W., and Riddihough, Robin P., 1989. The crustal structure of the western continental margin of North America. In: Pakiser, L.C., and Mooney, W.D., Geophysical framework of the continental United States. Geol. Soc. Amer., Memoir 172, pp.103-128.

Darienzo, Mark E., and Peterson, Curt D., 1990. Episodic tectonic subsidence of late Holocene salt marshes, northern Oregon central Cascadia margin. Tectonics, v.9, no.1, pp.1-22.

Duncan, Robert A., 1982. A captured island chain in the Coast Range of Oregon and Washington. Jour. Geophys. Res., v.87, pp.10,827-10837.

Heller, Paul L., and Ryberg Paul T., 1983. Sedimentary record of subduction to forarc transition in the rotated Eocene basin of western Oregon. Geology, v.11, p.380-383.

Kelsey, Harvey M., 1990. Late Quaternary deformation of marine terraces on the Cascadia subduction zone near Cape Blanco, Oregon. Tectonics, v.9, pp.983-1014.

Komar, Paul D., 1992. Ocean processes and hazards along the Oregon Coast. Oregon Geology, v.54, no.1, pp.3-19.

Kulm, L.D., and Fowler, G.A., 1974. Oregon continental margin structure and stratigraphy: a test of the imbricate thrust model. In: Burk, C.A., and Drake, M.J., eds., The geology of continental margins. New York, Springer-Verlag, pp.261-283.

Lund, Ernest H., 1973. Landforms along the coast of southern Coos County, Oregon. Ore Bin, v.35, no.12, pp.189-210.

------1974. Rock units and coastal landforms between Newport and Lincoln City, Oregon. Ore Bin, v.36, no.5, pp.69-90.

McInelly, Galan W., and Kelsey, Harvey M., 1990. Late Quaternary tectonic deformation in the Cape Arago-Bandon region of coastal Oregon as deduced from wave-cut platforms. Jour. Geophys. Res., v.95, no.B5, pp.6699-6713.

Molenaar, C.M., 1985. Depositional relations of Umpqua and Tyee Formations (Eocene), southwestern Oregon. Amer. Assoc. of Petroleum Geol., Bull.69, no.8, pp.1217-1229.

Muhs, Daniel R., et al., 1990. Age estimates and uplift rates for late Pleistocene marine terraces: southern Oregon portion of the Cascadia forearc. Jour. Geophys. Res., v.95, no.B, pp.6685-6698.

Niem, Alan R., Snavely, P.D., and Niem, W.A., 1990. Onshore-offshore geologic cross section from the Mist gas field, northern Oregon Coast Range, to the northwest Oregon continental shelf and slope. Oregon Dept. Geol. and Mineral Indus., Oil and Gas Inv. 17, 46p.

Peterson, Curt D., Gleeson, George W., and Wetzel, Nick, 1987. Stratigraphic development, mineral sources and preservation of marine placers from Pleistocene terraces in southern

Oregon, U.S.A. Sedimentary Geology, v.53, pp.203-229.

Pfaff, Virginia J., and Beeson, Marvin H., 1989. Miocene basalt near Astoria, Oregon; geophysical evidence for Columbia Plateau origin. In: Reidel, S.O., and Hooper, P.R., eds., Volcanism and tectonism in the Columbia River flood-basalt province. Geol. Soc. Amer., Special Paper 239, pp.143-146.

Reilinger, Robert, and Adams, J., 1982. Geodetic evidence for active landward tilting of the Oregon and Washington coastal ranges. Geophys. Res. Letters, v.9, no.4, pp.401-403.

Rytuba, Paul T., 1984. Sedimentation, structure and tectonics of the Umpqua Group (Paleocene to early Eocene) south-western Oregon. PhD., Univ. of Arizona, 180p.

Scholl, David W., Grantz, Arthur, and Vedder, John G., eds., 1987. Geology and resource potential of the continental margin of western North America and adjacent ocean basins-Beaufort Sea to Baja California. Circum-Pacific Council for Energy and Mineral Res., Earth Sci. Ser., v.6, 799pp.

Simpson, Robert W., and Cox, Allan, 1977. Paleomagnetic evidence for tectonic rotation of the Oregon Coast Range. Geology, v.5, pp.585-598.

Snavely, Parke D., 1987. Tertiary geologic framework, neotectonics, and petroleum potential of the Oregon-Washington continental margin. In:Scholl, David W., Grantz, Arthur, and Vedder, John G., eds. Geology and resource potential of the continental margin of western North America and adjacent ocean basins-Beaufort Sea to Baja California. Circum-Pacific Council for Energy and Mineral Res., Earth Sci. Series, v.6, pp.305-335.

The Willamette River floodplain just east of Corvallis displays abandoned meanders, old channels, and oxbow lakes typical of a mature stream [photo courtesy U.S. Department of the Interior].

Willamette Valley

Physiography

The Willamette Valley and Puget Sound physiographic province is a lowlands stretching from Cottage Grove, Oregon, to Georgia Strait in Washington. The smallest physiographic division in Oregon, the valley is a level, elongate alluvial plain which narrows at either end for 30 miles before it pinches out. Enclosed on the west by the Coast Range, on the east by the Cascade Mountains, and bordered on the north by the Columbia River, the main valley is 130 miles long and from 20 to 40 miles wide. From 400 feet at the southern end of the valley near Eugene, the elevation drops to sea level at Portland, an average of 3 feet per mile. The overall gradient is to the north and not from the margins toward the middle. The southern end of the valley is narrower but flatter than the northern hilly Salem and Portland areas. Salem is bordered by the Eola Hills to the west, the Ankeny Hills to the south, and the Waldo Hills to the east. The 1,000 feet high Tualatin Mountains are adjacent to Portland on the west, the Chehalem Mountains cross to the southwest, while to the east and southeast smaller volcanic buttes and peaks dot the landscape. Near the center of the valley, the 45th parallel, halfway between the equator and the North Pole, passes close to Salem.

With a watershed of 11,200 square miles, the Willamette River is the major waterway in the valley. Originating at the junction of the Coast and Middle forks near Eugene, the river runs north-northeast to its confluence with the Columbia. Flowing into the Willamette, sediment laden waters of the Coast and Middle forks from the south, the McKenzie, Calapooia, North and South Santiam, Pudding, Molalla, and Clackamas rivers from the Cascade Mountains, and the Long Tom, Marys, Luckiamute, Yamhill, and Tualatin rivers from the Coast Range drain the surrounding areas.

Although comparatively small, the Willamette Valley is the economic and cultural heart of Oregon. As the only natural lowland of any size, its moderate climate supports 70% of Oregon's population as well as intense and varied agriculture.

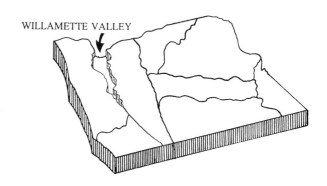

WILLAMETTE VALLEY

Geologic Overview

Physiologically the Willamette-Puget Sound lowland is similar to the Great Valley of California, but geologically the two are significantly different. The California valley was at one time an inland sea behind the Coast Range, whereas the Oregon province was part of a broad continental shelf of the ocean extending from the Cascades westward beyond the present coast. Structurally the Willamette Valley was more of a coastal marine environment than a true isolated basin or a valley cut by a river. Although subsurface geology of the Willamette Valley is closely related to that of the Coast Range, the later history of the valley is primarily one of glacial events. Thick layers of Late Pleistocene and Holocene alluvium cover all but a few areas of preTertiary rock from Eugene to Portland.

Older foundation rocks here are volcanics that erupted as part of a submarine oceanic island archipelago. Once the archipelago was attached or accreted to the western margin of North America, the volcanic rocks subsided, and a forearc basin formed on top. This basin was to become the focus of marine deposits from the Eocene through Pliocene. Fossils and sediments accumulating in the basin during the Oligocene,

203

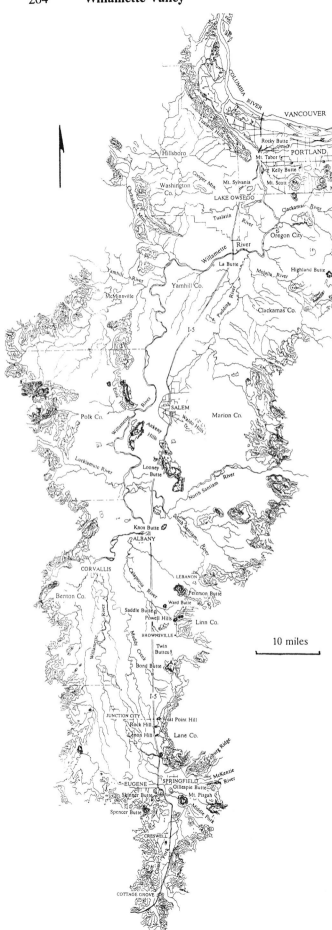

10 miles

Miocene, and Pliocene reflect shallowing as the ocean shoreline retreated northwestward. These marine sediments were covered in turn by Columbia River lavas that poured through the gorge from eastern Oregon during the middle and late Miocene to invade as far south as Salem.

Uplift and tilting of the Coast Range block and Western Cascades brought about the trough-like configuration of the Willamette Valley and the formation of a number of closed basins on the continental shelf. During the Pliocene and Pleistocene in the northern part of the province, a large lake received silts, muds, and gravels from the Willamette and Columbia rivers. The eruption of the Boring lavas from over 100 small volcanoes near Oregon City as well as east and west of Portland covered these earlier lake sediments, and today the vents project as small buttes.

The dominant signature on Willamette Valley geology resulted from a number of large-scale Pleistocene floods that scoured eastern Washington and the Columbia River gorge leaving deposits throughout the province. Enormous glacial lakes formed in Montana when the Clark Fork River was dammed by ice and debris. Once the ice blockage was breached, rushing flood waters carrying icebergs cascaded across Idaho, southeastern Washington, and down the Columbia gorge. Water backed up into the Willamette Valley creating temporary lakes and strewing a field of boulders in its wake. An unknown number of floods took place during a 2,500 year interval until the climate warmed, and glaciers retreated northward.

Because of its position close to the offshore subduction zone between Pacific Northwest plates, Oregon experiences a continual number of seismic events, and in the future the state could be the site of catastrophic earthquakes although details of time and place are uncertain. The coastal regions and Willamette Valley would be particularly vulnerable should a strong quake occur.

Geology

Geologically part of the eastern margin of the Coast Range block, foundation rocks of the Willamette Valley have played something of a passive role against the backdrop of moving tectonic plates. In Eocene time an undersea chain of volcanoes atop the Kula and Farallon plates collided with the westward moving North American plate where they were accreted. With a thickness of more than 2 miles, the volcanic rocks of the island chain form the basement of the Coast Range and Willamette Valley. After docking or making initial contact with North America, the island archipelago was rotated clockwise beginning in the early Eocene. With accretion, the old subduction zone east of the volcanic block was abandoned and a new one activated offshore to the west where it is today.

The slow subsidence of the block created a broad forearc trough along the western margin of North America. From Eocene through Pliocene time the basin was the recipient of deposits that blanketed the earlier volcanic platform. Rivers draining the Klamath Mountains and later the Idaho batholith provided abundant sediments that accumulated in the newly formed basin. During the early Eocene the eastern edge of the subsiding coastal block that was to become the Willamette Valley collected sandstones and siltstones of the Flournoy Formation near Lorane, Philomath, Falls City, the low hills around Camp Adair, and in the southern valley. Where Eocene rocks are exposed in the area north of Corvallis, rolling hills contrast sharply with the flat valley floor elsewhere that is covered with Pleistocene fill.

In the northern part of the valley these deposits were followed by middle Eocene Yamhill muds, sands, and silts, mixed with ash and lavas from the ancestral Cascades that were carried into the shallow seaway. Within the Yamhill, shoals of limestones around offshore banks formed the Rickreall and Buell limestones containing broken mollusc shells, foraminifera, and calcareous algae intermixed with volcanic debris. In the northern valley 2,000 feet of Nestucca Formation deposited in a deep water setting extended westward from McMinnville, while near-shore sands, silts, and muds of the shallow marine Spencer Formation produced deltas along the margin. Found along the western side of the valley from Eugene north to Gales Creek in Washington County, Spencer sands are covered by nonmarine tuffs and conglomerates of the late Eocene Fisher Formation. Fossil plants from the Fisher Formation southwest of Cottage Grove indicate a warm, moist tropical climate where broad leaf plants as the Aralia grew close to the shoreline. Beneath Eugene almost a mile of upper Eocene silts and sands of the Eugene Formation extend northward toward the Salem hills. Marine molluscs, crabs, and sharks in this formation suggest warm, semitropical seas. Sediments of the Spencer, Fisher, and Eugene formations were derived from the rapidly growing volcanics of the Western Cascades.

Oligocene

The Oligocene ocean in the Willamette Valley reached only as far south as Salem. The high-water mark on the western shoreline is recorded by marine sediments in the vicinity of Silverton and Scotts Mills in Marion and Clackamas counties. In the Scotts Mills Formation a transgressive, advancing seaway followed by a regressive, retreating ocean chronicles storm

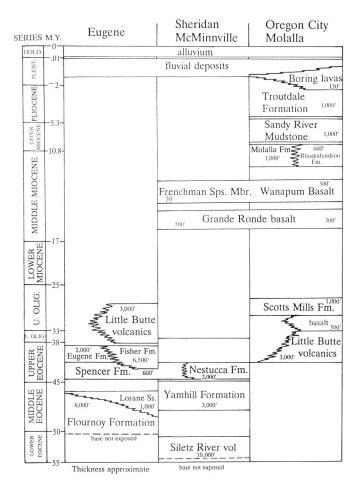

Willamette Valley stratigraphy (after COSUNA, 1983)

conditions, shallow water, and coastal swamps that gave rise to thin layers of low-grade coal. Coal beds at Wilhoit Springs and Butte Creek were deposited along the margins of the sea as it retreated. Prior to the arrival of the Columbia River lavas in the middle Miocene, the Scotts Mills sediments were tilted eastward and severely eroded.

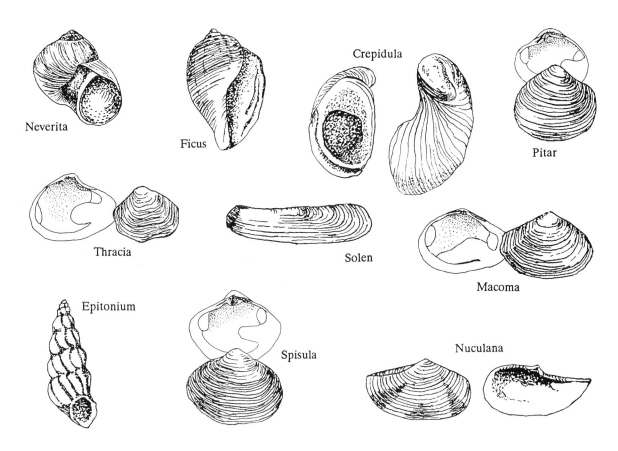

Invertebrate fossils of the late Eocene Eugene Formation

In the Nehalem Valley fine-grained shallow marine sediments of the Scappoose Formation are contemporaneous with the Scotts Mills Formation. Over 1,500 feet of Scappoose sandstones, mudstones, and conglomerates were deposited in an estuary or delta environment covering a dissected landscape.

Miocene

With slow uplift of the Coast Range, the sea withdrew from the region of the Willamette Valley. The deep shelf and slope environment became shallower as the basin filled. Little is known about the configuration and environment of the Willamette Valley following the regression of the late Oligocene seaway. A broad semitropical coastal plain, where lakes ponded in slight depressions, extended from the ancestral Cascades out to the present shoreline. Black, silty clays lying between weathered Pleistocene and late Eocene Spencer sandstones near Monroe reveal ancient lakebed sediments that contain fossil pollen from coniferous as well as broadleaf plants once living around the lacustrine basin. Much of the pollen from this lake is from plants now extinct in the Pacific northwest.

In middle and late Miocene time, voluminous sheets of basaltic lava from fissures and vents in northeastern Oregon poured through the Columbia gorge and into the Willamette Valley where they reached as far south as Salem. The fluid Columbia River lavas covered the region of the Portland Hills, most of the Tualatin Valley, as well as the slopes of the Chehalem, Eola, and Amity hills. The dark, finely crystalline, collumnar-jointed basalt ranges up to 1,000 feet in thickness. Near Portland the layers of lava produced a monotonous, flat landscape with only the tops of several higher hills projecting above the flows. After cooling and crystallizing, the lavas rapidly decomposed in western Oregon's wet climate so that almost all of the original volcanic landscape has been thoroughly dissected. Dark red soils around Dundee, the Eola Hills, and Silverton Hills are easily recognized as decomposed Columbia River basalts.

Interfingering with the Columbia River basalts in northern Marion and Clackamas counties, 1,000 feet of clastic sediments, mudflows, and volcanic tuffs of the Molalla Formation represent the first terrestrial sediments deposited after the withdrawal of the Oligo-

Pleistocene fluvial and flood deposits and structure
of the southern Willamette Valley (after Yeats, et
al., 1991; Graven, 1990)

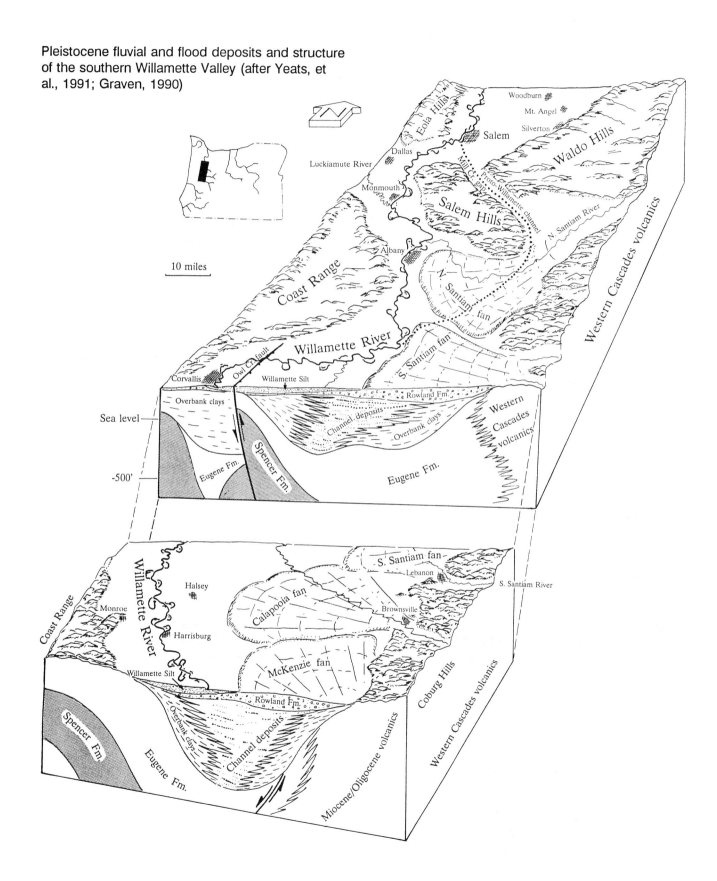

cene seaway. Mudflows here originated when heavy rains on the slopes of ancient volcanoes caused loose ash to flow for miles down valleys. Abundant fossil flora in these sediments suggest a hilly topography covered with Liquidambar, (sweet gum), Platanus, (sycamore), and Carya, (hickory) adjacent to swamps dominated by cypress. Ginkgo and Metasequoia commonly occured in the warm, rainy climate.

Around 1,000 feet of silt in the Portland and Tualatin basins, designated as the Helvetia Formation, sits directly above the Columbia River basalts. Pebbles of basalt quartzite, granite, along with abundant quartz and mica, suggest these sediments were deposited by streams of the surrounding mountains ranges as well as the ancestral Columbia River.

With margins around Portland and Tualatin in the north and west and Sandy on the east, a bowl-shaped structure gently sloped from 700 feet at Sandy to 250 feet at Portland and Vancouver. Flowing in a channel south of its present course, the Columbia River aided by the northerly flowing Willamette River emptied into the basin to form a lake and deposit silts and muds in a delta over 1,300 feet thick. More than 1,500 feet of the Sandy River Mudstone of sandstones, siltstones, and conglomerates filling the Portland basin underlie the city today. Later Pliocene Troutdale gravels from the Columbia River drainage washed into the lake, covering the mudstone to depths of 700 feet. Dissection and erosion by rivers and Ice Age floods removed portions of the Troutdale Formation from the Portland Hills and surrounding areas.

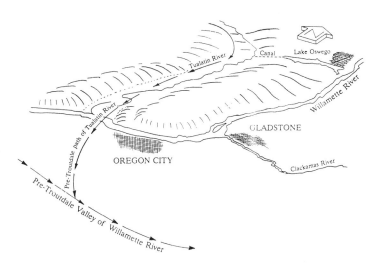

Pathways of the pre-Troutdale Willamette and Tualatin rivers in the northern Willamette Valley. Extrusion of the Boring lavas moved the Willamette River channel northward to Oregon City (after Baldwin, 1957).

Geomorphology of Lake Oswego (photo courtesy of Delano Photographics)

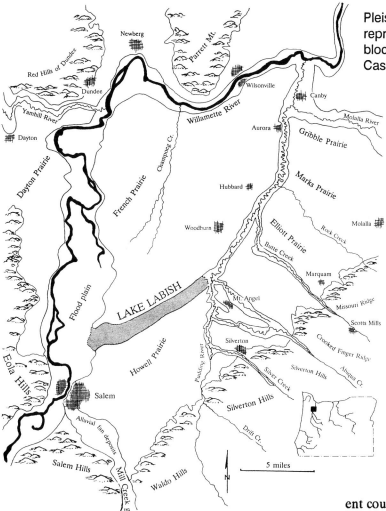

Pleistocene Lake Labish northeast of Salem may represent an older channel of the Willamette blocked by sediments from the Western Cascades (after Glenn, 1962).

Pliocene

During the Pliocene epoch, continued subduction of the Juan de Fuca plate and growth of the offshore accretionary wedge brought about the renewed uplift and tilting of the Coast Range and gentle folding of northern Willamette Valley rocks. The end of the wide coastal plain marked the beginning of the Willamette Valley as a separate physiographic feature once the sea had withdrawn to its present position. At this time major folds of the coastal mountain range formed, and outer continental shelf environments were raised over a mile from 1,000 to 2,000 feet below sea level to elevations of 3,000 feet above.

About 5 million years ago in the northern valley Boring Lavas erupted from over 100 small vents, cones, and larger shield volcanoes as Mt. Defiance, Larch Mt., Mt. Sylvania, and Highland Butte. Boring Lavas can be readily distinguished from the older Miocene Columbia River basalts by their fracture pattern. The fine-grained, light-weight Boring lavas break into large blocks and are seldom found in the small columns so common to the Columbia River Basalt.

The Pliocene lavas covered the gravels in the old lake bed, and today they make up many of the buttes near Gresham and Boring as well as capping a number of hills in the vicinity of Oregon City. Rocky Butte, Mt. Scott, Mt. Tabor, Kelly Butte, and Mt. Sylvania contain cinders and lavas of these flows, the most westerly of which is along the slopes of the Tualatin Valley near Beaverton and Metzger. La Butte southwest of Wilsonville in Marion County may be the most southern eruption of these lavas. The landscape in the valley and around Portland shows clear evidence of these late eruptions. Flows from Highland Butte southeast of Oregon City pushed the Willamette River ten miles to the northwest near its present course, whereas in Portland lavas filled old streambeds where Burnside Street and Canyon Road are now located. Rocky Butte, a volcanic vent that erupted approximately 1.2 million years ago, and Mt. Tabor, with a small cinder cone projecting from its north side, are presently in downtown Portland. Across the river in Washington, Bobs Mountain northeast of Portland and Battle Ground crater north of Vancouver still have intact cones.

Pleistocene Rivers

Just prior to deposition of the Willamette Formation, up to 150 feet of coarse gravel and sand of the Rowland Formation spread over a broad area in the southern Willamette Valley. Divided into two members, the Rowland represents glacial outwash from the Cascades that was flushed into the valley by the North and South Santiam, Calapooia, and McKenzie rivers. This massive outwash unit thins toward the northwest and forms a complex series of coalescing alluvial fans that have bulldozed the Willamette channel off to the western margin of the valley between Eugene and Corvallis.

The valley beneath the Rowland glacial-alluvial fan sequence is filled with almost 300 feet of proto-

With uplift and intensive erosion of the land during glacial periods when sea level was lowered, rejuvenated streams began to cut deeply into flood plains. Terrace gravels, recording the many changes in stream levels, can be seen along the Sandy and Clackamas rivers and in eastern Portland where the streets rise in elevation toward Gresham. At that time the ancient Willamette River ran southeast of Oregon City. Here it was joined by the ancestral Tualatin River flowing along the western margin of the Portland Hills. Both rivers wandered over a wide floodplain resulting from thick alluvial deposits which first filled the Portland and Tualatin basins, then backed up the stream valleys, and finally, in places, covered the divides between the streams. As sea level continued to drop, the Tualatin flowed through the channel at Lake Oswego while the Willamette established its present pathway in the Columbia River basalts at Oregon City where cutting action of the river produced the falls. In the final stage, flood waters flushed out the sands of the previously abandoned channel, diverting the Tualatin back into its original southward pathway to merge again with the Willamette. Today water from the Tualatin River is channelled into Lake Oswego by means of a low dam and ditch.

The rich dark soils of what may be a former channel of the Willamette River, now called Lake Labish, can be seen in a straight strip extending for almost 10 miles northeast of Salem in Marion County. The former course of the river was cut off during the Pleistocene when a natural dam of sand from Silver, Abiqua, and Butte creeks blocked the channel. The resulting shallow lake slowly filled with silt and organic

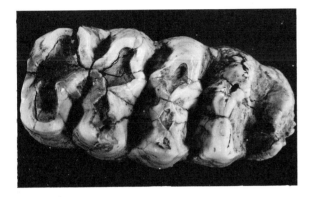

Crown views of elephant teeth clearly distinguish the low cusps of the browsing mastodon (top) from those of the grazing mammoth (bottom). (Fossils courtesy of the Thomas Condon collection, Univ. of Oregon)

Willamette River sands and gravels. Bordered on both sides by bluish clay overbank deposits, these coarse pebbles trace the path of the ancient river as it began filling the rugged erosional surface on top of the Columbia River basalts and older Tertiary units in the valley. Prior to the emplacement of the alluvial fans, the Willamette River may have followed a more easterly pathway in the valley. There is evidence that the old channel between the southern and northern valleys ran through a narrow gap at Mill Creek between Salem and the Waldo Hills.

About the size of a modern Indian elephant, mastodon roamed in herds throughout the Willamette Valley during the Pleistocene.

debris to become a marsh. Thick peat deposits in the old lake reflect a long period as a swamp and bog. In this organic layer, bones of Ice Age mammals such as mammoths, mastodon, giant sloth, and bison are frequently found. Unlike the Pleistocene LaBrea tar pits of Los Angeles, the animals were probably not mired in the peat, but carcasses were washed in and covered allowing the remains to be preserved in the oxygen-poor bog away from the attentions of scavengers. Today the fertile soils of Lake Labish support a thriving onion industry.

Pleistocene

Covering the Boring lavas and Troutdale gravels, a gritty, structureless, yellowish-brown sediment called the Portland Hills Silt was deposited within the last 700,000 years. Commonly 25 to 100 feet thick, the layer mantles much of the Portland area from the Tualatin and Chehalem mountains on the west all the way to the Gresham Hills and Ross Mountain on the east. Microscopically the silt is remarkably uniform, and the identified minerals include many which must have been derived from terrains as far north as Canada. This unique formation has long been something of a geologic puzzle and has been designated as both a wind-blown and water-laid deposit. The physical characteristics of the silt are similar to the palouse

strata in southwest Washington indicating that the sediments are wind blown in origin. The Yellow River in China derives its name from comparable Pleistocene loess deposits that blanket vast areas in the northern part of the country. Ground up rock flour, produced by the crushing and milling action of glacial ice, was transported by water to be deposited along flood plains of the Columbia River. Strong Pleistocene winds, collecting the fine dust, carried it aloft in enormous clouds to cover the Portland Hills. Four different layers of silt are separated by three soil horizons. Interglacial, warm intervals are represented by the silt deposits, whereas soil horizons reflect times of glacial advance.

Ice Age Floods

Beginning about 2 million years in the past, the Ice Ages mark the advance and retreat of continental glaciers, an event that triggered one of the most catastrophic episodes in Oregon's geologic history. When first proposed in the 1920s by J. Harlen Bretz, the theory of an enormous flood washing across Washington and through the Columbia River gorge was not readily accepted. Careful work by Bretz, however, built up a body of evidence that could not be ignored. Between 15,500 and 13,000 years ago, the Columbia River drainage experienced a series of spectacular floods from ruptured ice dams along its

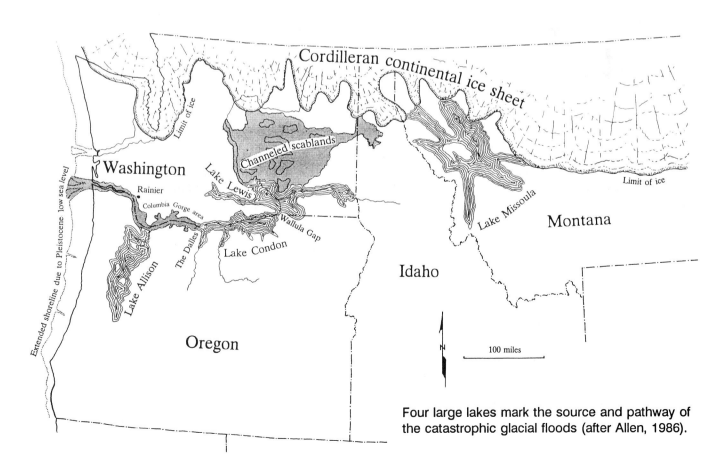

Four large lakes mark the source and pathway of the catastrophic glacial floods (after Allen, 1986).

Prior to each flood, an ice lobe from northern Idaho stretched southwest to dam up the Clark Fork River that flows northward to join the Columbia across the Canadian border. Old shorelines visible today high above the city of Missoula, Montana, are evidence that the ice dam backed up a vast lake covering a large area of western Montana. As the ice dam was breached, water, ice, and sedimentary debris poured out at a rate exceeding 9 cubic miles per hour for 40 hours. Flushing through the Idaho panhandle and scouring the area now known as the channeled scablands of southeast Washington, the lake drained in about 10 days.

After crossing eastern Washington, the water collected briefly at the narrows of Wallula Gap on the Oregon border where blockage produced the 1,000 foot deep Lake Lewis. Ponding up a second time at The Dalles to create Lake Condon, the rushing water stripped off gravels and picked up debris, steepening the walls of the Columbia gorge. Near Rainier the river channel was again constricted causing flood waters to back up all the way into the Willamette Valley. At Crown Point flood waters spilled south into the Sandy River drainage and across the lowland north of Vancouver taking over the Lacamas Creek channel. Most of the water exited through the gorge to the ocean, but as much as a third spread over the Portland region to depths of 400 feet. Only the tops of Rocky Butte, Mt. Tabor, Kelly Butte, and Mt. Scott would have been visible above the floodwaters. Surging up the ancestral Tualatin River, the waters covered the present day site of Lake Oswego to depths over 200 feet, while Beaverton, Hillsboro, and Forest Grove would have been under 100 feet of water.

Skulls and skeletal elements of Ice Age bison up to 8 feet high at the shoulder are common in Willamette Valley swamp deposits (specimen from the Thomas Condon Collection, Univ. of Oregon).

canyon and tributary streams high in the upper watershed. The amount of water in a single flood, estimated at up to 400 cubic miles, is more than the annual flow of all the rivers in the world. The natural reservoir of Lake Missoula filled and emptied repeatedly at regular intervals suggesting that natural processes were regulating the timing of the floods. Once the lake had filled to a certain level, it may have floated the ice lobe or glacial plug that jammed the neck of the valley which, in turn, released enough water to allow the flooding process to begin.

Swampy lowlands in the Willamette Valley yield the bones of the enormous Ice Age ground sloth.

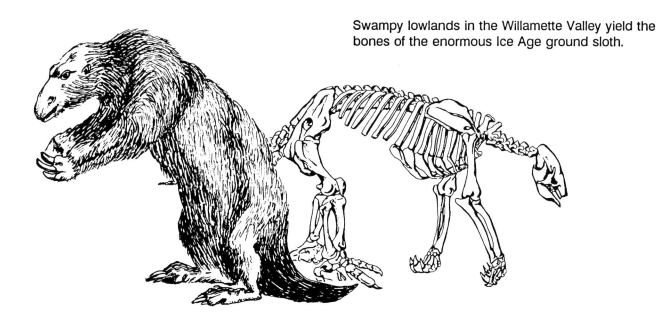

Along with floating icebergs, the rushing waters carried enormous amounts of gravel, sand, silt, and clay in suspension as well as rolling and tumbling along the bottom. In the valley this mass of flood borne sediment was segregated by the narrows at Lake Oswego leaving much of the coarse material in the vicinity of Portland and the Tualatin Valley. A huge sand and gravel bar dammed the Tualatin River for a period to form a small lake 5 miles west of Oswego. Sometime after the flood receded, the river cut through the bar, and the lake drained, but the basin which had been scoured and deepened by the flooding remains. Sweeping through Lake Oswego, large volumes of the finer sands and silts along with ice rafted boulders that had been collected upstream washed southward into the lower Willamette Valley where they were deposited over the valley floor and lower slopes of the surrounding hills. Sands predominate at Canby and Aurora, and silts, tens of feet thick, were spread as far south as Harrisburg.

The muddy waters filled the central valley temporarily creating Lake Allison. Extending from the Lake Oswego and Oregon City gaps southward almost to Eugene, the surface of this large body of water was over 350 feet above present sea level. The lake formed when the valley was dammed at its north end by ice jams or overwhelmed by the amount of water coming south. Repeated surging as new water came through the gap at Oregon City and ebbing as it drained out kept the lake level in a constant state of fluctuation. The tops of lake silt deposits are commonly 180 to 200 feet in elevation throughout the Tualatin, Yamhill, and Willamette valleys. After a brief interval the water drained back out to the Columbia and the ocean.

The multiple floods had a lasting impact on the channel of the Columbia River and the Willamette Valley. Because the flow of flood water into the Willamette Valley was opposite to the normal northward drainage, the river was disrupted for periods up to two weeks until the waters receded. Distinct banded layers of Willamette Silt brought into the valley indicate the flood waters must have invaded many times. As flood waters entered the valley they were quickly stripped of coarse sand and gravel when ponding took place. Silt and clay particles, however, remainded suspended in the turbid waters and covered the valley with a layer up to 100 feet deep exposed today along the banks of the Willamette and its tributaries. Surface deposits of these silts are best developed in the southern Willamette Valley where they are subdivided into four members of the Willamette Formation on the basis of subtle mineral and textural differences. Within the Willamette Silts, the Irish Bend Member has been identified as the primary

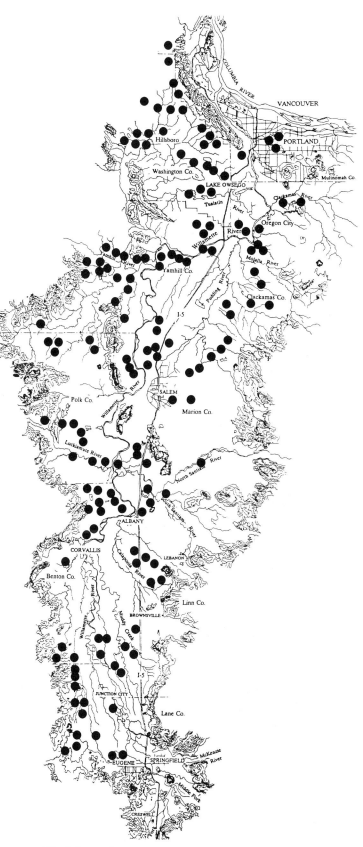

Glacial erratics carried into the Willamette Valley atop icebergs during large Pleistocene floods are scattered from Eugene to Portland (after Allison, 1935).

sediment of a large-scale flood. Extending over 300 square miles of the southern valley, the Irish Bend silt reaches a maximum thickness of nearly 50 feet just south of Corvallis.

With the separate silt layers in the Willamette Valley suggesting multiple periods of flooding, the precise number of floods is in doubt. Figures proposed are "many floods", "35 floods", or "7 to 8 floods". Whether the floods occurred in "two cycles", "annually", "every 175 years", "over an extended period of time", or were "short-lived" is still conjectural. These floods continued over the 2,500 year interval until the ice sheets permanently retreated northward with a warming climate.

Erratics

Flood waters spilling into the Willamette Valley carried large blocks of ice borne on the torrent. Atop and within the icebergs, rocks and sediment were transported all the way from Montana. Once the ice melted, the stones were dropped as glacial erratics in a wide pattern across the valley. Although more than 300 occurrences of these erratics have been recorded, thousands more lay unrecognized. More than 40 boulders over 3 feet in diameter have been located, and many smaller stones as well as chips and pebbles of foreign material have been noted in farm fields, road-cuts, and along old river terraces.

The Willamette meteorite from near West Linn may have been carried into the valley by an iceberg during flooding (photo courtesy of Oregon Historical Society).

Varying in composition and size, the erratics are granite, granodiorite, quartzite, gneiss, slate, and a few of basalt. With the exception of basalt, these rocks are common to central Montana and not the Willamette Valley. Because of the exotic composition of the erratic material, the path along which they were rafted into the valley can be traced down the Columbia River channel. Erratics were deposited through Wallula Gap to The Dalles where they are found up to 1,000 feet above sea level. In the Willamette Valley erratic fragments are imbedded in the top of the Willamette Silt.

The largest known erratic lies in the valley between McMinnville and Sheridan in Yamhill County. Composed of the metamorphic rock argillite, the boulder originally weighed about 160 tons, but over 70 tons have been removed by tourists. Perhaps the most famous erratic is the Willamette meteorite which may have fallen in Montana only to be transported by floodwaters to where it came to rest near West Linn in Clackamas County. The meteorite was subsequently purchased for $20,000 in 1905 by Mrs. William E. Dodge who donated it to the American Museum of Natural History in New York.

Structure

Uplift, tilting, and folding in the Coast Range and Willamette Valley are attributed to the continued subduction of the Juan de Fuca plate. Gentle folding accompanied by faulting with as much as 1,000 feet of vertical displacement produced deep valleys. Columbia River basalts, that had filled the bowl-shaped Tualatin, Wilsonville, and Newberg valleys in the middle Miocene, were depressed to 1,300 feet below sea level by steady eastward tilting of the Coast Range block. Within the Willamette Valley, broad northwest trending anticlinal folds are interspersed with parallel, subdued synclines. The dominant folds are the Portland Hills, Cooper Mountain, and Bull Mountain, as well as Parrett and Chehalem mountains. Separating the anticlinal hills, wide valleys of Tualatin, Newberg, and Wilsonville are gentle synclines or downfolds.

The Portland Hills anticline that formed during the late Miocene and Pliocene is steepened on the east side by a large fault that stretches for over 60 miles northwest and southeast of Portland. This fault system extends southeastward to join the Clackamas River lineament, an alignment of surface fractures on a grand scale that traverses Oregon all the way to Steens Mountain as part of the Brothers Fault zone. Running parallel to this immense feature, the Mt. Angel fault trends northwestward under Woodburn to project into the Gales Creek fault zone of the Coast Range. Northeast of Salem, the butte at Mt. Angel

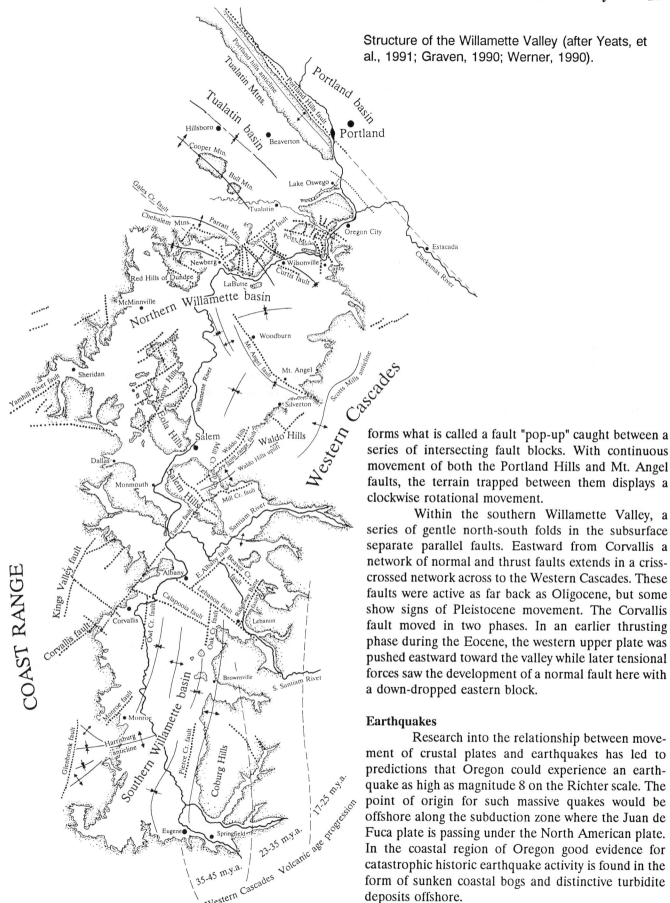

Structure of the Willamette Valley (after Yeats, et al., 1991; Graven, 1990; Werner, 1990).

forms what is called a fault "pop-up" caught between a series of intersecting fault blocks. With continuous movement of both the Portland Hills and Mt. Angel faults, the terrain trapped between them displays a clockwise rotational movement.

Within the southern Willamette Valley, a series of gentle north-south folds in the subsurface separate parallel faults. Eastward from Corvallis a network of normal and thrust faults extends in a criss-crossed network across to the Western Cascades. These faults were active as far back as Oligocene, but some show signs of Pleistocene movement. The Corvallis fault moved in two phases. In an earlier thrusting phase during the Eocene, the western upper plate was pushed eastward toward the valley while later tensional forces saw the development of a normal fault here with a down-dropped eastern block.

Earthquakes

Research into the relationship between movement of crustal plates and earthquakes has led to predictions that Oregon could experience an earthquake as high as magnitude 8 on the Richter scale. The point of origin for such massive quakes would be offshore along the subduction zone where the Juan de Fuca plate is passing under the North American plate. In the coastal region of Oregon good evidence for catastrophic historic earthquake activity is found in the form of sunken coastal bogs and distinctive turbidite deposits offshore.

Portland is on the down-thrown side of the Portland Hills fault shown here with the Tualatin Mountains rising off to the west (left) (photo courtesy Delano Photographics).

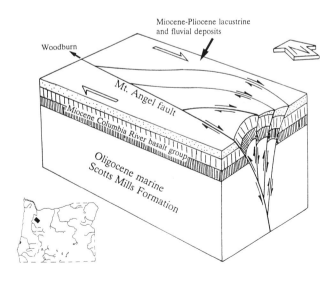

Fault pop-up or flower structure beneath Mt. Angel, Oregon (after Werner, 1990; Yeats, et al., 1991).

The lack of ongoing strong seismic activity along the boundaries of Pacific Northwest plates may be due to an unlocked subduction zone with virtually frictionless movement. However, this seems unlikely when comparing Oregon's continental margin and subduction zone to those of southern Chile where very young oceanic crust is being subducted in both locations. Beneath Oregon the 10 million year old crust is still warm and consequently very buoyant making it difficult to subduct. Rocks of the warm descending crust also tend to adhere to the overlying North American plate, thus the boundary between the two readily binds up and locks. As the plate collision process continues at the rate of about 1 inch per year, the bond will eventually break with catastrophic results. In 1960 Chile experienced a 9.5 magnitude earthquake, the largest of the 20th century.

In contrast to China where the record of moderate and major quakes goes back 3,000 years or California where Franciscan fathers began records in 1800, there is only scanty historical data on seismic activity in Oregon. Indian legends of more than 200 years ago tell about massive earthquakes as well as destructive tidal waves, but these stories cannot be

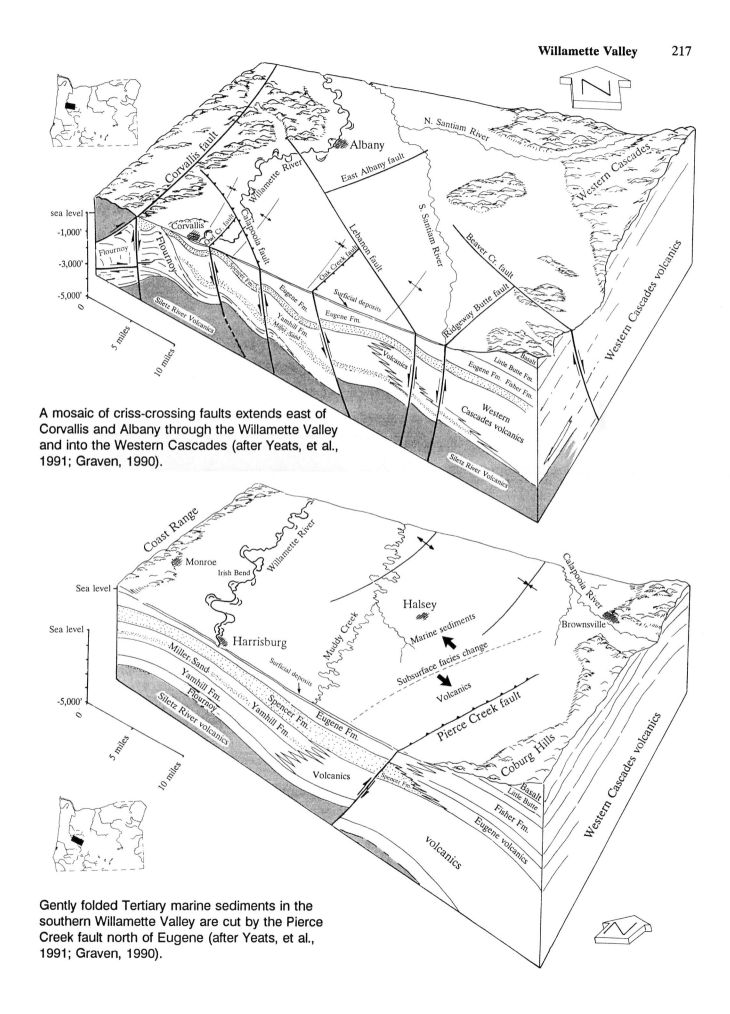

A mosaic of criss-crossing faults extends east of Corvallis and Albany through the Willamette Valley and into the Western Cascades (after Yeats, et al., 1991; Graven, 1990).

Gently folded Tertiary marine sediments in the southern Willamette Valley are cut by the Pierce Creek fault north of Eugene (after Yeats, et al., 1991; Graven, 1990).

The Corvallis fault, up-thrown on its northern side, passes diagonally between Corvallis and a low range of hills (Vineyard Mtn.) (photo by Western Ways of Corvallis).

pinpointed in time nor evaluated with respect to their magnitude. A catalog kept at Oregon State University lists a total of 1,286 known seismic occurrences from 1841 to 1986 in Oregon and southern Washington. Of these, approximately 440 earthquakes were associated with the eruption of Mount St. Helens in 1980, and 250 occurred before 1962 when records were first kept by modern instruments. The seven largest earthquakes in Oregon during this period began with the November 23, 1873, shaking of Port Orford and Crescent City at an estimated magnitude of 6.3 on the Richter scale. The outward limit of this shock was felt from Portland to San Francisco. Of the Oregon cities, Port Orford suffered the most with damage to chimneys. Other powerful tremors shook Portland in 1877, Umatilla in 1893, Richland, Washington, in 1916, Milton-Freewater in 1936, Sauvie Island in 1964, and the Adel-Warner lakes area in 1968.

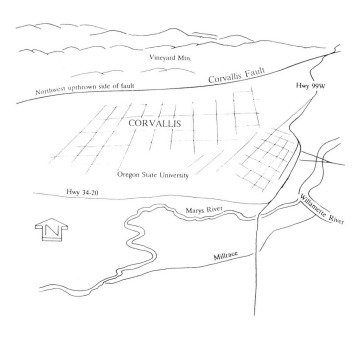

Although it may seem that the Willamette Valley and Portland have not had recent earthquakes, the contrary is true. Beneath Portland and the northern Willamette Valley a network of faults exist which could become active at any time. In 1990 earthquakes record-ed beneath Mt. Hood were similar to quakes occurring there in 1974, 1982, and 1989. The Mt. Angel fault in the central valley was the apparent source of dozens of small quakes in 1990. After 1882 Portland experienced a recorded quake on the average of once every 5 years. Brick buildings swayed, windows rattled, and terrified inhabitants rushed to the street during the tremor of February 3, 1892, although there was no damage. In 1962 and again in 1968 magnitude 3.7 and 5.0 quakes repeatedly struck Portland from epicenters on the eastern edge of the city. Shocks were due to movement of a fault as much as 10 to 15 miles deep. Centered in the Gorda plate off the southern coast, an earthquake of July 12, 1991, registering 6.6 was felt in many communities of western Oregon. A series of quakes on July 22, 1991, measuring 3.5 were focused in the Tualatin Mountains.

Because of the lack of recorded quake activity, few of the larger buildings in the Willamette Valley and Portland were designed to sustain powerful seismic shaking. Many of the building foundations in the northern valley are on water saturated sediment. These soils continue to shake even after the quake has stopped somewhat like a bowl of jello on a table. This type of shaking accentuates the destructive effects. Additionally, the lack of cohesion of many of the Portland and northern valley soils would cause them to flow during a quake. This slope failure, or soil liquifac-tion, not only destroys structural foundations, but the motion triggers disasterous mud slides and flows that spread destruction laterally. Combined with the inevita-ble movement along the subduction zone and the dense population of the valley, there is the potential for a disaster of unprecedented size when the predicted strong earthquake occurs.

Recorded earthquake activity is low in the eastern sections of the state. In May and June of 1968 earthquakes caused rockslides and building damage in southcentral Oregon where the epicenter was located between Crump Lake and Hart Lake in the Warner Valley. The largest quake on May 29 registered 5.1 on the Richter scale. Although not well understood, seismic processes in central Oregon are related to stresses within the Basin and Range province. A slightly milder earthquake of magnitude 4.8 centered northwest of Maupin in the Deschutes Valley, shaking central Oregon on April 12, 1976. Inhabitants stated that their houses shook and swayed although there was no serious damage. The Blue Mountain province, an area of little known earthquake activity, experienced a shock north of Baker registering 3.6 on August 14, 1969. Earthquake history of this region is very brief with only two other noted shocks in 1906 and 1916.

The proximity of subduction faults along with loose sediments and unconsolidated soils make the Willamette Valley particularly susceptible to quakes. Below the Cascades and eastern Oregon, the great depth of the subduction zone tends to reduce risk, whereas the coastal area close to the quake sources is perhaps at greatest peril.

Mining and Mineral Industry

Bauxite

The residual iron and aluminum ores in the Willamette Valley developed as deposits on top of basalts of the Columbia River series where much of the surface of the Miocene lava is deeply weathered. As rock decomposition proceeds, soluble chemicals are leached out leaving a clay soil with the highly insoluble bauxite enriched with iron and aluminum. Limited ferruginous bauxite deposits in the Pacific Northwest are located in scattered sections of Columbia, Washing-ton, Multnomah, Yamhill, Clackamas, and Marion counties and in Cowlitz and Wahkiakum counties in adjoining Washington state. The thickest deposits are in Washington and Columbia counties thinning toward Salem.

Iron

Within the sequences of Columbia River basalts, the iron oxide mineral limonite developed in bogs along undrained depressions. Several attempts were made to mine and process the limonite near Scappoose and Lake Oswego before 1867 when furnac-es of the Oregon Iron and Steel Company produced the first iron west of the Rocky Mountains. Unfortu-nately financial problems and difficulties with stock-holders hampered the operation throughout its history, and even the construction of an adjacent pipe foundry utilizing pig iron failed to keep the furnace going. Despite its stormy history, the company produced 83,400 tons of pig iron from 1867 to 1894. Several thousand feet of old tunnels still exist under the Lake Oswego Country Club and nearby Iron Mountain. The furnace used for smelting the iron ore, located near the outlet of Lake Oswego, is now part of the Lake Oswe-go City park.

Features of Geologic Interest

Buttes

Although most of the Willamette Valley is flat

where the deposition of alluvium has formed a broad flood plain, there are many local topographic features that project above the valley floor. South of Salem 14 isolated, rounded buttes are aligned in a north-south direction along the east side of the valley. Volcanic in origin, the buttes are composed of 30 to 35 million year old basalts. The mineral and chemical composition of the basalts as well as the age does not suggest any association with the younger Columbia River lavas.

The most northerly of these features, Knox Butte east of Albany rises 634 feet above the valley floor. To the south near Lebanon, Peterson Butte, with the highest elevation at 1,434 feet, differs from the others in that it is a central volcanic vent with 12 dikes radiating from the cone. Between Lebanon and Halsey, Ward Butte, the small Saddle Butte, Powell Hills, Twin Buttes, Bond Butte, as well as West Point Hill, Rock Hill, and Lenon Hill near the Coburg ridge have an average elevation just over 600 feet. West Point Hill, Rock Hill, and Lenon Hill are spurs of resistant rock from Western Cascade volcanics that extend into the valley. The west face of Skinner Butte in downtown Eugene with an elevation of 682 feet displays jointed columnar basalt as does the 602 foot high Gillespie Butte across the river. Spencer Butte dominates south Eugene at 2,065 feet in height.

The origin of these buttes is still the subject of controversy. In the past they have been interpreted as volcanic necks, lava flow remnants, and laccoliths, but in all probability they are sills. These are lense-shaped bodies that intruded in the subsurface parallel to layers of existing sediments where they cooled and hardened. As the less resistant surrounding sediments are dissected by erosion, these rounded hills of basalt are exposed as remnants.

Silver Falls

Silver Falls State Park, the largest in the state, covers 8,700 acres and is located 30 miles east of Salem. Thirty million years ago the region around Silver Falls was covered by coastal waters of a shallow warm Oligocene ocean. Once the ocean had receded westward 15 million years ago, flows of basalt from fissures in eastern Oregon repeatedly covered the area with more than 600 feet of Columbia River lavas. In the thousands of years between the volcanic events, thick soils developed atop the flows. Eruptions of ash and andesite lavas of the Fern Ridge Tuffs from adjacent Cascade volcanoes later blanketed the region before streams began to cut down through the layers. The softer ash beds were quickly removed although limited fragments of the layers remain in low spots

North Falls at Silver Falls State Park east of Salem drops 136 feet from a lip of Miocene basalt to a plunge pool cut into Oligocene marine rocks (photo courtesy Oregon State Highway Department).

around the park.

The 15 falls of the park were of more recent origin. Whereas the large streams easily cut through the basalts, the smaller creeks with less volume cannot wear down through the rock as quickly. As larger streams erode deep canyons, the smaller ones are left high above to spill over the uncut basalt as waterfalls. The largest cataract is South Falls, dropping 177 feet into a beautiful, deep plunge pool. At North Falls, with a height of 136 feet, the fossiliferous sandstone layer of the Scotts Mills Formation under the basalt wears away easily to create a 300 foot ampitheater behind the falls. The chimney-like holes in the overhanging rock are lava-cast tree molds where hot flowing lava surrounded and engulfed standing tree trunks. Once the lava had cooled, the tree decomposed, leaving only a mold of the trunk.

Table Rock

Standing out on the skyline about 35 miles east of Salem, Table Rock towers over the Western Cascades. The steep cliffs of the peak, shaped like a cardinal's hat, are easily visible and readily identified from far out in the Willamette Valley. Reached by an easy drive from Molalla, access to the crest of the monolith is by a well-maintained hiking trail through a dense forest that begins on the north face and winds almost completely around the mountain to ascend to

Offering a panorama of the Willamette Valley, Table Mountain east of Salem is a series of buttes capped by basalts.

the top on the south face. A dissected remnant of late Miocene or early Pliocene lava flows, Table Rock is capped by several isolated pieces of basalt. Columnar fractures in the lava and adjacent isolated fragments on surrounding peaks demonstrate the volcanic extrusive nature of the features. Flow rocks here represent a later sequence in development of the Western Cascades that took place before the extrusion of the High Cascades.

Suggested Readings

Allen, John E., Burns, Marjorie, and Sargent, Sam C., 1986. Cacaclysms on the Columbia. Portland, Oregon, Timber Press, 211p.
-------1982. The volcanic story of the Columbia River gorge. Jour. Geol. Ed., v.30, p.156-162.
Allison, Ira S., 1935. Glacial erratics in Willamette Valley. Geol. Soc. Amer., Bull., v.46, p.615-632.
-------1978. Late Pleistocene sediments and floods in the Willamette Valley. Ore Bin, v.40, no.11, p.177-191; 193-202.
Baldwin, Ewart M., 1957. Drainage changes of the Willamette River at Oregon City and Oswego, Oregon. Northwest Sci., v.31, no.3, pp.29-30.
Balsillie, J.H., and Benson, G.T., 1971. Evidence for the Portland Hills fault. Ore Bin, v.33, no.6, p.109-118.
Couch, Richard W., and Johnson, Stephen, 1968. The Warner Valley earthquake sequence: May and June, 1968. Ore Bin, v.30, no.10, pp.191-204.
Freed, M., 1979. Silver Falls State Park. Oregon Geology, v.41, no.1, p.-3-10.
Glenn, Jerry L., 1956. Late Quaternary sedimentation and geologic history of the north Willamette Valley, Oregon. PhD., Oregon State Univ., 231p.
Graven, Eric P., 1990. Structure and tectonics of the southern Willamette Valley, Oregon. Ms., Oregon State Univ., 119p.
Hanson, Larry G., 1986. Scenes from ancient Portland. Oregon Geology, v.48, no.11, p.130-131.
Heaton, Thomas H., and Hartzell, Stephen H., 1987. Earthquake hazards on the Cascadia subduction zone. Science, v.236, pp.119-168.
Jacobson, Randall S., et al., 1986. Map of Oregon seismicity, 1841-1986. Oregon Dept. Geol. and Mineral Indus., Geological Map Series,

GMS 49.

Lentz, Rodney T., 1981. The petrology and stratigraphy of the Portland Hills Silt - a Pacific Northwest loess. Oregon Geology, v.41, no.1, p.3-10.

Luedke, Robert G., and Smith, Robert L., 1982. Map showing distribution, composition, and age of late Cenozoic volcanic centers in Oregon and Washington. U.S. Geol. Survey, Misc. Inv. Series., I-1091-D.

Madin, Ian P., 1989. Evaluating earthquake hazards in the Portland, Oregon, metropolitan area: mapping potentially hazardous soils. Oregon Geology, v.51, no.5, p.106-110;118.

McDowell, Patricia F., and Roberts, Michael C., 1987. Field guidebook to the Quaternary stratigraphy, geomorphology and soils of the Willamette Valley, Oregon. Assoc. Amer. Geogr., Ann. Mtg., Portland, Oregon, Field Trip 3, 75p.

Miller, Paul R., and Orr, William N., 1985. Mid-Tertiary trans-gressive rocky coast sedimentation: central Western Cascade range, Oregon. Jour. Sed. Pet., v.58, no.6, pp.959-968.

-----and Orr, William N., 1986. The Scotts Mills Formation: Mid-Tertiary geologic history and paleogeography of the central Western Cascade range, Oregon. Oregon Geology, v.48, no.12, pp.139-151.

Newton, Vernon C., 1969. Subsurface geology of the lower Columbia and Willamette basins, Oregon. Oregon Dept. Geol. and Mineral Industries, Oil and Gas Inv., no.2, 121p.

Qamar, Anthony, et al., 1986. Earthquake hypocenters in Washington northern Oregon. Wash. Div. Geology and Earth Res., Info. Circular 82, 64p.

Roberts, M.C., 1984. The late Cenozoic history of an alluvial fill: the southern Willamette Valley, Oregon, pp.491-503. In: Mahaney, W.C., ed. Correlation of Quaternary chronologies. Geo Books, Norwich England.

-----and Whitehead, Donald R., 1984. The palynology of a non-marine Neogene deposit in the Willamette Valley, Oregon. Rev. of Palaeobotany and Palynology, v.41, pp.1-12.

Schlicker, Herbert G., and Deacon, Robert J., 1967. Engineering Geology of the Tualatin Valley region, Oregon. Oregon. Dept. Geol. and Mineral Indus., Bull.60, 103p.

Significant Oregon earthquakes in 1991. Oregon Geology, v.53, no.5, p.107.

Snavely, Parke D., and Wagner, Holly C., 1963. Tertiary geologic history of western Oregon and Washington. Wash., Div. of Mines and Geology, Report of Investig., no.22, 25p.

-----and Wells, Ray E., 1991. Ceonzoic evolution of the continental margin of Oregon and Washington. U.S. Geol. Survey, Open-File Report 91-441-B, 34p.

Thayer, Thomas P., 1939. Geology of the Salem Hills and the North Santiam River basin, Oregon. Oregon. Dept. Geol. and Mineral Indus., Bull.15, 40p.

Tolan, Terry L., and Beeson, M.H., 1984. Intracanyon flows of the Columbia River Basalt Group in the lower Columbia River gorge and their relationship to the Troutdale Formation. Geol. Soc. Amer., Bull., v.95, pp.463-477.

Trimble, Donald E., 1963. Geology of Portland, Oregon, and adjacent areas. U.S. Geol. Survey, Bull.1119, 119p.

Werner, Kenneth Stefan, 1990. I. Direction of maximum horizontal compression in western Oregon determined by borehole breakouts. II. Structure and tectonics of the northern Willamette Valley, Oregon. Ms., Oregon State Univ., 159p.

Wong, Ivan G., Silva, Walter J., and Madin, Ian P., 1990. Preliminary assessment of potential strong earthquake ground shaking in the Portland, Oregon, metropolitan area. Oregon Geology, v.52, no.6, p.131-134.

Yeats, Robert S., 1989. Current assessment of earthquake hazard in Oregon. Oreogn Geology, v.51, no.4, pp.90-92.

-----et al., 1991. Tectonics of the Willamette Valley, Oregon. U.S. Geol. Survey Open-file Report 91-441-P, 47p.

Bibliography

Adams, John, 1984. Active deformation of the Pacific Northwest continental margin. Tectonics, v.3, no.4, pp.449-472.

-----1990. Paleoseismicity of the Cascadia subduction zone: evidence from turbidites off the Oregon-Washington margin. Tectonics, v.9, no.4, pp.569-583.

Abbott, William H., 1970 Micropaleontology and paleoecology of Miocene diatoms from the Harper district, Malheur County, Oregon. M.S., Northeast Louisiana Univ., 86p.

Addicott, Warren O., 1964. A Late Pleistocene invertebrate fauna from southwestern Oregon. Jour. Paleo., v.38, no.4, pp.650-661.

-----1980. Miocene stratigraphy and fossils, Cape Blanco, Oregon. Oregon Geology, v.42, no.5, pp.87-98.

-----1983. Biostratigraphy of the marine Neogene sequence at Cape Blanco, southwestern Oregon. U.S. Geol. Survey, Prof. Paper 774-G, pp.G1-G20.

Aguirre, Martin R., and Fisk, Lanny H., 1987. Age of "Goose Rock Conglomerate," Wheeler and Grant Counties, north-Central Oregon. Amer. Assoc. Petrol. Geol., v.77 (8), p.999.

Ahmad, Raisuddin, 1986. Eocene geology of the Agness-Illahe area, southwest Oregon. Oregon Geology, v.48, no.2, pp.15-21.

Allen, John E., 1946. Reconnaissance geology of limestone deposits in the Willamette Valley, Oregon. Oregon Dept. Geol. and Mineral Indus., GMI Short Paper 15, 15p.

-----1947. Bibliography of the geology and mineral resources of Oregon (supplement), July 1, 1936 to December 31, 1945. Oregon Dept. Geol. and Mineral Indus., Bull.33, 108p.

-----1975. The Wallowa "Ice cap" of northeastern Oregon; an exercise in the interpretation of glacial landforms. Ore Bin, v.37, no.12, pp.189-202.

-----1979. The Magnificent gateway; a layman's guide to the geology of the Columbia River gorge. Portland, Timber Press, 144p.

-----1979. Speculations on Oregon calderas. Oregon Geology, v.41, no.2, pp.31-31.

-----1982. The volcanic story of the Columbia River Gorge. Jour. Geol. Education, v.30, p.156-162.

-----1986. Time travel in Oregon; a geology scrapbook. Portland, Ore., J. Allen Publ., various pagings.

-----1984. Oregon lakes and their origins. Oregon Geology, v.46, no.12, pp.143-146.

-----1984. The caves of Crater Lake National Park. Oregon Geology, v.46, no.1, pp.3-6.

-----1989. Ice-Age glaciers and lakes south of the Columbia River gorge. Oregon Geology, v.51, no.1, pp.12-14.

-----and Baldwin, Ewart M., 2944. Geology and coal resources of the Coos Bay Quadrangle, Oregon. Oregon, Dept. Geol. and Mineral Indus., Bull.27, 153p.

-----and Beaulieu, John D., 1976. Plate tectonic structures in Oregon. Ore Bin, v.38, no.6, pp.87-99.

-----and Mason, Ralph S., 1949. Brick and tile industry in Oregon. Oregon Dept. Geol. and Mineral Indus., GMI Short Paper 19, 28p.

-----Burns, Marjorie, and Sargent, Sam C., 1986. Cataclysms on the Columbia. Portland, Ore., Timber Press, 211p.

Allen, Victor T., 1948. Formation of bauxite from basaltic rocks of Oregon. Economic Geology, v.43, no.8, pp.619-626.

Allison, Ira S., 1935. Glacial erratics in Willamette Valley. Geol. Soc. Amer., v.46, pp.615-632.

-----1953. Geology of the Albany Quadrangle, Oregon. Oregon

Dept. of Geol. and Mineral Indus., Bull.37, 18p.

-----1978. Late Pleistocene sediments and floods in the Willamette Valley. Ore Bin, v.40, no.11, pp.177-191; pt.2 p.193-202.

-----1979. Pluvial Fort Rock Lake, Lake County, Oregon. Oregon, Dept. Geol. and Mineral Indus., Special Paper 7, 72p.

-----1982. Geology of pluvial Lake Chewaucan, Lake County, Oregon. Oregon State Univ., Studies in Geology, 11, 78p.

Anderson, Frank M., 1932. Pioneers in the geology of California. Calif. Div. Mines, Bull.104, 24p.

Anderson, J. Lawford, ed., 1990. The nature and origin of Cordilleran magmatism. Geol. Soc. Amer., Memoir 174, 414p.

Anderson, James Lee, and Vogt, Beverly F., 1987. Intracanyon flows of the Columbia River Basalt Group and adjacent Cascade Range, Oregon and Washington. Wash. Div. of Geology and Earth Res., Bull.77, pp.249-267.

-----et al., 1987. Distribution maps of stratigraphic units of the Columbia River basalt group. Wash. Div. of Geology and Earth Res., Bull.77, pp.183-195.

Appling, Richard N., 1956. Manganese deposits of southwestern Oregon. U.S. Bureau of Mines, Rept. of Investig. 5369, 56p.

-----1959. Manganese deposits of northeastern Oregon. U.S. Bureau of Mines, Rept. of Investig. 5472, 23p.

Armentrout, John M., 1980. Cenozoic stratigraphy of Coos Bay and Cape Blanco, southwestern Oregon. Oregon Dept. Geol. and Mineral Indus., Bull.101, pp.175-216.

----- ed., 1981. Pacific Northwest Cenozoic biostratigraphy. Geol. Soc. Amer., Special Paper 184, pp.137-148.

-----1987. Cenozoic stratigraphy, unconformity-bounded sequences and tectonic history of southwestern Washington. Wash. Div. Geology and Earth Sci., Bull.77, pp.291-317.

-----and Suek, David H., 1985. Hydrocarbon exploration in western Oregon and Washington. Amer. Assoc. Petroleum Geol., Bull.69, no.4, pp.627-643.

-----Cole, Mark R., and TerBest, Harry, eds., 1979. Cenozoic paleogeography of the Western United States. Soc. Econ. Paleont. and Mineralog., Pacific Coast Paleogr. Symp.3, 335p.

-----et al., 1983. Correlation of Cenozoic stratigraphic units of western Oregon and Washington. Oregon Dept. Geol. and Mineral Indus., Oil and Gas Inv. 7, 90p.

Armstrong, Richard Lee, 1978. Cenozoic igneous history of the U.S. Cordillera from lat 42 to 49 N. Geol. Soc. Amer., Memoir 152, pp.265-280.

Ash, Sidney R., 1991. A new Jurassic flora from the Wallowa terrane in Hells Canyon, Oregon and Idaho. Oregon Geology, v.53, no.2, pp.27-33.

Ashwill, Melvin S., 1979. An exposure of limestone at Gray Butte, Jefferson County, Oregon. Oregon Geology, v.41, no.7, pp.107-109.

----- 1982. Thermal springs near Madras, Oregon. Oregon Geology, v.44, no.l, pp.8-9.

-----1987. Paleontology in Oregon; workers of the past. Oregon Geology, v.49, no.12, pp.147-153.

Atwater, Brian F., 1987. Evidence for great Holocene earthquakes along the outer coast of Washington state. Science, v.236, pp.942-944.

-----1988. Comment and reply on "Coastline uplift in Oregon and Washington and the nature of Cascadia subduction-zone tectonics". Geology, v.16, no.10, pp.952-953.

Atwater, Tanya, 1970. Implications of plate tectonics for the Cenozoic tectonic evolution of Western North America. Geol. Soc. Amer., Bull., v.81, pp.3513-1536.

-----1989. Plate tectonic history of the northeast Pacific and western North America. Geol. Soc. Amer., The Geology of North America, v.N, The eastern Pacific and Hawaii, pp.21-72.

Ave Lallemant, Hans G., 1976. Structure of the Canyon Mountain (Oregon) ophiolite complex and its implication for sea-floor spreading. Geol. Soc. Amer., Memoir 173, pp.1-49.

-----Schmidt, W.J., and Kraft, J.L., 1985. Major late-Triassic strike-slip displacement in the Seven Devils terrane, Oregon and Idaho: a result of left-oblique plate convergence? Tectonophysics, v.119, pp.299-328.

Avramenko, Walter, 1981. Volcanism and structure in the vicinity of Echo Mountain, central Oregon Cascade Range. Ms, Univ. of Oregon, 156p.

Babcock, R.S., et al., in press. A rifted margin origin for the Crescent Basalts and related rocks in the northern Coast Range volcanic province, Washington and British Columbia. Jour. Geophys. Res., Solid Earth.

Bacon, Charles R., 1983. Eruptive history of Mount Mazama and Crater Lake, Cascade Range, U.S.A. Jour. of Volcanology and Geothermal Res., v.18, pp.57-115.

Bailey, Michael M., 1989. Evidence for magma recharge and assimilation in the Picture Gorge Basalt subgroup, Columbia River Basalt group. In: Reidel, S.P., and Hooper, P.R., eds., 1989. Volcanism and tectonism in the Columbia flood-basalt province. Geol. Soc. Amer., Special Paper 239, pp. 332-355.

-----1989. Revisions to stratigraphic nomenclature of the Picture Gorge Basalt subgroup, Columbia River basalt group. In: Reidel, S.P., and Hooper, P.R., eds., 1989. Volcanism and tectonism in the Columbia flood-basalt province. Geol. Soc. Amer., Special Paper 239, pp.67-84.

Baker, Victor R., and Bunker, Russell, 1985. Cataclysmic late Pleistocene flooding from Glacial Lake Missoula: a review. Quaternary Sci. Rev., v.4, pp.1-41.

Baksi, Ajoy K., 1989. Reevaluation of the timing and duration of extrusion of the Imnaha, Picture Gorge, and Grande Ronde basalts, Columbia River Basalt group. In: Reidel, S.P., and Hooper, P.R., eds., 1989. Volcanism and tectonism in the Columbia River flood-basalt province. Geol. Soc. Amer., Special Paper 239, pp.105-110.

Baldwin, Ewart M., 1945. Some revisions of the late Cenozoic stratigraphy of the southern Oregon Coast. Jour. Geol., v.52, pp.35-46.

-----1946. Diatomite. Ore Bin, v.8, no.1, pp.1-7.

-----1950. Pleistocene history of the Newport, Oregon, region. Geol. Soc. Oregon Country, Newsl., v.16, no.10, pp.77-81.

-----1952. The geology of Saddle Mountain Clatsop County, Oregon. Geol. Soc. Oregon Country, Newsl., v.18, no.4, pp.29-30.

-----1957. Drainage changes of the Willamette River at Oregon City and Oswego, Oregon. Northwest Sci., v.31, no.3, pp.109-117.

-----1964. Geology of the Dallas and Valestz quadrangles, Oregon. Oregon, Dept. Geol. and Mineral Indus., Bull.35, 52p.

-----1966. Geology of the Columbia River gorge. Northwest Sci., v.40, no.4, pp.121-128.

-----1969. Thrust faulting along the lower Rogue River, Klamath Mountains, Oregon. Geol. Soc. Amer., Bull., v.80, pp.2047-2052.

-----1972. Thrusting of the Rogue Formation near Marial on the lower Rogue River, Oregon. Ore Bin, v.34, no.4, pp.57-66.

-----1974. Eocene stratigraphy of southwestern Oregon. Oregon Dept. Geol. and Mineral Indus., Bull.83, 40p.

-----1976. Geology of Oregon. Third ed., Dubuque, Kendall/Hunt Publ., 170p.

-----1980. The gold dredge at Whisky Run north of the mouth of the Coquille River, Oregon. Oregon Geology, v.42, no.8, pp.145-146.

-----1984. The origin of olistostromes in the Roseburg Formation in southwestern Oregon. Oregon Geology, v.46, no.7, pp. 75-76, 82.

-----1989. Eocene unconformities in the Camas Valley quadrangle, Oregon. Oregon Geology, v.51, no.1, pp.3-8.

-----and Boggs, Sam Jr., 1969. Study of heavy metals and heavy minerals of the southern Oregon coast. Office of Marine Geology and Hydrology, U.S. Geol. Survey, Technical Rept. 2, 126p.

-----and Howell, P.W., 1949. The Long Tom, a former tributary of the Siuslaw River. Northwest Sci., v.23, p.112-124.

-----et al., 1973. Geology and mineral resources of Coos County, Oregon. Oregon Dept. Geol. and Mineral Indus., Bull.80, 82p.

Balsillie, J.H., and Benson, G.T., 1971. Evidence for the Portland hills fault. Ore Bin, v.33, no.6, pp.109-118.

Barksdale, Julian D., 1984. Memorial to Elliot Bakes McKee, Jr., 1934-1982. Geol. Soc. Amer., Memorials, v.14, pp.1-3.

Barnes, Melanie Ames Weed, 1981. The geology of Cascade Head, an Eocene volcanic center. Ms., Univ. of Oregon, 94p.

Barrash, Warren, and Venkatakrishnan, Ramesh, 1982. Timing of late Cenozoic volcanic and tectonic events along the western margin of the North American plate. Geol. Soc. Amer., Bull., v.93, pp.977-989.

-----Bond, John, and Venkatakrishnan, Ramesh, 1983. Structural evolution of the Columbia Plateau in Washington and Oregon. Amer. Jour. Sci., v.283, pp.897-935.

----- et al., 1980. Geology of the LaGrande area, Oregon Oregon, Dept. Geol. and Mineral Indus., Special Paper 6, 47p.

Beaches and dunes of the Oregon coast, 1975. U.S. Dept. Agriculture, Oregon Coastal Coastal Conservation and Development Commission, 161p.

Beaulieu, John D., 1971. Geologic formations of western Oregon. Oregon. Dept. of Geol. and Mineral Indus., Bull.70, 72p.

-----1972. Geologic formations of the southern half of the Huntington quadrangle, Oregon. Bull.73, 80p.

-----1974. Environmental geology of western Linn County, Oregon. Oregon Dept. Geol. and Mineral Indus., Bull.84, 117p.

-----1976. Geologic hazards in Oregon. Ore Bin, v.38, no.5, pp.67-83.

-----and Hughes, Paul W., 1975. Environmental geology of western Coos and Douglas counties, Oregon. Oregon Dept. Geol. and Mineral Indus., Bull.87, 148p.

-----and Hughes, Paul W., 1976. Land use geology of western Curry County, Oregon. Oregon Dept. Geol. and Mineral Indus., Bull.90, 148p.

-----and Hughes, Paul W., 1977. Land use geology of central Jackson County, Oreogn. Oregon Dept. Geol. and Mineral Indus., Bull.94, 83p.

Beck, Myrl E., and Burr, Cynthia D., 1979. Paleomagnetism and tectonic significance of the Goble Volcanic series, southwestern Washington. Geology, v.7, pp.175-179.

-----and Plumley, Peter W., 1980. Paleomagnetism of intrusive rocks in the Coast Range of Oregon: microplate rotations in middle Tertiary time. Geology, v.8, p.573-577.

Beeson, Marvin H., and Moran, Michael R., 1979. Columbia River Basalt group stratigraphy in western Oregon. Oregon Geology, v.41, no.1, pp.11-14.

-----and Tolan, Terry L., 1985. Regional correlations within the Frenchman Springs Member of the Columbia River Basalt group: New insights into the middle Miocene tectonics of north-western Oregon. Oregon Geology, v.47, no.8, pp.87-96.

-----and Tolan, Terry L., 1987. Columbia River gorge: The geologic evolution of the Columbia River in northwestern Oregon and southwestern Washington. In: Hill, Mason L., ed., Cordilleran Section of the Geol. Society of America, Centennial Field Guide, 1987, pp.321-326.

-----and Tolan, Terry L., 1990. The Columbia River Basalt Group in the Cascade Range: a middle Miocene reference datum for structural anaylsis. Jour. Geophys. Res., v.95, no.B12, pp.19,547-19,559.

-----Perttu, Rauno, and Perttu, Janice, 1979. The origin of the Miocene basalts of coastal Oregon and Washington: an alternative hypothesis. Oregon Geology, v.41 no.10, pp.159-166.

-----Tolan, Terry L., and Anderson, James Lee, 1989. The Columbia River Basalt Group in western Oregon; geologic structures and other factors that controlled flow emplacement patterns. In: Reidel, S.P., and Hooper, P.R., eds. Volcanism and tectonism in the Columbia River flood-basalt province. Geol. Soc. Amer., Special Paper 239, pp.223-246.

-----Tolan, Terry L., and Maden, Ian P., 1989. Geologic map of the Lake Oswego Quadrangle, Clackamas, Multnomah, and Washington counties, Oregon. Oregon Dept. Geol. and Mineral Indus., Geol. Map Series, GMS 59.

-----Tolan, Terry L., and Maden, Ian P., 1991. Geologic map of the Portland Quadrangle, Multnomah and Washington counties, Oregon, and Clark County, Washington. Oregon Dept. Geol. and Mineral Indus., Geol. Map Series, GMS 75.

Bela, James L., 1979. Geologic hazards of eastern Benton County, Oregon. Oregon Dept. Geol. and Mineral Indus., Bull.98, 122p.

Bentley, Elton B., 1974. The glacial morphology of eastern Oregon uplands. PhD., Univ. of Oregon, 250p.

Berg, Joseph W., and Baker, Charles D., 1963. Oregon earthquakes, 1841 through 1958. Seismological Soc. America, Bull., v.53, no.1, pp.95-108.

Berggren, W.A., and Van Couvering, J.A., 1974. The late Neogene, biostratigraphy, geochronology, and paleoclimatology of the last 15 m.y. in marine and continental sequences. Paleogeog., Paleoclimatology, and Paleoecology, v.16, no.1 and 2.

Bestland, Erick A., 1987. Volcanic stratigraphy of the Oligocene Colestin Formation in the Siskiyou Pass area of southern Oregon. Oregon Geology, v.49, no.7, pp.79-86.

Bishop, Ellen Morris, 1989. Smith Rock and the Gray Butte complex. Oregon Geology, v.51, no.4, pp.75-80.

-----1990. A field guide to the geology of Cove Palisades State Park and the Deschutes basin in central Oregon. Oregon Geology, v.52, no.1, pp.3-12, 16.

Black, Gerald L., 1983. Heat flow in the Oregon Cascades. In: Priest, George R., and Vogt, Beverly F., Geology and geothermal resources of the central Oregon Cascade Range. Oregon Dept. of Geol. and Mineral Indus., Special Paper 15, pp.69-76.

Blackwell, David D., et al., 1978. Heat flow of Oregon. Oregon Dept. Geol. and Mineral Indus., Special Paper 4, 42p.

-----et al., 1982. Heat flow, arc volcanism, and subduction in northern Oregon. Jour. Geophys. Res., v.87, no.B10, pp.8735-8754.

-----1990. Heat flow in the Oregon Cascade range and its correlation with regional gravity, Curie Point depths, and geology. Jour. Geophys. Res., v.95, no. B12, pp.19,475-19,493.

Blake, M.C. jr., et al., 1985. Tectonostratigraphic terranes in south-west Oregon. In: Howell, David G., Techtonostratigraphic terranes of the circum-Pacific region. Earth Sci. Ser. v.1, Circumpacific Council for Energy and Mineral Resources, pp.147-157.

Blakely, Richard J., 1990. Volcanism, isostatic residual gravity, and regional tectonic setting of the Cascade volcanic province. Jour. Geophys. Res., v.95, no.B12, pp.19,439-19,451.

----- et al., 1985. Tectonic setting of the southern Cascade Range as interpreted from its magnetic and gravity fields. Geol. Soc. Amer., Bull., v.96, pp.43-48.

Blome, Charles D., 1984. Upper Triassic radiolaria and radiolarian zonation from western North America. Bull. of Amer. Paleo. 318, 88p.

----1991. Evolution of a Permo-Triassic sedimentary melange, Grindstone terrane, east-central Oregon. Geol. Soc. Amer. Bull., v.103, pp.1280-1296.

-----and Irwin, William P., 1983. Tectonic significance of late Paleozoic to Jurassic radiolarians from the North Fork terrane, Klamath Mountains, California. In: Stevens, Calvin, ed., PreJurassic rocks in western North American suspect terranes. Soc. Econ. Paleont. and Mineral. Pacific Sect., pp.77-89.

-----and Nestell, Merlynd K., 1991. Evolution of a Permo-Triassic sedimentary melange, Grindstone terrane, east-central Oregon. Geol. Soc. Amer., Bull., v.103, pp.1280-1296.

-----et al., 1986. Geologic implications of radiolarian-bearing Paleozoic and Mesozoic rocks from the Blue Mountains province, eastern Oregon. U.S. Geol. Survey, Prof. Paper 1435, pp.79-93.

Bogen, Nicholas L., 1986. Paleomagnetism of the Upper Jurassic Galice Formation, southwestern Oregon: evidence for differential rotation of the eastern and western Klamath Mountains. Geology, v.14, no.3, pp.335-338.

Bones, T.J., 1979. Atlas of fossil fruits and seeds from north central Oregon. Oregon Mus. of Science and Indus., Occasional Papers in Natural Sci., no.1, 23p.

Bostwick, David A., and Koch, George S., 1962. Permian and Triassic rocks of northeastern Oregon. Geol. Soc. Amer. Bull., v.73, pp.419-422.

Bourgeois, Joanne, 1980. A transgressive shelf sequence exhibiting hummocky stratification: the Cape Sebastian sandstone (upper Cretaceous), southwestern Oregon. Jour. Sed. Petrology, v.50, no.3, pp.0681-0702.

-----1984. Late Cretaceous transgressive sedimentation: a comparison of the basal Hornbrook Formation and the Cape Sebastian Sandstone, northern California and southwestern Oregon. In: Nilsen, Tor H., ed., 1984. Geology of the upper Cretaceous Hornbrook Formation, Oregon and California: Pacific Sect., Soc. Econ. Mineralog. and Paleont., v.42, pp.149-158.

-----and Leithold, Elana L., 1983. Sedimentation, tectonics and sea-level change as reflected in four wave-dominated shelf sequences in Oregon and California, pp.1-16. In: Larue, D.K., ed., et al. 1983. Cenozoic marine sedimentation, Pacific margin, U.S.A., Soc. Econ. Paleont. and Mineral., Pacific Sect., Los Angeles.

Bowen, Richard G., Blackwell, David D., and Hull, Donald A., 1977. Geothermal exploration studies in Oregon. Oregon Dept. Geol. and Mineral Indus., Misc. Paper 19, 50p.

-----Peterson, N.V., and Riccio, J.F., 1978. Low-to intermediate-temperature thermal springs and wells in Oregon. Oregon Dept. Geol. and Mineral Indus., Geol. Map Ser., GMS-10.

-----et al., 1976. Progress report on heat-flow study of the Brothers fault zone, central Oregon. Ore Bin, v.38, no.3, pp.39-46.

Brandon, Alan D., and Goles, Gordon G., 1988. A Miocene sub-
continental plume in the Pacific Northwest: geochemical
evidence. Earth and Planet. Sci. Letters, 88, pp.273-283.

Bretz, J.H., 1923. The channeled scablands of the Columbia
Plateau. Jour. Geol., v.77, p.505-543.

Brikowski, Tom H., 1983. Geology and petrology of Gearhart
Mountain: a study of calc-alkaline volcanism east of the
Cascades in Oregon. Univ. of Oregon, Ms., 157p.

Brittain, R.C., 1986. Eagle-Pitcher diatomite mine and processing
plant, eastern Oregon. Oregon Geology, v.48, no.9, pp.
108-109.

Brogan, Phil F., 1964. East of the Cascades. Binfords and
Mort, Portland, Ore., 304p.

Brooks, Howard C., 1957. Oregon's opalite mining district
active again. Ore Bin, v.19, no.10, pp.83-88.

-----1963. Quicksilver in Oregon. Oregon Dept. Geol. and
Mineral Indus., Bull.55, 223p.

-----1979. Plate tectonics and the geologic history of the Blue
Mountains. Oregon Geol., v.41, no.5, pp.71-80.

-----1990. Limestone deposits in Oregon. Oregon Dept. Geol.
and Mineral Indus., Special paper 23, pp.8-10.

-----1990. Mining and exploration in Oregon in 1989. Oregon
Geology, v.52, no.2, pp.37-42.

-----and Ramp, Len, 1968. Gold and silver in Oregon. Oregon
Dept. Geol. and Mineral Indus., Bull.61, 337p.

-----and Vallier, Tracy L., 1967. Progress report on the geology
of part of the Snake River Canyon, Oregon and Idaho.
Oregon Dept. Geol. and Mineral Indus., v.29, no.12,
pp.233-266.

-----and Vallier, Tracy L., 1978. Mesozoic rocks and tectonic
evolution of eastern Oregon and western Idaho. In: Howell,
D., and McDougall, K., eds., Mesozoic Paleogeography of
the western United States. Pacific Coast Paleogeography
Symp. 2, Soc. Econ. Paleont. and Mineral., Pacific Sect.,
pp.133-145.

Brown, D.E., and Petros, J.R., 1985. Geochemistry, geochronology, and magnetostratigraphy of a measured section
of the Owyhee basalt, Malheur County, Oregon. Oregon
Geology, v.47, no.2, pp.15-20.

-----Peterson, N.V., and McLean, G.D., 1980. Preliminary geology
and geothermal resource potential of the Lakeview area,
Oregon. Oregon Dept. Geol. and Mineral Indus., Open-
file Report 0-80-9, 108p.

-----et al., 1980. Preliminary geology and geothermal resource
potential of the Willamette Pass area, Oregon. Oregon
Dept. Geol. and Mineral Indus., Open-file Rept. 0-80-3, 65p.

Brownfield, Michael E., 1981. Oregon's coal and its economic
future. Oregon Geology, v.43, no.5, pp.59-67.

Bruer, Wesley G., 1984. Correlation Section 24; northwest Oregon.
Amer. Assoc. Petroleum Geologists, Pacific Section, chart.

Buddenhagen, H.J., 1967. Structure and orogenic history of the
southwestern part of the John Day uplift, Oregon. Ore Bin,
v.29, no.7, pp.129-138.

Burns, L. K., and Ethridge, Frank G., 1979. Petrology and
diagenetic effects of lithic sandstones: Paleocene and
Eocene Umpqua Formation, southwest Oregon. In: Scholle,
Peter, and Schluger, Paul R., eds., 1979. Aspects of
diagenesis. Soc. Econ. Paleont. and Mineralog., Special
Publ. 26, pp.307-317.

Byerly, Perry, 1952. Pacific coast earthquakes. Oregon State
System of Higher Ed., Eugene, Condon Lectures, 38p.

Byrne, John V., 1962. Geomorphology of the continental terrace
off the central coast of Oregon. Ore Bin, v.24, no.5,
pp.65-74.

-----1963. Geomorphology of the Oregon continental terrace south
of Coos Bay. Ore Bin, v.25, no.9, pp.149-155.

-----1963. Geomorphology of the continental terrace off the
northern coast of Oregon. Ore Bin, v.25, no.12, pp.201-
209.

Callaghan, Eugene, and Buddington, A.F., 1938. Metalliferous
mineral deposits of the Cascade Range in Oregon. U.S.
Geol. Survey, Bull.893, 141p.

Cameron, Kenneth Allan, 1980. Geology of the southcentral
margin of the Tillamook highlands; southwest quarter of
the Enright Quadrangle, Tillamook County, Oregon. Ms.,
Portland State Univ., 87p.

-----1991. Prehistoric buried forests of Mount Hood. Oregon
Geology, v.53, no.2, pp.34-43.

Camp, V.E., and Hooper, P.R., 1981. Geologic studies of the
Columbia Plateau: Part I. Late Cenozoic evolution of the
southeast part of the Columbia River basalt province.
Geol. Soc. Amer., Bull., Part I, v.92, pp.659-688.

-----1981. Geologic studies of the Columbia Plateau: Part II.
Upper Miocene basalt distribution, reflecting source
locations, tectonism, and drainage history in the Clearwater embayment, Idaho. Geol. Soc. Amer., Bull.,
Part I, v.92, pp.669-678.

Carlson, Paul R., and Nelson, C. Hans., 1987. Marine geology and
resource potential of Cascadia Basin. In: Scholl, David W.,
Grantz, Arthur, and Vedder, John G., eds., Geology and
resource potential of the continental margin of western
North America and adjacent ocean basins-Beaufort Sea to
Baja California. Cirvum-Pacific Council for Energy and
Mineral Res., Earth Sci., Ser., v.6, pp.523-535.

Carlson, Richard W., 1981. Late Cenozoic rotations of the Juan
de Fuca ridge and the Gorda rise: a case study. Tectonophysics, v.77, pp.171-188.

-----1984. Isotopic constraints on Columbia River flood basalt
genesis and the nature of the sub-continental mantle.
Geochemica et Cosmochim. Acta, v.48, pp.2357-2372.

-----and Hart, William K., 1983. Geochemical study of the
Steens Mountain flood basalt. Carnegie Inst., Wash.,
Yearbook 82, pp.475-480.

-----and Hart, William K., 1987. Crustal genesis on the Oregon
plateau. Jour. Geophys. Res., v.92, no.B7, pp.6191-6206.

-----Lugmair, G.W., and MacDougall, J.D., 1981. Columbia River
volcanism: the question of mantle heterogeneity or crustal
contamination. Geochemica et Cosmochim. Acta, v.45,
pp.2482-2499.

Carpenter, Paul K., 1983. Origin of comb layering in the Willow
Lake intrusion, N.E. Oregon. Univ. of Oregon, Ms., 118p.

Carson, Bobb, 1973. Acoustic stratigraphy, structure, and history
of Quaternary deposition in Cascadia Basin. Deep-Sea
Res., v.20, pp.387-396.

Carson, M.A., and MacLean, P.A., 1985. Storm-controlled
oblique dunes of the Oregon coast: discussion and reply.
Geol. Soc. Amer., Bull., v.96, pp.409-410.

Castor, Stephen B., and Berry, Michael R., 1981. Geology of
the Lakeview uranium district, Oregon. In: Goodell,
Philip, and Waters, Aaron, Uranium in volcanic and
volcaniclastic rocks. Amer. Assoc. Petroleum Geologists,
Studies in Geology, v.13, pp.55-62.

Catchings, R.D., and Mooney, W.D., 1988. Crustal structure of
the Columbia Plateau: evidence for continental rifting.
Jour. Geophys. Res., v.93, no.B1, pp.459-474.

-----and Mooney, W.D., 1988. Crustal structure of east central
Oregon: relation between Newberry Volcano and regional
crustal structure. Jour. Geophys. Res., v.93, no.B9,

pp.10,081-10,094.

Cavender, Ted., 1969. An Oligocene mudminnow (family Umbridae) in Oregon with remarks on relationships within the Escoidei. Museum of Zoology, Univ. of Michigan, Occ. Papers no.660, pp.1-28.

Chan, Marjorie A., 1985. Correlations of diagenesis with sedimentary facies in Eocene sandstones, western Oregon. Jour. Sed. Pet., v.55, no.3, pp.0322-0333.

-----and Dott, R.H., 1983. Shelf and deep-sea sedimentation in Eocene forearc basin, western Oregon - fan or non-fan? Amer. Assoc. Petroleum Geol., Bull., v.67, pp.2100-2116.

-----and Dott, R.H., jr., 1986. Depositional facies and progradational sequences in Eocene wave-dominated deltaic complexes, southwestern Oregon. Amer. Assoc. Petroleum Geol., Bull., v.70, pp.415-429.

-----and Dott, R.H., 1986. Wave-dominated deltaic complexes, southwestern Oregon. Amer. Assoc. Petroleum Geol., Bull., v.70, p.415-429.

Chaney, Ralph W., 1944. Pliocene floras of California and Oregon. Carnegie Inst., Wash., Contrib. to Paleo., 553p.

-----1956. The Ancient forests of Oregon: a study of earth history in western America. Condon Lectures, Eugene, Oregon. 56p.
    Also: Carnegie Inst., Wash., Publ.501, pp.631-648.

-----Condit, Carlton, and Axelrod, D.I., 1959. Miocene floras of the Columbia Plateau. Carnegie Inst., Wash., Publ. 617, 237p.

Chaplet, Michel, Chorowicz, Jean, and Roure, Francoise, 1986/87. Fault patterns by space remote sensing and the rotation of western Oregon during Cenozoic times. Earth and Planet. Sci. Letters, no.81, pp.425-233.

Chitwood, Lawrence A., and McKee, Edwin H., 1981. Newberry Volcano, Oregon. U.S. Geol. Survey, Circular 838, pp.85-91.

-----Jensen, Robert A., and Groh, Edward A., 1977. The age of Lava Butte. Ore Bin, v.39, no.10, pp.157-164.

Choiniere, Stephen R., and Swanson, Donald A., 1979. Magnetostratigraphy and correlation of Miocene basalts of the northern Oregon coast and Columbia Plateau, southeast Washington. Amer. Jour. Sci., v.279, pp.755-777.

Christiansen, Robert L., and McKee, Edwin H., 1978. Late Cenozoic volcanic and tectonic evolution of the Great Basin and Columbia intermontane regions. In: Smith, R., and Eaton, Gordon, eds., Cenozoic tectonic and regional geophysics of the western Cordillera. Geol. Soc. Amer., Memoir 152, pp.283-311.

Church, S.E., et al., 1986. Lead-isotopic data from sulfide minerals from the Cascade Range, Oregon and Washington. Geochimica et Cosmochim. Acta., v.50, pp.317-328.

Ciesiel, Robert F., and Wagner, Norman S., 1969. Lava-tube caves in the Saddle Butte area of Malheur County, Oregon. Ore Bin, v.31, no.8, pp.153-171.

Clark, Robert D., 1989. The odyssey of Thomas Condon. Portland, Oregon, Oregon Historical Soc. Press, 569p.

Cleary, John, et al., 1981. Gravity, isotope, and geochemical study of the Alvord Valley geothermal area, Oregon: summary. Geol. Soc. Amer., Bull., part 1, v.92, pp.319-322; 934-962.

Clifton, H.E., and Luepke, G., 1987. Heavy-mineral placer deposits of the continental margin of Alaska and the Pacific coast states. In: Scholl, David W., Grantz, Arthur, and Vedder, John G., eds. Geology and resource potential of the continental margin of western North America and adjacent ocean basins-Beaufort Sea to Baja California. Circum-

Pacific Council for Energy and Mineral Res., Earth Sci., Ser., v.6, pp.691-738.

Cohn, Lisa, 1991. The big one. Old Oregon, v.71, no.1, pp.18-22.

Colbath, G. Kent, and Steele, Matthew J., 1982. The geology of economically significant lower Pliocene diatomites in the Fort Rock basin near Christmas Valley, Lake County, Oregon. Oregon Geology, v.44, no.10, pp.111-118.

Cole, Mark R., and Armentrout, John M., 1979. Neogene paleogeography of the western United States. In: Armentrout, John M., Cole, M. R., and TerBest, H.J., eds., 1979. Cenozoic paleogeography of the western United States. Pacific Coast Paleogeogr. Symposium 3, Soc. Econ. Paleont. and Mineralog., pp.297-323.

Coleman, R.G., and Irwin, W.P., eds., 1977. North American ophiolites. Oregon Dept. Geol. and Mineral Indus., Bull.95, pp.74-105.

Condon, Thomas, 1902. The two islands. J.K. Gill, Portland, Oregon, 211p.

Coney, Peter J., Jones, David L., and Monger, James W.H., 1980. Cordilleran suspect terranes. Nature, v.288, pp.329-333.

Connard, G., Couch, Richard M., and Gemperle, M., 1983. Analysis of aeromagnetic measurements from the Cascade Range in central Oregon. Geophysics, v.48, no.3, pp.376-390.

Conrey, Richard M., 1985. Volcanic stratigraphy of the Deschutes Formation, Green Ridge to Fly Creek, north-central Oregon. Ms., Oregon State Univ., 349p.

Cooper, William S., 1958. Coastal sand dunes of Oregon and Washington. Geol. Soc. Amer., Memoir 72, 169p.

Cope. Edward D., 1888. Sketches of the Cascade Mountains of Oregon. American Naturalist, v.22, pp.996-1003.

-----1889. The Silver Lake of Oregon and its region. Amer. Naturalist, v.23, no.275, pp.970-982.

Corcoran, R.E., 1965. Geology of Lake Owyhee State Park and vicinity, Malheur County, Oregon. Ore Bin, v.27, no.5, pp.81-98.

-----1969. Geology of the Owyhee Upland province. Oregon Dept. Geol. and Mineral Indus., Bull.64, pp.80-88.

-----and Libbey, F.W., 1955. Investigation of Salem hills bauxite deposits. Ore Bin, v.17, no.4, pp.23-27.

Correlation of stratigraphic units of North America (COSUNA) project, 1983. Northwest Region F. Alan Lindberg, ed. Amer. Assoc. Petroleum Geol., chart.

Couch, Richard W., and Braman, David E., 1980. Geology of the continental margin near Florence, Oregon. Oregon Geology, v.41, on.11, pp.171-179.

-----and Foote, R., 1985. The Shukash and Lapine basins: Pleistocene depressions in the Cascade Range of central Oregon. EOS, v.66, no.3, p.24.

-----and Johnson, Stephen, 1968. The Warner Valley earthquake sequence: May and June, 1968. Ore Bin, v.30, no.10, pp.191-204.

-----and Pitts, Stephen G., 1980. The structure of the continental margin near Coos Bay, Oregon. In: Newton, V., et al., Prospects for oil and gas in the Coos, Douglas, and Lane counties, Oregon. Oregon, Dept. Geol. and Mineral Indus., Oil and Gas Inv.6, pp.23-28.

-----and Riddihough, Robin P., 1989. The crustal structure of the western continental margin of North America. In: Pakiser, L.C., and Mooney, W.D. Geophysical framework of the continental United States. Geol. Soc. Amer., Memoir. 172, pp.103-128.

-----and Whitsett, Robert, 1969. The North Powder earthquake of August 14, 1969. Ore Bin, v.31, no.12, pp.239-244.

-----Thrasher, Glenn, and Keeling, Kenneth, 1976. The Deschutes valley earthquake of April 12, 1976. Ore Bin, v.38, no.10, pp.151-154.

Crandall, Dwight R., 1965. The glacial history of western Washington. In: Wright, H.E., and Frey, D., eds., 1965. The Quaternary history of the United States. Princeton Univ. Press, Princeton, N.J., pp.341-353.

-----1967. Glaciation at Wallowa Lake, Oregon. U.S. Geol. Survey, Prof. Paper 575-C, pp.C145-C153.

-----1980. Recent eruptive history of Mount Hood, Oregon, and potential hazards from future eruptions. U.S. Geol. Survey, Bull.1492, 81p.

Crowell, John C., 1985. The recognition of transform terrane dispersion within mobile belts. In: Howell, David G., ed. Tectonostratigraphic terranes of the circum-Pacific region. Pacific Council for Energy and Mineral Resources, Earth Science Ser., no.1, pp.51-61.

Cummings, Michael L., and Growney, Lawrence P., 1988. Basalt hydrovolcanic deposits in the Dry Creek arm area of the Owyhee Reservoir, Malheur County, Oregon: stratigraphic relations. Oregon Geology, v.50, no.7/8, pp.75-82.

-----et al., 1990. Stratigraphic development and hydrothermal activity in the central Western Cascade range, Oregon. Jour. Geophys. Res., v.95, no.B12, pp.19,601-19,610.

Curray, J.R., 1965. Late Quaternary history, continental shelves of the United States. In: Wright, H.E., and Frey, D.G., eds., The Quaternary history of the United States. Princeton Univ. Press, Princeton, N.J., 723-735.

Dake, H.C., 1938. The gem minerals of Oregon. Oregon Dept. Geol. and Mineral Indus., Bull.7, 16p.

Danner, Wilbert R., 1977. Recent studies of the Devonian of the Pacific Northwest. In: Murphy, M.A., Berry, W.B.N., and Sandberg, C.A., eds. Westerm North American: Devonian. Univ. of Calif., Riverside. Campus Museum, Contrib.4, pp.220-225.

Darienzo, Mark E., and Peterson, Curt D., 1990. Episodic tectonic subsidence of late Holocene salt marshes, northern Oregon central Cascadia margin. Tectonics, v.9, no.1, pp.1-22.

Davidson, George, 1869. Coast pilot of California, Oregon, and Washington territory. U.S. Coast Survey, Pacific Coast, 262p.

Davis, Gregory A., 1979. Problems of intraplate extensional tectonics, western United States, with special emphasis on the Great Basin. In: Newman, Gary W., et al., eds., Basin and Range symposium and Great Basin field conference, pp.41-54.

-----Monger, J.W.H., Burchfiel, B.C., 1978. Mesozoic construction of the Cordilleran "collage", central British Columbia to central California. In: Howell, D.G., and McDougall, Kristin, Mesozoic paleogeography of the western United States, Soc. Econ. Paleont. and Mineralog., Pacific Coast Paleogeography Symp. 2, p.1-32.

Derkey, Robert E., 1980. The Silver Peak volcanogenic massive sulfide, northern Klamath Mountains, Oregon. Geol. Soc. Amer., Abstr. with prog., v.12, no.3, p.104.

Dethier, David P., 1980. Reconnaissance study of Holocene glacier fluctuations in the Broken Top area, Oregon. Geol. Soc. Amer., Abstr. with prog., v.12, no.3, p.104.

Dick, Henry J.B., 1977. Partial melting in the Josephine peridotite, I: the effect on mineral composition and its consequence for geobarometry and geothermometry. Amer. Jour. Sci., v.277, pp.801-832.

Dicken, Samuel N., 1973. Oregon geography. Eugene, Ore., Univ. of Oregon Bookstore, 147p.

-----1980. Pluvial lake Modoc, Klamath County, Oregon, and Modoc

and Siskiyou counties, California. Oregon Geology, v.42, no.11, pp.179-187.

-----and Dicken, Emily F., 1985. The legacy of ancient Lake Modoc: a historical geography of the Klamath Lakes basin Oregon and California. Eugene, Ore., Univ. of Oregon Bookstore, distr., various pagings.

Dickinson, William R., 1977. Paleozoic plate tectonics and the evolution of the Cordilleran continental margin. In: Stewart, J.H., et al., eds., 1977. Paleozoic paleogeography of the western United States. Soc. Econ. Paleont. and Mineralog., Pacific Coast Paleogeogr. Symp.1, pp.137-155.

-----1979. Cenozoic plate tectonic settings of the Cordilleran region in the United States. In: Armentrout, J.M., Cole, M.R., and TerBest, H.J., eds, Cenozoic paleogeography of the western United States. Soc. Econ. Paleontologists and Mineralog., Pacific Coast Paleogeography Symp. 3, pp.1-13.

-----1979. Mesozoic forearc basin in central Oregon. Geology (Boulder), v.7, no.4, pp.166-170.

-----and Thayer, Thomas P., 1978. Paleogeographic and paleo-tectonic implications of Mesozoic stratigraphy and structure in the John Day inlier of central Oregon. In: Howell, D.G., and McDougall, K.A., eds. Mesozoic paleogeography of the western United States. Soc. Econ. Paleont. and Mineral. Pacific Sect., Pacific Coast Paleogr. Symp.2, pp.147-161.

-----and Vigrass, Laurence W., 1965. Geology of the Suplee-Izee area Crook, Grant, and Harney counties, Oregon. Oregon Dept. Geol. and Mineral Indus., Bull.58, 109p.

Dignes, T.W., and Woltz, D., 1981. West coast. Amer. Assoc. Petroleum Geol., Bull.65, no.10, pp.1781-1791.

Diller, J.S., 1896. Geological reconnaissance in northwestern Oregon. U.S. Geol. Survey, 17 Ann. rept., p.1-80.

-----1903. Description of the Port Orford quadrangle. U.S. Geol. Survey, Bull. 196, 69p.

Donato, Mary Margaret, 1975. The geology and petrology of a portion of the Ashland pluton, Jackson County, Oregon. Ms., Univ. of Oregon, 89p.

-----Coleman, R.G., and Kays, M.A., 1980. Geology of the Condrey Mountain schist, northern Klamath Mountains, California and Oregon. Oregon Geology, v.42, no.7, pp.125-129.

Dott, R.H., 1962. Geology of the Cape Blanco area, southwest Oregon. Ore Bin, v.24, no.8, pp.121-133.

-----1966. Eocene deltaic sedimentation at Coos Bay, Oregon. Jour. Geol., v.74, pp.373-420.

-----1971. Geology of the southwestern Oregon coast west of the 124th meridian. Oregon Dept. Geol. and Mineral Indus., Bull.69, 63p.

-----and Bird, Kenneth J., 1979. Sand transport through channels across an Eocene shelf and slope in southwestern Oregon, U.S.A. Soc. Econ. Paleont. and Mineralog., Special Publ. 27, pp.327-342.

Downs, Theodore, 1956. The Mascall fauna from the Miocene of Oregon. Univ. Calif. Publ. Geol. Sci., v.31, pp.199-354.

Drake, Ellen T., 1982. Tectonic evolution of the Oregon continental margin. Oregon Geology, v.44, no.2, pp.15-21.

-----et al., 1990. Crater Lake; an ecosystem study. Amer. Assoc. for the Advancement of Science, Pacific Division, 69th Ann. Mtg., Oregon State Univ., Corvallis, Oregon, 221p.

Driedger, Carolyn L., and Kennard, Paul M., 1984. Ice volumes on Cascade volcanoes: Mount Rainier, Mount Hood, Three Sisters, and Mount Shasta. U.S. Geol. Suvery, Open-file Report 84-581, 42p.

Duncan, Robert A., 1982. A captured island chain in the Coast Range of Oregon and Washington. Jour. Geophys. Res., v.87, no.B13, pp.10,827-10,837.

-----and Kulm, LaVerne D., 1989. Plate tectonic evolution of the Cascades arc-subduction complex. In: Winterer, E.L., Hussong, D.M., and Decker, R.W., eds. The Eastern Pacific Ocean and Hawaii. Geol. Soc. Amer., the Geology of North America, v.N, pp.413-438.

-----Fowler, Gerald, and Kulm, L.D., 1970. Planktonic foraminiferan-radiolarian ratios and Holocene-Late Pleistocene deep-sea stratigraphy off Oregon. Geol. Soc. Amer., Bull., v.81, pp.561-566.

Dymond, Jack, et al., 1987. Hydrothermal activity in Crater Lake: evidence from sediments. EOS, v.68, no.50, p.1771.

Easterbrook, Don J., 1986. Stratigraphy and chronology of Quaternary deposits of the Puget lowland and Olympic Mountains of Washington and the Cascade Mountains of Washington and Oregon. In: Sibrava, W., et al., eds. Quaternary glaciations in the northern hemisphere. Quaternary Science Reviews 5, pp.145-159.

Eaton, Gordon P., 1979. Regional geophysics, Cenozoic tectonics, and geologic resources of the Basin and Range province and adjoining regions. In: Newman, Gary W., et al., eds., Basin and Range Symposium and Great Basin field conference, pp.11-39.

-----1982. The Basin and Range province: origin and tectonic significance. Ann. Rev. Earth Planet. Sci., v.10, pp.409-440.

-----1984. The Miocene Great Basin of western North America as an extending back-arc region. Tectonophysics, 102, pp.275-295.

Edmiston, R.C., and Benoit, W.R., 1984. Characteristics of Basin and Range geothermal systems with fluid temperatures of 150 C to 200 C. Geothermal Res. Council, Trans., v.8, pp.417-424.

Ehlen, Judi, 1967. Geology of state parks near Cape Arago, Coos County, Oregon. Ore Bin, v.29, no.4, pp.61-82.

Engebretson, David C., Cox, Allan, and Gordon, Richard G., 1985. Relative motions between oceanic and continental plates in the Pacific Basin. Geol. Soc. Amer., Special Paper 206, 59p.

Enlows, Harold E., 1976. Petrography of the Rattlesnake Formation the type area, central Oregon. Oregon, Dept. Geol. and Mineral Indus., Short Paper 25, 34p.

Erdman, James A., and Harrach, George H., 1981. Uranium in big sagebrush from Western U.S. and evidence of possible mineralization in the Owyhee Mountains of Idaho. Jour. Geochem. Explor., v.14, pp.83-94.

Evans, James G., 1984. Structure of part of the Josephine peridotite, northwestern California and southwestern Oregon. U.S. Geol. Survey, Bull. 1546 A-D, pp.7-37.

Farooqui, Saleem M., et al., 1981. Dalles Group: Neogene formations overlying the Columbia River Basalt group in north-central Oregon. Oregon Geol., v.43, no.10, pp.131-140.

Farr, Leonard Carl., 1989. Stratigraphy, diagenesis, and depositional environment of the Cowlitz Formation (Eocene) northwest, Oregon. Ms., Portland State Univ., 168p.

Fassio, Joseph Michael, 1990. Geochemical evolution of ferruginous bauxite deposits in northwestern Oregon and southwestern Washington. Ms., Portland State Univ., 103p.

Faul, Henry, and Faul, Carol, 1983. It began with a stone: a history of geology from the stone age to the age of plate tectonics. New York, Wiley, 270p.

Fecht, Karl R., Reidel, Stephen P., and Tallman, Ann M., 1987. Paleodrainage of the Columbia River system on the Columbia Plateau of Washington state - a summary. Wash. Div. of Geol. and Earth Res., Bull.77, pp.238-245.

Ferguson, Denzel, and Ferguson, Nancy, 1978. Oregon's Great Basin country. Burns, Ore., Gail Graphics, 178p.

Ferns, Mark L., 1990. Talc and soapstone in Oregon. Oregon Dept. Geol. and Mineral Indus., Special Paper 23, pp.11-13.

-----and Huber, Donald F., 1984. Mineral resources map of Oregon. Oregon Dept. Geol. and Mineral Indus., Geological Map Series, GMS 36.

Fisher, Richard V., 1963. Zeolite-rich beds of the John Day Formation, Grant and Wheeler Counties, Oregon. Ore Bin, v.25, no.11, pp.185-195.

-----and Rensberger, John M., 1972. Physical stratigraphy of the John Day Formation, central Oregon. Univ. Calif., Publ. Geol. Sci., v.101, pp.1-45.

Fitterman, David V., 1988. Overview of the structure and geothermal potential of Newberry Volcano, Oregon. Jour. Geophys. Res., v.93, no.B9, pp.10,059-10,066.

Flanagan, Guy, and Williams, David L., 1982. A magnetic investigation of Mount Hood, Oregon. Jour. Geophys. Res., v.87, no.B4, pp.2804-2814.

Fouch, Thomas D., 1983. Petroleum potential of wilderness lands in Oregon. U.S. Geol. Survey, Circular 902-J, pp.J1-J5.

Fowler, Don and Koch, David, 1982. The Great Basin. In: Bender, Gordon L., ed., 1982. Reference handbook on the deserts of North America, Greenwood, Conn., Westport, 594p.

Fox, T.P., 1987. Stratigraphy of the Standard Kirkpatrick No.l. Gilliam County, Oregon: New insight into Tertiary tectonism of the Blue Mountains. Oregon Geology, v.49, no.2, pp.15-22.

Foxworthy, Bruce, and Hile, Mary, 1982. Cascade Range volcanoes. In: Volcanoes of the Cascade Range in North America. U.S. Geol. Survey, Prof. Paper 1249, pp.3-8.

Freed, Michael, 1979. Silver Falls State Park. Oregon Geology, v.41, no.1, pp.3-10.

Fryberger, John Summers, 1959. The geology of Steens Mountain, Oregon. Univ. of Oregon, Ms., 65p.

Gamer, Robert L., 1981. Gold deposits of the Baker area. Geol. Soc. Oregon Country, Newsl., v.47, no.9, pp.107-108.

-----1982. Mt. Hood - Portland's own volcano. Geol. Soc. Oregon Country, Newsl., v.48, no.10, pp.226-228.

Gandera, William E., 1977. Stratigraphy of the middle to late Eocene formations of southwestern Willamette Valley. Univ. of Oregon, Ms., 71p.

Gaona, M.T., 1984. Stratigraphy and sedimentology of the Osburger Gulch sandstone member of the upper Cretaceous Hornbrook Formation, northern California and southern Oregon. In: Nilsen, Tor H., ed. Geology of the upper Cretaceous Hornbrook Formation, Oregon and California. Soc. Econ. Paleont. and Mineralogists, Pacific Sect., v.42, pp.141-148.

Garcia, Michael O., 1982. Petrology of the Rogue River island-arc complex, southwest Oregon. Amer. Jour. Sci., v.282, pp.783-807.

-----1989. Industrial minerals in Oregon. Oregon Geology, v.51, no.6, pp.123-129.

Gehr, Keith D., and Newman, Thomas M., 1978. Preliminary note on the late Pleistocene geomorphology and archaeology of the Harney Basin, Oregon. Ore Bin, v.40, no.10, pp.165-170.

Geitgey, Ronald P., 1987. Oregon sunstones. Oregon Geology, v.49, no.2, pp.23-24.

-----and Vogt, Beverly F., 1990. Industrial rocks and minerals of the Pacific northwest. Oregon Dept. Geol. and Mineral

Indus., Special Paper 23, 110p.

Gentile, John R., 1982. The relationship of morphology and material to landslide occurrence along the coastline in Lincoln County, Oregon. Oregon Geology, v.44, no.9, pp. 99-102.

Geologic evolution of the northernmost Coast Ranges and western Klamath Mountains, California, 1989. Field trip guidebook T308, Galice, Oregon, to Eureka, California. Fieldtrips for the 28th International Geological Congress. American Geophysical Union, 58p.

George, Anthony J., 1985. The Santiam mining district of the Oregon Cascades; a cultural property inventory and historical survey. Lyons, Oregon, Shiny Rock Mining Corp., 336p.

Gerlach, David C., Ave Lallemant, H.G., and Leeman, William P., 1981. An island arc origin for the Canyon Mountain ophiolite complex, eastern Oregon, U.S.A. Earth and Planet. Sci. Letters, 53, pp.255-265.

Getahun, Aberra, and Retallack, Gregory J., 1991. Early Oligocene paleoenvironment of a paleosol from the lower part of the John Day Formation near Clarno, Oregon. Oregon Geology, v.53, no.6, pp.131-136.

Gilluly, James, 1933. Copper deposits near Keating, Oregon. U.S. Geol. Survey Bull.830, pp.1-32.

-----1937. Geology and mineral resources of the Baker Quadrangle, Oregon. U.S. Geol. Survey, Bull.879, 119p.

Glenn, Jerry Lee, 1962. Gravel deposits in the Willamette Valley between Salem and Oregon City, Oregon. Ore Bin, v.24, no.3, pp.33-47.

-----1965. Late Quaternary sedimentation and geologic history of the north Willamette Valley, Oregon. PhD., Oregon State Univ., 231p.

Goles, Gordon G., 1986. Miocene basalts of the Blue Mountains province in Oregon. I: Compositional types and their geological settings. Jour. Petrology, v.27(2), pp.495-520.

-----Brandon, Alan D., and Lambert, R.St J., 1989. Miocene basalts of the Blue Mountains province in Oregon; Part 2, Sr isotopic ratios and trace element features of little-known Miocene basalts of central and eastern Oregon. Geol. Soc. Amer., Special Paper 239, pp.357-365p.

Goodge, John W., 1989. Evolving early Mesozoic convergent margin deformation, central Klamath Mountains, northern California. Tectonics, v.8, no.4, pp.845-864.

Graven, Eric P., 1990. Structure and tectonics of the southern Willamette Valley, Oregon. Oregon State Univ., Ms., 119p.

Gray, Gary G., 1977. A geological field trip guide from Sweet Home, Oregon, to the Quartzville mining district. Ore Bin, v.39, no.6, pp.93-108.

-----1978. Overview of the Bohemia mining district. Ore bin, v.40, no.5, pp.77-91.

-----1986. Native terranes of the central Klamath Mountains, California. Tectonics, v.5, no.7, pp.1043-1054.

Gray, Jerry J., 1980. Forecasting rock material demand: an overview of several techniques and detailed review of two. Oregon Geology, v.42, no.1, pp.3-15.

-----1990. Oregon bentonite. Oregon Dept. Geol. and Mineral Indus., Special Paper 23, pp.14-18.

Greeley, Ronald, 1971. Geology of selected lava tubes in the Bend area, Oregon. Oregon, Dept. Geol. and Mineral Indus., Bull.71, 47p.

Gresens, Randall L., and Stewart, Richard J., 1981. What lies beneath the Columbia Plateau? Oil & Gas Jour., v.79, no.31, pp.257-164.

Griggs, A.B., 1945. Chromite-bearing sands of the southern part of the coast of Oregon. U.S. Geol. Survey Bull., 945-E, 150p.

Griggs, G.B., et al., 1970. Deep-sea gravel from Cascadia Channel. Jour. Geol., v.78, no.5, pp.611-619.

-----et al., 1970. Holocene faunal stratigraphy and paleo-climatic implications of deep-sea sediments in Cascadia Basin. Paleo., Paleo., Paleo., v.7, pp.5-12.

Gromme, C. Sherman, et al., 1986. Paleomagnetism of the Tertiary Clarno Formation of central Oregon and its significance for the tectonic history of the Pacific Northwest. Jour. of Geophys. Res., v.91, no.B14, pp.14,089-14,130.

Grose, L.T., and Keller, G.V., 1979. Geothermal energy in the Basin and Range province. In: Newman, Gary W., et al., eds., Basin and Range symposium and Great Basin field conference, pp.361-369.

Gross, M. Grant, 1966. Distribution of radioactive marine sediment derived from the Columbia River. Jour. Geophys. Res., v.71, no.8, pp.2017-2021.

Gutmanis, J.C., 1989. Wrench faults, pull-apart basins, and volcanism in central Oregon: a new tectonic model based on image interpretation. Geol. Jour., v.24, pp.183-192.

Hamilton, Warren, 1978. Mesozoic tectonics of the western United States. In: Howell, D.G., and McDougall, K., eds., Mesozoic paleogeography of the western United States, Soc. Econ. Paleont. and Mineralogists, Pacific Coast Paleogr. Symp. 2, p.33-70.

Hammond, Paul E., 1979. A tectonic model for evolution of the Cascade Range. In: Armentrout, J.M., Cole, M.R., and TerBest, H., eds., Cenozoic paleogeography of the western United States, Soc. Econ. Paleont. and Mineralogists, Pacific Coast Paleogeogr. Symp. 3, pp.219-237.

-----1989. Guide to the geology of the Cascade Range, Portland, Oregon, to Seattle, Washington. Field Trip Guidebook T306, Amer. Geophysical Union, 215p.

Hansen, Henry P., 1946. Postglacial forest succession and climate in the Oregon Cascades. Amer. Jour. Sci., v.244, pp.710-734.

-----1947. Postglacial vegetation of the northern Great Basin. Amer. Jour. Botany, v.31, pp.164-171.

Hanson, Larry G., 1986. Scenes from ancient Portland; sketches of major events in the city's geologic history. Oregon Geology, v.48, no.11, pp.130-131.

Harper, Gregory D., 1983. A depositional contact between the Galice Formation and a late Jurassic ophiolite in northwestern California and southwestern Oregon. Oregon Geology, v.45, no.1, pp.3-7,9.

----- 1984. The Josephine ophiolite, northwestern California. Geol. Soc. Amer., Bull., v.95, pp.1009-1026.

-----and Wright, James E., 1984. Middle to late Jurassic tectonic evolution of the Klamath Mountains, California-Oregon. Tectonics, v.3, no.7, pp.759-772.

-----Saleeby, Jason B., and Norman, Elizabeth A.S., 1985. Geometry and tectonic setting of sea-floor spreading for the Josephine ophiolite, and implications for Jurassic accretionary events along the California margin. In: Howell, David G., ed., Tectonostratigraphic terranes of the circum-Pacific region. Circum-Pacific Council for Energy and Mineral Resources, Earth Sci. Series no.l, pp.239-257.

-----Bowman, John R., and Kuhms, Roger, 1988. A field, chemical, and stable isotope study of subseafloor metamorphism of the Josephine ophiolite, California-Oregon. Jour. Geophys. Res., v.93, no.B5, pp.4625-4656.

Harris, Stephen L., 1988. Fire mountains of the west; the Cascade and Mono Lake volcanoes. Missoula, Mountain Press,

378p.

Hart, D.H., and Newcomb, R.C., 1965. Geology and ground water of the Tualatin Valley, Oregon. U.S. Geol. Survey, Water Supply Paper 1697, 172p.

Hart, William K., and Mertzman, Stanley A., 1983. Late Cenozoic volcanic stratigraphy of the Jordan Valley area, southeastern Oregon. Oregon Geology, v.45, no.2, pp.15-19.

Haskell, Daniel C., 1968. The United States Exploring Expedition, 1833-1842 and its publications 1844-1874. New York, Greenwood, 188p.

Hatton, Raymond R., 1988. Oregon's big country. Bend, Maverick Publ., 137p.

Heaton, Thomas H., and Hartzell, Stephen H., 1987. Earthquake hazards on the Cascadia subduction zone. Science, v.236, p.162-168.

Heiken, G.H., Fisher, R.V., and Peterson, N.V., 1981. A field trip to the maar volcanoes of the Fort Rock - Christmas Lake Valley basin, Oregon. U.S. Geol. Survey, Circular 838, pp.119-140.

Heller, Paul L., and Ryberg, Paul T., 1983. Sedimentary record of subduction to forearc transition in the rotated Eocene basin of western Oregon. Geology, v.11, pp.380-383.

-----Tabor, Rowland, W., and Suczek, Christopher A., 1987. Paleogeographic evolution of the United States Pacific Northwest during Paleogene time. Can. Jour. Earth Sci., v.24, pp.1652-1667.

Hering, Carl William, 1981. Geology and petrology of the Yamsay Mountain complex, south-central Oregon: a study of bimodal volcanism. Univ. of Oregon, PhD., 194p.

Hickman, Carole J.S., 1969. The Oligocene marine molluscan fauna of the Eugene Formation in Oregon. Univ. of Oregon, Mus. Natural History, Bull.16, 112p.

Higgins, Michael W., and Waters, Aaron C., 1967. Newberry Caldera, Oregon. Ore Bin, v.29, no.3, pp.37-60.

Hill, Brittain E., and Scott, William E., 1990. Field trip guide to the central Oregon High Cascades. Part 2 (conclusion): Ash-flow tuffs in the Bend area. Oregon Geology, v.52, pp.123-124.

-----and Taylor, Edward M., 1990. Paper connected with field trip Oregon High Cascade pyroclastic units in the vicinity of Bend, Oregon. Oregon Geology, v.52, no.6, pp.125-126, 139.

-----and Priest, George, 1991. Initial results from the 1990 geothermal drilling program at Santiam Pass, Cascade Range, Oregon. Oregon Geology, v.53, no.5, pp.101-103.

Hill, Mason, ed., 1987. Cordilleran Section of the Geological Society of America. Decade of North American Geology, Project Series, Centennial Field Guide, 490p.

Hines, Pierre R., 1969. Notes on the history of the sand and gravel industry in Oregon. Ore Bin, v.31, no.12, pp.225-237.

Hodge, Edwin T., 1925. Mount Multnomah; ancient ancestor of the Three Sisters. Univ. of Oregon Publ., v.2, no.10, 158p.

------1942. Geology of north central Oregon. Oregon State College, Studies in Geol., no.3, 76p.

Holmes, Kenneth L., 1968. New horizons in historical research: the geologists' frontier. Ore Bin, v.30, no.8, pp.151-159.

Holtz, Preston E., 1971. Geology of lode gold districts in the Klamath Mountains, California and Oregon. U.S. Geol. Survey, Bull.1290, 91p.

------1971. Plutonic rocks of the Klamath Mountains, California and Oregon. U.S. Geol. Survey, Prof. Paper 684-B, pp.B1-B20.

-----1979. Regional metamorphism in the Condrey Mountain Quadrangle, north-central Klamath Mountains, Califor-

nia. U.S. Geol. Survey, Prof. Paper 1086, 25p.

Hooper, Peter R., and Camp, Victor E., 1981. Deformation of the southeast part of the Columbia Plateau. Geology, v.7, pp.322-328.

-----and Conrey, R.M., 1989. A model for the tectonic setting of the Columbia River basalt eruptions. In: Reidel, S.P., and Hooper, P.R., eds. Volcanism and tectonism in the Columbia River flood-basalt province. Geol. Soc. Amer., Special Paper 239, pp.293-305.

-----and Swanson, D.A., 1987. Evolution of the eastern part of the Columbia Plateau. Wash. Div. Geol. and Earth Res., Bull.77, pp.197-214.

-----Knowles, C.R., and Watkins, N.D., 1979. Magnetostratigraphy of the Imnaha and Grande Ronde basalts in the southeast part of the Columbia Plateau. Amer. Jour. Sci., v.279, pp.737-754.

Howard, Donald G., 1978. Levyne in the Wallowas. Ore Bin, v.40, no.7, pp.117-126.

Howard, J.K., and Dott, R.H., jr., 1961. Geology of Cape Sebastian State Park and its regional relationships. Ore Bin, v.23, no.8, pp.75-81.

Howell, David G., ed., 1985. Tectonostratigraphic terranes of the circum-Pacific region. Circum-Pacific Council for Energy and Mineral Resources, Earth Science Series, No.1, 581p.

-----and McDougall, Kristin A., 1978. Mesozoic paleogeography of the Western United States. Pacific Coast Paleogeography Symp. 2. Soc. Econ. Paleont. and Mineral., Pacific Sect., 573p.

-----Jones, David L., and Schermer, Elizabeth R., 1985. Tectonostratigraphic terranes of the circum-Pacific region. In: Howell, David G., ed., Tectonostratigraphic terranes of the circum-Pacific region. Circum-Pacific Council for Energy and Mineral Resources, Earth Science Series, No.1, pp.3-30.

Hughes, Scott S., and Taylor, Edward M., 1986. Geochemistry, petrogenesis, and tectonic implications of central High Cascade mafic platform lavas. Geol. Soc. Amer., Bull., v.97, p.1024-1036.

Hunt, Charles B., 1979. The Great Basin, an overview and hypotheses of its history. In: Newman, Gary W., et al., eds., Basin and Range symposium and Great Basin field conference, pp.1-9.

Hunter, Ralph E., 1980. Depositional environments of some Pleistocene coastal terrace deposits, southwestern Oregon-case history of a progradational beach and dune sequence. Sedimentary Geology, v.27, pp.241-262.

-----and Clifton, H. Edward, 1982. Cyclic deposits and hummocky cross-stratification of probable storm origin in upper Cretaceous rocks of the Cape Sebastian area, southwestern Oregon. Jour. Sed. Pet., v.27, no.4, pp.241-262.

-----Clifton, H. Edward, and Phillips, R. Lawrence, 1970. Geology of the stacks and reefs off the southern Oregon coast. Ore Bin, v.32, no.10, pp.185-201.

-----Richmond, Bruce M., and Alpha, Tau Rho, 1983. Storm-controlled oblique dunes of the Oregon coast. Geol. Soc. Amer., v.94, pp.1450-1465.

Imlay, Ralph W., et al., 1959. Relations of certain upper Jurassic and lower Cretaceous formations in southwestern Oregon. Amer. Assoc. of Petroleum Geol., v.43, no.12, pp.2770-2785.

Ingersoll, Raymond V., 1982. Triple-junction instability as cause for late Cenozoic extension and fragmentation of the western United States. Geology, v.10, no.12, pp.621-624.

Irwin, William P., 1960. Geologic reconnaissance of the northern Coast Ranges and Klamath Mountains, California, with a summary of the mineral resources. Calif. Div. Mines

Bull., 179, 80p.

-----1964. Late Mesozoic orogenies in the ultramafic belts of northwestern California and south-western Oregon. U.S. Geol. Survey, Prof. Paper 501-C, pp.C1-C9.

-----1966. Geology of the Klamath Mountains Province. In: Bailey, E.N., ed., Geology of northern California. Calif. Div. Mines, Bull.190, pp.19-38.

-----1977. Review of Paleozoic rocks of the Klamath Mountains. In: Stuart, J.H., Stevens, C.H., and Frische, A.E., eds., Paleozoic paleogeography of the western United States. Soc. Econ. Paleont. and Mineralogists, Pacific Coast Paleogeography Symp.1, pp.441-454.

-----1985. Age and tectonics of plutonic belts in accreted terranes of the Klamath Mountains, California and Oregon. In: Howell, David G., ed., 1985. Techno-stratigraphic terranes of the circum-Pacific region. Circum-Pacific Council for Energy and Mineral Res., Earth Science Series, No.1, pp.187-198.

-----Jones, David L., and Kaplan, Terry A., 1978. Radiolarians from pre-Nevadan rocks of the Klamath Mountains, California and Oregon. In: Howell, D. G., and McDougall, K.A., eds. Mesozoic paleogeography of the western United States. Soc. Econ. Paleont. and Mineralogists, Pacific Coast Symposium 2, pp.303-310.

-----Wardlaw, Bruce R., and Kaplan, Terry A., 1983. Conodonts of the western Paleozoic and Triassic belt, Klamath Mountains, California and Oregon. Jour. Paleo., v.57, no.5, pp.1030-1039.

Jackson, Philip L., 1987. Erosional changes at Alsea Spit, Waldport, Oregon. Oregon Geology, v.49, no.5, pp.55-59.

Jackson, Ronald Laverne, 1974. A mineralogical and geochemi-cal study of the ferruginous bauxite deposits in Columbia County, Oregon and Wahkiakum County, Washington. Ms., Portland State Univ., 87p.

Jacobson, Randall S., 1985. The 1984 landslide and earthquake activity on the Baker-Homestead highway near Halfway, Oregon. Oregon Geology, v.74, no.5, pp.51-58.

-----et al., 1986. Map of Oregon seismicity, 1841-1986. Oregon Dept. Geol. and Mineral Indus., Geological Map Series, GMS 49.

Jensen, Robert A., 1988. Roadside guide to the geology of Newberry Volcano. Bend, CenOreGeoPub, 75p.

Johnson, H.Paul, and Holmes, Mark L., 1989. Evolution in plate tectonics; the Juan de Fuca Ridge. In: Winterer, E.L., Hussong, D.M., and Decker, R.W., eds., The eastern Pacific Ocean and Hawaii. Geol. Soc. Amer., Vol.N., pp.73-90.

Johnson, Keith E., and Ciancanelli, Eugene V., 1984. Geothermal exploration at Glass Buttes, Oregon. Oregon Geol., v.46, no.2, pp.15-18.

Johnston, David A., and Donnelly-Nolan, Julie, eds., 1981. Guides to some volcanic terranes in Washington, Idaho, Oregon, and northern California. U.S. Geol. Survey, Circular 838, 189p.

Jones, David L., Silberling, N.J., and Hillhouse, John, 1977. Wrangellia - A displaced terrane in northwestern North America. Can. Jour. Earth Sci., v.14, pp.2565-2577.

-----et al., 1982. The growth of western North America. Scientific American, v.247, no.5, pp.70-85.

-----et al., 1983. Recognition, character and analysis of tectonostratigraphic terranes in western North America. Jour. Geol. Educ., v.31, pp.295-303.

Jurdy, Donna M., 1984. The subduction of the Farallon plate beneath North America as derived from relative plate motions. Tectonics, v.3, no.2, pp.107-113.

Kadri, Moinoddin Murtuzamiya, 1982. Structure and influence of the Tillamook uplift on the stratigraphy of the Mist area, Oregon. Ms., Portland State Univ., 105p.

-----et al., 1983. Geochemical evidence for changing provenance of Tertiary formations in northwestern Oregon. Oregon Geology, v.45, no.2, pp.20-22.

Karl, H.A., Hampton, M.A., and Kenyon, N.H., 1989. Lateral migration of Cascadia Channel in response to accretionary tectonics. Geology, v.17, pp.144-147.

Katsura, Kurt T., 1988. The geology and epithermal vein mineralization at the Champion mine, Bohemia mining district, Oregon. Ms., Univ. of Oregon, 254p.

Kays, Allan M., 1970. Western Cascades volcanic series, south Umpqua Falls region, Oregon. Ore Bin, v.32, no.5, pp. 81-94.

-----and Ferns, M.L., 1980. Geologic field trip guide through the north-central Klamath Mountains. Oregon Geology, v.42, no.2, pp.23-35.

Kelsey, Harvey M., 1990. Late Quaternary deformation of marine terraces on the Cascadia subduction zone near Cape Blanco, Oregon. Tectonics, v.9, pp.983-1014.

Kimmel, Peter G., 1982. Stratigraphy, age, and tectonic setting of the Miocene-Pliocene lacustrine sediments of the western Snake River plain, Oregon and Idaho. In: Bonnichsen, Bill, and Breckenridge, R.M. Cenozoic geology of Idaho. Idaho Bur. of Mines and Geol., Bull. 26, pp.559-578.

Kittleman, Laurence R., 1965. Cenozoic stratigraphy of the Owyhee Region, southeastern Oregon. Mus. Natural Hist., Univ. of Oregon, Bull.1, 45p.

-----1973. Guide to the geology of the Owyhee region of Oregon.. Mus. Natural Hist., Univ. of Oregon, Bull.21, 61p.

-----et al., 1967. Geologic map of the Owyhee Region, Malheur County, Oregon. Univ. of Oregon, Mus. Natural Hist., Bull.8.

Kleck, Wallace D., 1972. The geology of some zeolite deposits in the southern Willamette Valley, Oregon. Ore Bin, v.34, no.9, pp.145-151.

Kleinhans, Lewis C., Balcells-Baldwin, E.A., and Jones, R.A., 1984. A paleogeographic reinterpretation of some middle Cretaceous units, north-central Oregon: evidence for a submarine turbidite system. In: Nilsen, Tor H., ed. Geology of the upper Cretaceous Hornbrook Formation, Oregon and California. Soc. Econ. Paleont. and Mineral., Pacific Sect., v.42, pp.239-257.

Koch, John G., 1966. Late Mesozoic stratigraphy and tectonic history, Port Orford-Gold Beach area, southwestern Oregon coast. Amer. Assoc. Petrol. Geol., v.50, no.1, pp.25-71.

-----and Camp., C.L., 1966. Late Jurassic ichthyosaur from Sisters Rocks, coastal southwestern Oregon, Ore Bin, v.28, no.3, pp.65-68.

Kohler, W.M., Healy, J.H., and Wegener, S.S., 1982. Upper crustal structure of the Mount Hood, Oregon, region as revealed by time term analysis. Jour. Geophys. Res., v.87, no.B1, pp.339-355.

Komar, Paul D., 1983. The erosion of Siletz Spit, Oregon. In: Komar, Paul D., ed., CRC Handbook of coastal processes and erosion. CRC Press, Boca Raton, Florida, pp.65-76.

-----1992. Ocean processes and hazards along the Oregon coast. Oregon Geology, v.54, no.1, pp.3-19.

-----and Rea, C. Cary, 1976. Beach erosion on Siletz Spit, Oregon. Ore Bin, v.38, no.8, pp.119-134.

-----and Schlicker, Herb, 1980. Beach processes and erosion

problems on the Oregon Coast. Oregon, Dept. Geol. and Mineral Indus., Bull.101, pp.169-173.

Koski, Randolph A. and Derkey, Robert E., 1981. Massive sulfide deposits in oceanic-crust and island-arc terranes of southwestern Oregon. Oregon Geology, v.43, no.9, pp.119-125.

Kulm, LaVerne D., 1980. Sedimentary rocks of the central Oregon continental shelf. In: Newton, Vern C., et al., Prospects for oil and gas in the Coos basin, western Coos, Douglas, and Lane counties, Oregon. Oregon, Dept. Geol. and Mineral Indus., Oil and Gas Inv. 6, pp.29-34.

----and Fowler, G.A., 1974. Oregon continental margin structure and stratigraphy: a test of the imbricate thrust model. In: Burk, C.A., and Drake, M.J., eds., The geology of continental margins. New York, Springer-Verlag, pp.261-283.

-----and Scheidegger, K.F., 1979. Quaternary sedimentation on the tectonically active Oregon continental slope. Soc. Econ. Paleont. and Mineralog., Special Publ., 27, pp.247-264.

-----et al., 1968. Evidence for possible placer accumulations on the southern Oregon continental shelf. Ore Bin, v.30, no.5, pp.81-104.

-----et al., 1975. Oregon continental shelf sedimentation interrelationships of facies distribution and sedimentary processes. Jour. Geol., v.83, pp.145-175.

-----et al., 1984. Ocean margin drilling program, regional atlas, western North American continental margin and adjacent ocean floor of Oregon and Washington. Marine Science International, Woods Hole, Mass.

Larson, Charlie, and Larson, Jo., 1987. Lava River Cave. Vancouver, Wash., ABC Printing, 24p.

Lawrence, Robert D., 1976. Strike-slip faulting terminates the Basin and Range province in Oregon. Geol. Soc. Amer., v.87, pp.846-850.

-----1980. Marys Peak field trip: structure of the eastern flank of the central Coast Range, Oregon. Oregon Dept. Geology and Mineral Indus., Bull.101, pp.121-167.

Leithold, Elana L., 1984. Characteristics of coarse-grained sequences deposited in nearshore, wave-dominated environments-examples from the Miocene of south-central Oregon. Sedimentology, v.31, pp.749-775.

-----and Bourgeois, Joanne, 1983. Sedimentology of the sandstone of Floras Lake (Miocene) transgressive, high-energy shelf deposition, SW Oregon. In: Larue, D.K., et al., eds., Cenozoic marine sedimentation, Pacific margin, U.S.A. Soc. Econ. Paleont. and Mineralogists, Pacific Section, pp.17-28.

Lentz, Rodney T., 1981. The petrology and stratigraphy of the Portland Hills Silt - a Pacific Northwest loess. Oregon Geology, v.43, no.1, pp.3-10.

Leppert, Dave, 1990. Developments in applications for southeast Oregon bentonites and natural zeolites. Oregon Dept. Geol. and Mineral Indus., Special Paper 23, pp.19-24.

Libbey, F.W., 1960. Boron in Alvord Valley, Harney County, Oregon. Ore Bin, v.22, no.10, pp.97-105.

-----1962. The Oregon King mine, Jefferson County, Oregon. Ore Bin, v.24, no.7, pp.101-115.

-----1967. The Almeda Mine Josephine County, Oregon. Oregon Dept. Geol. and Mineral Indus., Short Paper 24, 53p.

-----1976. Lest we forget. Ore Bin, v.38, no.12, pp.179-195.

Lillie, Robert J., and Couch, Richard W., 1979. Geophysical evidence of fault termination of the Basin and Range province in the vicinity of the Vale, Oregon, geothermal area. In: Newman, Gary W., et al., eds., Basin and Range symposium and Great Basin field conference, pp.175-184.

Lindgren, W., 1901. The gold belt of the Blue Mountains of Oregon. U.S. Geol. Survey, 22 Ann. Rept., pt.2, pp.551-776.

Lindsley-Griffin, Nancy and Kramer, J. Curtis, eds., 1977. Geology of the Klamath Mountains, northern California. Geol. Soc. Amer., Cordilleran Sect., 73rd Ann. Mtg. [fieldtrip guide], 156p.

Loam, Jayson, 1980. Hot springs and pools of the northwest. Santa Barbara, Capra Press, 159p.

Long, Michael T., and Leverton, Mark A., 1984. Pleistocene interglacial volcanism: Upper Salmon Creek drainage, Lane County, Oregon. Oregon Geology, v.46, no.11, pp.131-138.

Lowry, W.D., and Baldwin, E.M., 1952. Late Cenozoic geology of the lower Columbia River valley, Oregon and Washington. Geological Soc. Amer., Bull., v.63, p.1-24.

Luedke, Robert G., and Smith, Robert L., 1982. Map showing distribution, composition, and age of late Cenozoic volcanic centers in Oregon and Washington. U.S. Geol. Survey, Misc. Inv. Ser., I-1091-D.

-----Smith, Ribert L., and Russell-Robinson, Susan L., 1983. Map showing distribution, composition, and age of late Cenozoic volcanoes and volcanic rocks of the Cascade Range and vicinity, northwestern United States. U.S. Geol. Survey, Misc. Inv. Ser., I-1507.

Lund, Ernest H., 1971. Coastal landforms between Florence and Yachats, Oregon. Ore Bin, v.33, no.2, pp.21-44.

-----1972. Coastal landforms between Tillamook Bay and the Columbia River, Oregon. Ore Bin, v.34, no.11, pp.173-194.

-----1972. Coastal landforms between Yachats and Newport, Oregon. Ore Bin, v.34, no.5, pp.73-91.

-----1973. Landforms along the coast of southern Coos County, Oregon. Ore Bin, v.35, no.12, pp.189-210.

-----1973. Oregon coastal dunes between Coos Bay and Sea Lion Point. Ore Bin, v.35, no.5, pp.73-92.

-----1974. Coastal landforms between Roads End and Tillamook Bay, Oregon. Ore Bin, v.36, no.11, pp.173-194.

-----1974. Rock units and coastal landforms between Newport and Lincoln City, Oregon. Ore Bin, v.36, no.5, pp.69-90.

-----1977. Geology and hydrology of the Lost Creek glacial trough. Ore Bin, v.39, no.9, pp.141-156.

-----and Bentley, Elton, 1976. Steens Mountain, Oregon. Ore Bin, v.38, no.4, pp.51-66.

Lupton, J.E., 1987. Excess 3He in Crater Lake, Oregon: evidence for geothermal input. EOS, v.68, no.50, p.1722.

Lux, Daniel R., 1983. K-Ar and 40 Ar-39 Ar ages of mid-Tertiary volcanic rocks from the Western Cascade Range, Oregon. Isochron/West, no.33, p.27-32.

Mack, Gregory Stebbins, 1982. Geology, ground water chemistry, and hydrogeology of the Murphy area, Josephine County, Oregon. Ms., Oregon State Univ.

MacLeod, N.S., 1982. Newberry Volcano, Oregon: a Cascade range geothermal prospect. Oregon Geology, v.44, no.11, pp. 123-131.

Madin, Ian P., 1989. Evaluating earthquake hazards in the Portland, Oregon, metropolitan area: mapping potentially hazardous soils. Oregon Geology, v.51, no.5, pp.106-110, 118.

Magill, James and Cox, Allan, 1980. Tectonic rotation of the Oregon Western Cascades. Oregon, Dept. Geol. and Mineral Indus., Special Paper 10, 67p.

-----and Cox, Allan, 1981. Post-Oligocene tectonic rotation of the Oregon Western Cascade Range and the Klamath Mountains. Geology, v.9, pp.127-131.

Mahoney, Kathleen A., and Steere, M.L., 1983. Index to the Ore Bin (1939-1978) and Oregon Geology (1979-1982). Oregon

Dept. Geol. and Mineral Indus., Special Paper 16, 46p.

Mamay, S.H., and Read, C.B., 1956. Additions to the flora of the Spotted Ridge Formation in central Oregon. U.S. Geol. Survey, Prof. Paper 274-I, pp.211-226.

Mangan, Margaret T., et al., 1986. Regional correlation of Grande Ronde basalt flows, Columbia River Basalt group, Washington, Oregon, and Idaho. Geol. Soc. Amer., Bull. 97, pp.1300-1318.

Mangum, Doris, 1967. Geology of Cape Lookout State Park, near Tillamook, Oregon. Ore Bin, v.29, no.5, pp.85-109.

Mankinen, Edward A., Irwin, W.P., and Sherman, G., 1984. Implications of paleomagnetism for the tectonic hisotry of the eastern Klamath and realted terranes in California and Oregon. In: Nilsen, Tor, ed., Geology of the upper Cretaceous Hornbrook Formation, Oregon and California. Soc. Econ. Paleont. and Mineralogists, Pacific Sect., v.42, pp.221-229.

-----et al., 1987. The Steens Mountain (Oregon) geomagnetic polarity transition; 3, Its regional significance. Jour. Geophys. Res., v.92, no.B8, pp.8057-8076.

Marsh, S.P., Kropschot, S.J., and Dickinson, R.G., 1984. Wilderness mineral potential; assessment of mineral-resource potential in U.S. Forest Service lands studied 1964-1984. U.S. Geol. Survey, Prof. Paper 1300, v.2, 1183p.

Mason, Ralph S., 1952. Semiprecious gems in Oregon. Ore Bin, v.14, no.7, pp.45-46.

-----1956. Eden Ridge coal to be investigated. Ore Bin, v.18, no.8, pp.67-69.

-----1982. Finding formations fit for firing. Oregon Geology, v.44, no.7, pp.79-81.

Maxime, Misseri, and Boudier, Francoise, 1985. Structures in the Canyon Mountain ophiolite indicate an island-arc intrusion. Tectonophysics, v.120, pp.191-209.

Maynard, L.C., 1974. Geology of Mt. McLoughlin, Oregon. Univ. Oregon, Ms., 139p.

McArthur, Lewis A., 1982. Oregon Geographic Names. 5th ed. rev. and enl. Portland, Oregon Historical Society, 839p.

McBirney, Alexander R., 1968. Compositional variations of the climactic eruption of Mount Mazama. In: Dole, Hollis, Andesite conference guidebook - international mantle project; science report 16-S. Oregon, Dept. Geol. and Mineral Indus., Bull.62, pp.53-56.

-----1978. Volcanic evolution of the Cascade Range. Ann. Rev. Earth Planet. Sci., v.6, pp.437-456.

-----et al., 1974. Episodic volcanism in the central Oregon Cascade Range, Geology, v.2, pp.585-590.

McBride, James Noland, 1976. Science at a land grant college: the science controversy in Oregon, 1931-1942, and the early development of the college of science at Oregon State University. Ms., Oregon State Univ., 208p.

McChesney, Stephen M., 1987. Geology of the Sultana vein, Bohemia mining district, Oregon. Ms., Univ. of Oregon, 160p.

McCornack, Ellen Condon, 1928. Thomas Condon, pioneer geologist of Oregon. Eugene, Univ. of Oregon Press, 255p.

McDougall, Ian, 1976. Geochemistry and origin of basalt of the Columbia River group, Oregon and Washington. Geol. Soc. Amer., Bull.87, pp.777-792.

McDougall, Kristin A., 1980. Paleoecological evaluation of late Eocene biostratigraphic zonations of the Pacific Coast of North America. Jour. Paleo., v.54, no.4, Suppl., 75p.

McDowell, Patricia F., and Roberts, Michael C., 1987. Field guidebook to the Quaternary stratigraphy, geomorphology,

and soils of the Willamette Valley. Assoc. Amer. Geogr. 1987 Ann. Mtg., Portland, Ore., 75p.

McFadden, Jerome J., 1986. Fossil flora near Gray Butte, Jefferson County, Oregon. Oregon Geology, v.48, pp.51-55.

McInelly, Galan W., and Kelsey, Harvey M., 1990. Late Quaternary tectonic deformation in the Cape Arago-Bandon region of coastal Oregon as deduced from wave-cut platforms. Jour. Geophys. Res., v.95, no.B5, pp.6699-6713.

McKee, Bates, 1972. Cascadia, the geologic evolution of the Pacific Northwest. McGraw-Hill, New York, 394p.

McKee, Edwin H., Duffield, Wendell A., and Stern, Robert J., 1983. Late Miocene and early Pliocene basaltic rocks and their implications for crustal structure, northeastern California and southcentral Oregon. Geol. Soc. Amer., Bull., v.94, pp.292-304.

McKeel, Daniel R., 1983. Subsurface biostratigraphy of the east Nehalem Basin, Columbia County, Oregon. Oregon, Dept. Geol. and Mineral Indus., Oil and Gas Inv.9, 33p.

-----1984. Biostratigraphy of exploratory wells in western Coos, Douglas, and Lane counties, Oregon. Oregon Dept. Geol. and Mineral Indus., Oil and Gas Inv. 11, 19p.

-----1984. Biostratigraphy of exploratory wells, northern Willamette Basin, Oregon. Oregon Dept. Geol. and Mineral Indus., Oil and Gas Inv. 12, 19p.

-----1985. Biostratigraphy of exploratory wells, southern Willamette Basin, Oregon. Oregon Dept. Geol. and Mineral Indus., Oil and Gas Inv. 13, 17p.

-----and Lipps, Jere H., 1975. Eocene and Oligocene planktonic foraminifera from the central and southern Oregon Coast Range. Jour. Foram. Res., v.5, no.4, pp.249-269.

McMullen, D.E., 1975. Oregon under foot. Oregon Mus. Science and Indus., OMSI Press Guide, 60p.

McWilliams, Robert G., 1980. Eocene correlations in western Oregon-Washington. Oregon Geology, v.42, no.9, pp.151-157.

McKnight, Brian K., 1984. Stratigraphy and sedimentology of the Payne Cliffs Formation, southwestern Oregon. In: Nilsen, Tor, ed., Geology of the Upper Cretaceous Hornbrook Formation, Oregon and California. Soc. Econ. Paleont. and Mineralogists, Pacific Sect., v.42, pp.187-194.

Meinzer, Oscar E., 1922. Map of the Pleistocene lakes of the Basin-and-Range province and its significance. Geol. Soc. Amer., Bull., v.33, pp.541-552.

Merriam, C.W., and Berthiaume, S.A., 1943. Late Paleozoic formations of central Oregon. Geol. Soc. Amer., Bull. v.54, pp.145-171.

Merrill, George P., 1924. First hundred years of American geology. New Haven, Yale Univ. Press, 773p.

Michaelson, Caryl A., and Weaver, Craig S., 1986. Upper mantle structure from teleseismic P wave arrivals in Washington and northern Oregon. Jour. Geophys. Res., B91, no.2, pp.2077-2094.

Miles, Gregory A., 1981. Planktonic foraminifers of the lower Tertiary Roseburg, Lookingglass, and Flournoy formations (Umpqua group), southwest Oregon. Geol. Soc. Amer., Special Paper 184, pp.85-103.

Miller, M. Meghan, 1987. Dispersed remnants of a northeast Pacific fringing arc: upper Paleozoic terranes of Permian McCloud faunal affinity, western U.S. Tectonics, v.6, no.6, pp.807-930.

Miller, Paul R., 1984. Mid-Tertiary stratigraphy, petrology, and paleogeography of Oregon's central Western Cascade Range. Ms., Univ. of Oregon, 187p.

-----and Orr, William N., 1986. The Scotts Mills Formation: mid-

Tertiary geologic history and paleogeography of the central Western Cascade range, Oregon. Oregon Geology, v.48, no.12, pp.139-151p.

-----and Orr, William N., 1988. Mid-Tertiary transgressive rocky coast sedimentation: central Western Cascade range, Oregon. Jour. Sed. Pet., v.58, no.6, pp.959-968.

Minor, Scott A., et al., 1987. Mineral resources of the High Steens and Little Blitzen Gorge wilderness study areas, Harney County, Oregon. U.S. Geol. Survey, Bull.1740-A, pp.A1-A21.

Mitchell, Edward, 1966. Faunal succession of extinct North Pacific marine mammals. Norsk Hvalfangst-Tidende, no.3, pp.47-60.

Molenaar, C.M., 1985. Depositional relations of Umpqua and Tyee formations (Eocene), southwestern Oregon. Amer. Assoc. of Petroleum Geol., Bull., v.69, pp.1217-1229.

Mooney, Walter D., and Weaver, Craig S., 1989. Regional crustal structure and tectonics of the Pacific coastal states; California, Oregon, and Washington. In: Pakiser, L.C., and Mooney, W.D., 1989. Geophysical framework of the continental United States. Geol. Soc. Amer., Mem. 172, pp.129-161.

Moore, Bernard N., 1937. Nonmetallic mineral resources of eastern Oregon. U.S. Geol. Survey, Bull.875, 180p.

Moore, E.J., 1963. Miocene marine mollusks from the Astoria Formation in Oregon. U.S. Geol. Survey Prof. Paper 419, 109p.

-----1976. Oligocene marine mollusks from the Pittsburgh Bluff Formation in Oregon. U.S. Geol. Surv., Prof. Paper 922, 66p.

Moore, George W., 1991. Tectonic terranes. Geotimes, February, pp.45-46.

Mortimer, N., 1985. Structural and metamorphic aspects of middle Jurassic terrane juxtaposition, northeastern Klamath Mountains, California. In: Howell, D. G., ed., Techtonostratigraphic terranes of the circum-Pacific region, Circum-Pacific Council for Energy and Mineral Resources, Earth Sci., Series, no.l, pp.201-214.

-----and Coleman, R.G., 1984. A Neogene structural dome in Klamath Mountains, California and Oregon. In: Nilsen, Tor, ed., Geology of the upper Cretaceous Hornbrook Formation, Oregon and California. Soc. Econ. Paleont. and Mineralog., Pacific Sect., v. 42, pp.179-186.

Muhs, Daniel R., et al., 1990. Age estimates and uplift rates for late Pleistocene marine terraces: southern Oregon portion of the Cascadia forearc. Jour. Geophys. Res., v.95, no.B5, pp.6685-6698.

Mullen, Ellen D., 1983. Paleozoic and Triassic terranes of the Blue Mountains, northeast Oregon: Discussion and field trip guide. Part I. A new consideration of old problems. Oregon Geology, v.45, no.6, pp.65-68.

-----1983. Paleozoic and Triassic terranes of the Blue Mountains, northeast Oregon: discussion and field trip guide. Part II: Road log and commentary. Oregon Geol., v.45,no.7/8, pp.75-82.

Munts, Steven Rowe, 1973. Platinum in Oregon. Ore Bin, v.35, no.9, pp.141-152.

-----1978. Geology and mineral deposits of the Quartzville mining district, Linn County, Oregon. M.S., Univ. of Oregon, 213p.

-----1981. Geology and mineral deposits of the Quartzville mining district, Linn County, Oregon. Oregon Geology, v.43, no.2, p.18.

Neal, Victor T., Neshyba, Stephen J., and Denner, Warren W., 1972. Vertical temperature structure in Crater Lake, Oregon.

Limnology and Oceanogr., v.17, pp.451-454.

Nelson, C. Hans, 1984. The Astoria fan: an elongate type fan. Geo-Marine Letters, v.3, pp.65-70.

-----et al., 1970. Development of the Astoria Canyon-fan physiography and comparison with similar systems. Marine Geology, v.8, pp.259-291.

-----et al., 1980. Sedimentary history of the floor of Crater Lake, Oregon; restricted basin with non-channelized turbidites. Geol. Soc. Amer., Abstr. with Prog., v.12, no.7, p.491.

Nelson, Dennis O., and Shearer, Gerald B., 1969. The geology of Cedar Butte northern Coast Range of Oregon. Ore Bin, v.31, no.6, pp.113-130.

Neuendorf, Klaus K.E., ed., 1978 (GeoRef, comp.) Bibliography of the geology and mineral resources of Oregon (sixth supplement), January 1, 1971 to December 31, 1975. Oregon Dept. Geol. and Mineral Indus., Bull.97, 74p.

-----1981. (GeoRef, comp.) Bibliography of the geology and mineral resources of Oregon (seventh supplement) January 1, 1976 to December 31, 1979. Oregon Dept. Geol. and Mineral Indus., Bull.102, 68p.

-----1987. (GeoRef, comp.) Bibliography of the geology and mineral resources of Oregon (eight supplement), January 1, 1980 to December 31, 1984. Oregon Dept. Geol. and Mineral Indus., Bull.103, 176p.

Nevins, Allan, 1955. Fremont; pathmarker of the west. New York, Longmans, 689p.

Newcomb, R.C., 1966. Lithology and eastward extension of The Dalles Formation, Oregon and Washington. U.S. Geol. Survey, Prof. Paper 550-D, pp.D59-D63.

-----1967. The Dalles-Umatilla syncline, Oregon and Washington. U.S. Geol. Survey, Prof. Paper 575 B, pp.B88-B93.

Newton, Cathryn R., 1990. Significance of "Tethyan" fossils in the American Cordillera. Science, v.249, pp.385-390.

Newton, Vernon C., 1969. Subsurface geology of the lower Columbia and Willamette basins, Oregon. Oregon Dept. of Geol. and Mineral Indus., Oil and Gas Inv.2, 121p.

-----1979. Oregon's first gas wells completed. Oregon Geology, v.41, no.6, pp.89-90.

-----and Van Atta, Robert O., 1976. Prospects for natural gas production and underground storage of pipe-line gas in the upper Nehalem River basin Columbia-Clatsop counties, Oregon. Oregon Dept. Geol. and Mineral Indus., Oil and Gas Investig.5, 56p.

-----et al., 1980. Prospects for oil and gas in the Coos basin, western Coos, Douglas, and Lane counties, Oregon. Oregon Dept. Geol. and Mineral Indus., Oil and Gas Inv.6, 74p.

Niem, Alan R., 1974. Wright's Point, Harney County, Oregon. An example of inverted topography. Ore Bin, v.36, no.3, pp.33-49.

-----1975. Geology of Hug Point state park northern Oregon coast. Ore Bin, v.37, no.2, pp.17-36.

-----and Niem, Wendy A., 1985. Oil and gas investigations of the Astoria basin, Clatsop and northernmost Tillamook counties, northwest Oregon. Oregon Dept. Geol. and Mineral Indus., OGI-14.

-----and Niem, Wendy A., 1990. Geology and oil, gas, and coal resources, southern Tyee basin, southern Coast Range, Oregon. Oregon Dept. Geol. and Mineral Indus., Open-file Report O-89-3, various pagings.

-----and Van Atta, Robert O., 1973. Ceonozoic stratigraphy of northwestern Oregon and adjacent southwestern Washington. Oregon Dept. Geol. and Mineral Indus., Bull.77, pp.75-132.

-----Snavely, Parke D., and Niem, Wendy A., 1990. Onshore-offshore geologic cross section from the Mist gas field, northern Oregon Coast Range, to the northwest Oregon continental shelf. Oregon Dept. Geol. and Mineral Indus., Oil and Gas Inv.17, 46p.

-----et al., 1985. Correlation of exploration wells, Astoria basin, N.W. Oregon. Oregon Dept. of Geology and Mineral Indus., Oil and Gas Inv.14, 8p.

Nilsen, Tor H., ed., 1984. Geology of the upper Cretaceous Hornbrook Formation, Oregon and California. Soc. Econ. Paleont. and Mineral., Pacific Sect., v.42, pp.239-257.

-----1984. Stratigraphy, sedimentology, and tectonic framework of the upper Cretaceous Hornbrook Formation, Oregon and California. In: Nilsen, Tor H., ed., Geology of the upper Cretaceous Hornbrook Formation, Oregon and California. Soc. Econ. Paleont. and Mineralog., Pacific Sect., v.42, pp.51-88.

Noblett, Jeffrey B., 1981. Subduction-related origin of the volcanic rocks of the Eocene Clarno Formation near Cherry Creek, Oregon. Oregon Geol., v.43, no.7, pp.91-99.

Nolf, Bruce, 1966. Broken Top breaks: flood released by erosion of glacial moraine. Ore Bin, v.28, no.10, pp.182-188.

North, William B., and Byrne, John V., 1965. Coastal landslides of northern Oregon. Ore Bin, v.27, no.11, pp.217-241.

Noson, Linda L., Quamar, Anthony, and Thorsen, Gerald W., 1988. Washington State earthquake hazards. Wash. Div. Geol. and Earth Resources, Info. Circular 85, 77p.

Nur, Amos, 1983. Accreted terranes. Reviews of Geophys. and Space Physics, v.21, no.8, pp.1779-1785.

Obermiller, Walter A., 1987. Geologic, structural and geo-chemical features of basaltic and rhyolitic volcanic rocks of the Smith Rock/Gray Butte area, central Oregon. Ms., Univ. of Oregon, 189p.

Oles, Keith F., and Enlows, Harold E., 1971. Bedrock geology of the Mitchell Quadrangle, Oregon. Oregon Dept. Geol. and Mineral Indus., Bull.72, 62p.

-----et al., 1980. Geologic field trips in western Oregon and southwestern Washington. Oregon Dept. Geol. and Mineral Indus., Bull.101, 232p.

Olmstead, D.L., 1988. Hydrocarbon exploration and occurrences in Oregon. Oregon Dept. Geol. and Mineral Indus., Oil and Gas Inv. 15, 78p.

Oregon metal mines handbook, 1951. Oregon Dept. Geol. and Mineral Indus., Bull.14-D, 166p.

Orr, Elizabeth, and Orr, William, 1984. Bibliography of Oregon paleontology, 1792-1983. Oregon Dept. Geol. and Mineral Indus., Special Paper 17, 82p.

Orr, William N., 1986. A Norian (late Triassic) ichthyosaur from the Martin Bridge Limestone, Wallowa Mountains, Oregon. U.S. Geol. Survey, Prof. Paper 1435, pp.41-47.

-----and Katsura, Kurt T., 1985. Oregon's oldest vertebrates [Ichthyosauria (Reptilia)]. Oregon Geology, v.47, no.7, pp.75-77.

-----and Orr, Elizabeth, 1981. Handbook of Oregon plant and animal fossils, Eugene, Ore., 285p.

Otto, Bruce R., and Hutchinson, Dana A., 1977. The geology of Jordan Craters, Malheur County, Oregon. Ore Bin, v.39, no.8, pp.125-140.

Parker, Garald G., Shown, Lynn M., and Ratzlaff, Karl W., 1964. Officer's Cave, a pseudokarst feature in altered tuff and volcanic ash of the John Day Formation in eastern Oregon. Geol. Soc. Amer., Bull. v.75, pp.393-402.

Parmenter, Tish, and Bailey, Robert, 1985. The Oregon ocean-book; an introduction to the Pacific Ocean...off Oregon.

Oregon Dept. of Land Conservation and Development, 85p.

Parsons, Roger B., 1969. Geomorphology of the Lake Oswego area, Oregon. Ore Bin, v.31, no.9, pp.186-192.

-----1969. Late Pleistocene stratigraphy, southern Willamette Valley, Oregon. Northwest Sci., v.43, no.3, pp.116-123.

Patterson, Peter V., 1969. Sawtooth Ridge: a northeast Oregon volcanic crater. Ore Bin, v.31, no.9, pp.173-183.

Peck, Dallas L., et al., 1964. Geology of the central and northern parts of the Western Cascade Range in Oregon. U.S. Geol. Survey, Prof. Paper 449, 56p.

Pegasus Gold Corporation, 1990. Heap leach technology workshop, Portland, Oregon, July 9, July 16, 1990. Presented by Dirk Van Zyl and William M. Schafer. Various pagings.

Perttu, R.K., and Benson, G.T., 1980. Deposition and deformation of the Eocene Umpqua group, Sutherlin area, southwestern Oregon. Oregon Geology, v.42, no.8, pp. 135-140,146.

Pessagno, Emile A., and Blome Charles D., 1986. Faunal affinities and tectonogenesis of Mesozoic rocks in the Blue Mountains province of eastern Oregon and western Idaho. U.S. Geol. Survey, Prof. Paper 1435, pp.65-78.

-----1990. Implications of new Jurassic stratigraphic, geochron-ometric, and paleo-latitudinal data from the western Klamath terrane (Smith River and Rogue Valley sub-terranes). Geology, v.18, pp.665-668.

Peterson, Carolyn P., Kulm, LaVerne D., and Gray, Jerry J., 1986. Geologic map of the ocean floor off Oregon and the adjacent continental margin. Oregon Dept. Geol. and Mineral Indus., Geological Map Series, GMS-42.

Peterson, Curt D., and Darienzo, M.E., 1988. Coastal neotectonic field trip guide for Netarts Bay, Oregon. Oregon Geology, v.50, no.9/10, pp.99-106,117.

-----Gleeson, George W., and Wetzel, Nick, 1987. Stratigraphic development, mineral sources and preservation of marine placers from Pleistocene terraces in southern Oregon, U.S.A. Sedimentary Geology, v.53, pp.203-229.

-----Scheidegger, Kenneth F., and Schrader, Hans J., 1984. Holocene depositional evolution of a small active-margin estuary of the northwestern United States. Marine Geology, v.59, pp.51-83.

Peterson, Norman V., 1959. Lake County's new continuous geyser. Ore Bin, v.21, no.9, pp.83-88.

-----1962. Geology of Collier State Park area, Klamath County, Oregon. Ore Bin, v. 24, no.6, pp.88-97.

-----1972. Oregon "sunstones". Ore Bin, v.34, no.12, pp.197-215.

-----and Groh, Edward A., 1961. Hole-in-the-Ground. Ore Bin, v.23, no.10, pp.95-100.

-----and Groh, Edward A., 1963. Maars of south-central Oregon. Ore Bin, v.25, no.5, pp.73-88.

-----and Groh, Edward A., 1963. Recent volcanic landforms in central Oregon. Ore Bin, v.25, no.3, pp.33-34.

-----and Groh, Edward A., 1964. Crack-in-the-Ground, Lake County, Oregon. Ore Bin, v.26, no.9, pp.158-166.

-----and Groh, Edward A., 1964. Diamond Craters, Oregon. Ore Bin, v.26, no.2, pp.17-34.

-----and Groh, Edward A., 1969. The ages of some Holocene volcanic eruptions in the Newberry Volcano area, Oregon. Ore Bin, v.31, no.4, pp.73-87.

-----and Groh, Edward A., 1970. Geologic tour of Cove Palisades State Park near Madras, Oregon. Ore Bin, v.32, no.8, pp. 141-168.

-----and Groh, Edward A., 1972. Geology and origin of the

Metolius Springs Jefferson County, Oregon. Ore Bin, v.34, no.3, pp.41-51.

-----and Ramp, Len, 1978. Soapstone industry in southwest Oregon. Ore Bin, v.40, no.9, pp.149-157.

-----et al., 1976. Geology and mineral resources of Deschutes County Oregon. Oregon Dept. Geol. and Mineral Indus., Bull.89, 66p.

Pfaff, Virginia J., and Beeson, Marvin H., 1989. Miocene basalt near Astoria, Oregon; geophysical evidence for Columbia Plateau origin. In: Reidel, S.P., and Hooper, P.R., eds., Volcanism and tectonism in the Columbia River flood-basalt province. Geol. Soc. Amer., Special Paper 239, pp.143-156.

Phelps, David W., 1978. Petrology, geochemistry, and structural geology of Mesozoic rocks in the Sparta Quadrangle and Oxbow and Brownlee Reservoir areas, eastern Oregon and western Idaho. Rice Univ., PhD., 229p.

-----and Ave Lallemant, Hans G., 1980. The Sparta ophiolite complex, northeast Oregon: a plutonic equivalent to low K20 island-arc volcanism. Amer. Jour. Sci., v.280-A, pp.345-358.

Piper, A.M., Robinson, T.W., and Park, C.F., jr., 1939. Geology and ground-water resources of the Harney Basin, Oregon. U.S. Geol. Survey, Water Supply Paper 841, 189p.

Pollock, J. Michael., 1985. Geology and geochemistry of hydro-thermal alteration, eastern portion of the North Santiam mining area. M.S., Portland State Univ.

-----and Cummings, Michael L., 1985. North Santiam mining area, Western Cascades - relations between alteration and volcanic stratigraphy: discussion and field trip guide. Oregon Geology, v.47, no.12, pp.139-145.

Popenoe, W.P., Imlay, R.W., and Murphy, M.A., 1960. Correlation of the Cretaceous formations of the Pacific Coast (United States and northwestern Mexico). Geol. Soc. Amer., Bull., v.71, pp.1491-1540.

Porter, Stephen C., Pierce, Kenneth L., and Hamilton, Thomas D., 1983. Late Wisconsin mountain glaciation in the western United States. In: Wright, H.E., ed. Late Quaternary environments of the U.S., Minneapolis, Univ. Minnesota Press, v.1, pp.71-108.

Potter, Miles F., and McCall, Harold, 1978. The golden years of eastern Oregon. Ore Bin, v.40, no.4, pp.57-64.

Power, Sara Glen, 1984. The "Tops" of porphyry copper deposits -Mineralization and plutonism in the Western Cascades, Oregon. PhD., Oregon State Univ., 234p.

Priest, George R., 1990. Geothermal exploration in Oregon, 1990. Oregon Geol., v.52, no.3, pp.81-86.

-----1990. Volcanic and tectonic evolution of the Cascade volcanic arc, central Oregon. Jour. of Geophys. Res., v.95, no.B12, pp.19,583-19,599.

-----and Vogt, Beverly F., 1983. Geology and geothermal resources of the central Oregon Cascade Range. Oregon Dept. Geol. and Mineral Indus., Special Paper 15, 123p.

-----et al., 1983. Overview of the geology of the central Oregon Cascade Range. In: Priest, George R., and Vogt, B.F., Geology and geothermal resources of the central Oregon Cascade Range. Oregon Dept. Geol. and Mineral Indus., Special Paper 15, pp.3-28.

Purcell, David, 1978. Guide to the lava tube caves of central Oregon. Corvallis, Ore., High Mountain Press, 55p.

Purdom, William B., 1966. Fulgurites from Mount Thielsen, Oregon. Ore Bin, v.28, no.9, pp.153-159.

Quamar, Anthony, et al., 1986. Earthquake hypocenters in Washington and northern Oregon. Wash. Div. Geology and Earth Res., Info. Circular 82, 64p.

Ramp, Len, 1953. Structural data from the Chrome Ridge area Josephine County, Oregon. Ore Bin, v.18, no.3, pp.19-25.

-----1961. Chromite in southwestern Oregon. Oregon Dept. Geol. and Mineral Indus., Bull.52, 169p.

-----1962. Jones marble deposit, Josephine County, Oregon. Ore Bin, v.24, no.10, pp.153-158.

-----1975. Geology and mineral resources of Douglas County, Oregon. Oregon Dept. Geol. and Mineral Indus., Bull. 75, 106p.

-----1975. Geology and mineral resources of the upper Chetco drainage area, Oregon. Oregon Dept. Geol. and Mineral Indus., Bull.88, 195p.

-----1978. Investigations of nickel in Oregon. Oregon Dept. Geol. and Mineral Indus., Misc. Paper 20, 68p.

-----and Peterson, Norman V., 1979. Geology and mineral resources of Josephine County, Oregon. Oregon Dept. Geol. and Mineral Indus., Bull.100, 45p.

-----Schlicker, Herbert G., Gray Jerry J., 1977. Geology, mineral resources, and rock material of Curry County, Oregon. Oregon Dept. Geol. and Mineral Indus., Bull. 93, 79p.

Rand, Helen B., 1959. Gold and Oregon's settlement. Ore Bin, v.21, no.5, pp.41-47.

Rankin, David Karl., 1983. Holocene geologic history of the Clatsop Plains foredune ridge complex. Ms., Portland State Univ., 189p.

Reidel, Stephen P., and Hooper, Peter R., 1989. Volcanism and tectonism in the Columbia River flood-basalt province. In: Reidel, S.P., and Hooper, P.R., eds. Volcanism and tectonism in the Columbia River flood-basalt province. Geol. Soc. Amer., Special Paper 239, 386p.

-----et al., 1989. The Grande Ronde Basalt, Columbia River Basalt group; stratigraphic descriptions and correlations in Washington, Oregon, and Idaho. In: Reidel, S.P., and Hooper, P.R., eds. Volcanism and tectonism in the Columbia River flood-basalt province. Geol. Soc. Amer., Special Paper 239, pp.21-52.

Renne, Paul R., 1988. Structural chronology, oroclinal deformation and tectonic evolution of the southweastern Klamath Mountains, California. Tectonics, v.7, no.6, pp.1223-1242.

Reilinger, Robert, and Adams, J., 1982. Geodetic evidence for active landward tilting of the Oregon and Washington coastal ranges. Geophys. Res. Letters, v.9, no.4, pp. 402-403.

Retallack, Gregory J., 1981. Preliminary observations on fossil soils in the Clarno Formation (Eocene to early Oligocene) near Clarno, Oregon. Oregon Geology, v.43, no.11, pp.147-150.

-----1991. A field guide to mid-Tertiary paleosols and paleoclimatic changes in the high desert of central Oregon. Part 1. Oregon Geology, v.53, no.3, pp.51-59.

-----1991. A field guide to mid-Tertiary paleosols and paleo-climatic changes in the high desert of central Oregon. Part 2. Oregon Geol., v.53, no.4, pp.75-80.

Ritchie, Beatrice, 1987. Tholeiitic lavas from the Western Cascade Range, Oregon. Ms., Univ. of Oregon, 94p.

Roberts, Michael C., 1984. The late Cenozoic history of an alluvial fill: the southern Willamette Valley, Oregon. In: Mahaney, W.C., ed. Correlation of quaternary chronologies. Geo Books, Norwich, England, pp.491-503.

-----and Whitehead, Donald R., 1984. The palynology of a non-marine Neogene deposit in the Willamette Valley, Oregon. Rev. of Palaeobotany and Palynology, v.41, pp.1-12.

Roberts, Miriam S., comp., 1970. Bibliography of the geology and mineral resources of Oregon (fourth supplement), January 1, 1956 to December 31, 1960. Oregon Dept. Geol. and Mineral Indus., Bull.67, 88p.

-----Steere, Margaret L., and Brookhyser, Caroline S., comp., 1973. Bibliography of the geology and mineral resources of Oregon (fifth supplement). January 1, 1961 to December 31, 1970. Oregon Dept. Geol. and Mineral Indus., Bull.78, 198p.

Robertson, Richard D., 1982. Subsurface stratigraphic correlations of the Eocene Coaledo Formation, Coos Bay basin, Oregon. Oregon Geology, v.44, no.7, pp.75-78,82.

Robinson, Paul T., 1987. John Day Fossil Beds National Monument, Oregon: Painted Hills unit. Geol. Soc. Amer., Cordilleran Sect. Centennial Field Guide, pp.317-310.

-----Brem, Gerald F., and McKee, Edwin H., 1984. John Day Formation of Oregon: A distal record of early Cascade volcanism. Geology, v.12, p.229-232.

-----Walker, George W., and McKee, Edwin H., 1990. Eocene(?), Oligocene, and lower Miocene rocks of the Blue Mountains region. U.S. Geol. Survey, Prof. Paper 1437, pp.29-61.

Robyn, Thomas L., and Hoover, James D., 1982. Late Cenozoic deformation and volcanism in the Blue Mountains of central Oregon: microplate interactions? Geology (Boulder), v.10, no.11, pp.572-576.

Roche, Richard Louis, 1987. Stratigraphic and geochemical evolution of the Glass Buttes complex, Oregon. Ms., Portland State Univ., 99p.

Rodgers, David W., Hackett, William R., and Ore, H. Thomas, 1990. Extension of the Yellowstone plateau, eastern Snake River plain, and Owyhee plateau. Geology, v.18, pp.1138-1141.

Rogers, John J.W., and Novitsky-Evans, Joyce M., 1977. The Clarno Formation of central Oregon, U.S.A. - volcanism on a thin continental margin. Earth and Planet. Sci. Letters 34, pp.56-66.

-----and Ragland, Paul C., 1980. Trace elements in continental-margin magmatism: Part I. Trace elements in the Clarno Formation of central Oregon and the nature of the continental margin on which eruption occurred: summary. Geol. Soc. Amer., Bull., Part I, v.91, pp.196-198.

Rollins, Anthony, 1976. Geology of the Bachelor Mountain, area, Linn and Marion counties, Oregon. Ms., Oregon State Univ., 83p.

Ross, C.P., 1938. The geology of part of the Wallowa Mountains. Oregon Dept. Geol. and Mineral Indus., Bull.3, 74p.

Ross, Martin E., 1980. Tectonic controls of topographic development within Columbia River basalts in a portion of the Grande Ronde River-Blue Mountains region, Oregon and Washington. Oregon Geol., v.42, no.10, pp.167-174.

Roure, Francois and Blanchet, Rene, 1983. A geological transect between the Klamath Mountains and the Pacific Ocean (southwestern Oregon): a model for paleosubductions. Tectonophysics, v.91, pp.53-72.

Ruddiman, W.F., and Wright, H.E., 1987. North America and adjacent oceans during the last deglaciation. Geol. Soc. Amer., The Geology of North America, v.K-3, 501p.

Russell, I.C., 1883. A geological reconnaissance in southern Oregon. U.S. Geol. Survey, 4th Ann. Rept., pp.431-464.

-----1905. Preliminary report on the geology and water-resources of central Oregon. U.S. Geol. Survey Bull.252, 138p.

Ryberg, Paul Thomas, 1978. Lithofacies and depositional environments of the Coaledo Formation, Coos County, Oregon. Ms., Univ. of Oregon, 159p.

-----1984. Sedimentation, structure and tectonics of the Umpqua Group (Paleocene to early Eocene) southwestern Oregon. PhD., Univ. of Arizona, 180p.

Rytuba, James J., 1989. Volcanism, extensional tectonics, and epithermal mineralization in the northern Basin and Range province, California, Nevada, Oregon, and Idaho. U.S. Geol. Survey, Circular 1035, pp.59-61.

-----and McKee, Edwin H., 1984. Perkaline ash flow tuffs and calderas of the McDermitt volcanic field, southeast Oregon and north central Nevada. Jour. Geophys. Res., v.89, no.B10, pp.8616-8628.

-----Glanzman, Richard K., and Conrad, W.K., 1979. Uranium thorium, and mercury distribution through the evolution of the McDermitt caldera complex. In: Newman, Gary W., et al., eds. Basin and Range Symposium and Great Basin field conference, Utah Geological Assoc., pp.405-412.

-----et al., 1990. Field guide to hot-spring gold deposits in the Lake Owyhee volcanic field, eastern Oregon. Geol. Soc. Nevada and United States Geological Survey 1990, Spring field trip guidebook; field trip no.10, 15p.

Saleeby, Jason B., et al., 1982. Time relations and structural-stratigraphic patterns in ophiolite accretion, west central Klamath Mountains, California. Jour. Geophys. Res., v.87, no. B5, pp.3831-3848.

Sanborn, E.I., 1937. The Comstock flora of west central Oregon; Eocene flora of western America. Carnegie Inst. Wash., Publ.465, pp.1-28.

Sarewitz, Daniel, 1983. Seven Devils terrane: is it really a piece of Wrangellia? Geology, v.11, pp.634-637.

Scharf, D.W., 1935. A Miocene mammalian fauna from Sucker Creek, southeastern Oregon. Carnegie Inst., Wash. Publ.435, pp.97-118.

Schermer, Elizabeth R., Howell, David G., and Jones, David L., 1984. The origin of allochthonous terranes: perspectives on the growth and shaping of continents. Ann. Rev. Earth Planet. Sci., v.12, pp.107-131.

Schlicker, Herbert G., 1961. Geology of the Ecola State Park landslide area, Oregon. Oregon Dept. Geol. and Mineral Indus., Ore Bin, v.23, no.9, pp.85-88.

-----and Deacon, Robert J., 1967. Engineering geology of the Tualatin Valley region, Oregon. Oregon Dept. Geol. and Mineral Indus., Bull.60, 103p.

-----and Finlayson, Christopher T., 1979. Geology and geologic hazards of northwestern Clackamas County, Oregon. Oregon Dept. Geol. and Mineral Indus., Bull.99, 79p.

-----Deacon, R.J., and Twelker, N.H., 1964. Earthquake geology of the Portland area, Oregon. Ore Bin, v.29, no.12, pp.206-230.

-----et al., 1973. Environmental geology of Lincoln County, Oregon. Oregon Dept. Geol. and Mineral Indus., Bull.81, 171p.

-----et al., 1974. Environmental geology of coastal Lane County Oregon. Oregon Dept. Geol. and Mineral Indus., Bull.85, 116p.

Schmela, Ronald J., and Palmer, Leonard A., 1972. Geologic analysis of the Portland Hills-Clackamas River alignment, Oregon. Ore Bin, v.34, no.6, pp.93-103.

Schneer, Cecil J., ed., 1979. Two hundred years of geology in America. Hanover, New Hampshire, Univ. of New Hampshire, 385p.

Scholl, David W., Grantz, Arthur, and Vedder, John G., eds., 1987. Geology and resource potential of the continental margin of western North America and adjacent ocean basins-Beaufort Sea to Baja California. Circum-Pacific Council for Energy and Mineral Res., Earth Sci. Ser., v.6, 799p.

Scholle, Peter, and Schluger, Paul., eds., 1979. Aspects of diagenesis. Soc. Econ. Paleont. and Mineralogists, Special Publ. 26, 443p.

Schroeder, Nanci A.M., Kulm, LaVerne D., and Muehlberg, Gary E., 1987. Carbonate chimneys on the outer continental shelf: evidence for fluid venting on the Oregon margin. Oregon Geology, v.49, no.8, pp.91-98.

Schuster, J. Eric, ed., 1987. Selected papers on the geology of Washington. Wash. Div. Geology and Earth Res., Bull.77, 395p.

Scott, William E., 1990. Geologic map of the Mount Bachelor volcanic chain and surrounding area, Cascade Range, Oregon. U.S. Geol. Survey Misc. Inv. Map I-1967.

-----1990. Temporal relations between eruptions of the Mount Bachelor volcanic chain and fluctuations of late Quaternary glaciers. Oregon Geology, v.52, no.5, pp.114-117.

-----and Mullineaux, D.R., 1981. Late Holocene eruptions, South Sister volcano, Oregon. U.S. Geol. Survey, Geological Research, 1981. p.217.

-----Gardner, Cynthia A., and Johnston, David A., 1990. Field trip guide to the central Oregon High Cascades. Part I: Mount Bachelor-South Sister area. Oregon Geology, v.52, no.5, pp.99-114.

Seiders, V.M., and Blome, C.D., 1987. Stratigraphy and sedimentology of Upper Cretaceous rocks in coastal southwest Oregon: evidence for wrench-fault tectonics in a postulated accretionary terrane: alternative interpretation and reply. Geol. Soc. Amer., Bull., 98, pp.739-744.

Shaffer, Leslie L.D., and Hashimoto, Steve T., 1969. The semi-precious gem industry of Oregon. Oregon Business Review, v.XXVII, no.7, pp.1-4.

Sheppard, R.A., 1986. Field trip guide to the Durkee zeolite deposit, Durkee, Oregon. Oregon Geology, v.48, no.ll, pp.127-132.

Sherrod, David R., and Smith, James G., 1990. Quaternary extrusion rates of the Cascade Range, northwestern United States and southern British Columbia. Jour. Geophys. Res., v.95, no.B12, pp.19,465-19,474.

Sibrava, W., et al., eds., 1986. Quaternary glaciations in the northern hemisphere. Quaternary Sci. Reviews 5, 510p.

Significant Oregon earthquakes in 1991. Oregon Geology, v.53, no.5, p.107.

Silberling, N.J., and Jones, D.L., eds., 1984. Lithotectonic terrane maps of the North American Cordillera. U.S. Geol. Survey, Open File Rept., 84-523. various pagings.

Silver, Eli A., 1978. Geophysical studies and tectonic development of the continental margin off the western United States, lat 34 to 48 N . In: Smith, R.B., and Eaton, G.P., eds. Ceonzoic tectonic and regional geophysics of the western Cordillera. Geol. Soc. Amer., Memoir 152, pp.251-262.

Simpson, Robert W., and Cox, Allan, 1977. Paleomagnetic evidence for tectonic rotation of the Oregon Coast range. Geology, v.5, pp.585-589.

Skinner, Craig, 1982. The Squaw Ridge lava field: new central Oregon territory for the desert-loving lava tube hunter. The Speleograph, August, pp.83-85.

Sliter, W.V., Jones, David L., and Throckmorton, C.K., 1984. Age and correlation of the Cretaceous Hornbrook For-mation, California and Oregon. In: Nilsen, Tor., ed., Geology of the upper Cretaceous Hornbrook Formation, Oregon and California. Soc. Econ. Paleont. and Mineral. Pacific Sect., v.42, pp.89-98.

Smith, Gary Allen, 1986. Simtustus Formation: paleogeographic and stratigraphic significance of a newly defined Miocene unit in the Deschutes basin, central Oregon. Oregon Geology, v.48, no.6, pp.63-72.

-----1986. Stratigraphy, sedimentology, and petrology of Neogene rocks in the Deschutes basin, central Oregon: a record of continental-margin volcanism and its influence on fluvial sedimentation in an arc-adjacent basin. Ph.D., Oregon State Univ., 467p.

-----1987. The influence of explosive volcanism on fluvial sedimentation: the Deschutes Formation (Neogene) in central Oregon. Jour. Sed. Pet., v.57, no.4, pp.613-629.

-----1991. A field guide to depositional processes and facies geometry of Neogene continental volcaniclastic rocks, Deschutes basin, central Oregon. Oregon Geology, v.53, no.1, pp.3-20.

-----Bjornstad, Bruce N., and Fecht, Karl R., 1989. Neogene terrestrial sedimentation on and adjacent to the Columbia Plateau; Washington, Oregon, and Idaho. In: Reidel, S.P., and Hooper, P.R., eds. Volcanism and tectonism in the Columbia River flood-basalt province. Geol. Soc. Amer., Special Paper 239, pp.187-197.

Smith Robert B., 1978. Seismicity, crustal structure, and intraplate tectonics of the interior of the western Cordillera. Geol. Soc. Amer., Memoir 152, pp.111-142.

-----and Eaton, G.P., eds. 1978. Cenozoic tectonic and regional geophysics of the western Cordillera. Geol. Soc. America, Memoir 152, 388p.

-----Snee, Lawrence W., and Taylor, Edward M., 1987. Stratigraphic, sedimentologic, and petrologic record of late Miocene subsidence of the central Oregon High Cascades. Geology, v.15, pp.389-392.

Smith, Warren Du Pre, and Allen, J.E., 1941. Geology and physiography of the northern Wallowa Mountains, Oregon. Oregon Dept. Geol. and Mineral Indus., Bull.12, 64p.

-----and Ruff, Lloyd L., 1938. The Geology and mineral resources of Lane County, Oregon. Oregon Dept. Geol. and Mineral Indus., Bull.11, 65p.

Snavely, Parke D., 1987. Depoe Bay, Oregon. In: Hill, Mason L., ed., Cordilleran Section of the Geological Society of America, Centennial Field Guide, pp.307-310.

-----1987. Tertiary geologic framework, neotectonics, and petroleum potential of the Oregon-Washington continental margin. In: Scholl, David W., Grantz, Arthur, and Vedder, John G., eds. Geology and resource potential of the continental margin of western North America and adjacent ocean basins-Beaufort Sea to Baja California. Circum-Pacific Council for Energy and Mineral Res., Earth Sci. Ser., v.6, pp.305-335.

-----and Baldwin, E.M., 1948. Siletz River Volcanic series, northwestern Oregon. Amer. Assoc. Petroleum Geol., v.32, no.5, pp.805-812.

-----and MacLeod, Norman S., 1971. Visitor's guide to the geology of the coastal area near Beverly Beach State Park, Oregon. Ore Bin, v.33, no.5, pp.85-106.

-----and Wagner, Holly C., 1963. Tertiary geologic history of western Oregon and Washington. Washington, Div. of Mines and Geol., Report of Inv.22, 25p.

-----and Wells, Ray E., 1991. Cenozoic evolution of the continental margin of Oregon and Washington. U.S. Geol. Survey, Open-File Report 91-441-B, 34p.

-----MacLeod, Norman S., and Wagner, Holly C., 1973. Miocene tholeiitic basalts of coastal Oregon and Washington and their relations to coeval basalts of the Columbia Plateau. Geol. Soc. Amer., Bull., v.84, pp.387-424.

-----Pearl, J.E., and Lander, D.L., 1977. Interim report on petroleum resources potential and geologic hazards in the outer continental shelf - Oregon and Washington Tertiary province. U.S. Geol. Surv., Open-file Rept. 77-282, 57p.

-----Rau, W.W., and Wagner, Holly C., 1964. Miocene stratigraphy of the Yaquina Bay area, Newport, Oregon. Ore Bin, v.26, no.8, pp.133-151.

-----Wagner, Holly C., and Lander, Diane L., 1980. Interpretation of the Cenozoic history, central Oregon continental margin: cross-section summary. Geol. Soc. Amer., Bull., pt.I, v.91, pp.143-146.

-----Wagner, Holly C., and MacLeod, Norman S., 1965. Preliminary data on compositional variations of Tertiary volcanic rocks in the central part of the Oregon Coast range. Ore Bin, v.27, no.6, pp.101-117.

-----et al., 1975. Alsea Formation-an Oligocene marine sedimentary sequence in the Oregon Coast Range. U.S. Geol. Survey, Bull. 1395-F, pp.F1-F21.

-----et al., 1980. Geology of the west-central part of the Oregon Coast Range. Geologic field trips in western Oregon and southwestern Washington. Oregon Dept. Geol. and Mineral Indus., Bull.101, pp.39-76.

Socolow, Arthur A., ed., 1988. The state geological surveys; a history. [n.p.] Amer. Assoc. of State Geologists, 499p.

Sorenson, Mark Randall, 1983. The Queen of Bronze copper deposit, southwestern Oregon; an example of sub-sea floor massive sulfide mineralization. Ms., Univ. of Oregon, 205p.

Stanley, George D., 1986. Late Triassic coelenterate faunas of western Idaho and northeastern Oregon: implications for biostratigraphy and paleogeography. U.S. Geol. Survey, Prof. Paper 1435, pp.23-36.

-----1987. Travels of an ancient reef. Natural History, v.87, no.11, pp.36-42.

-----1991. Geologic basis for petroleum resource assessment of onshore western Oregon and Washington (Province 72). U.S. Geol. Survey, Open-File Report 88-450-X, 29p.

-----amd Beauvais, Louise, 1990. Middle Jurassic corals from the Wallowa terrane, west-central Idaho. Jour. Paleo., v.64, no.3, pp.352-362.

Staples, Lloyd W., 1948. Oregon's new minerals and discredited species. Mineralogist, v.16, no.10, pp.470-476.

-----1962. The discoveries of new minerals in Oregon. Ore Bin, v.24, no.6, pp.81-86.

-----1965. Origin and history of the thunder egg. Ore Bin, v.27, no.10, pp.195-204.

-----Evans, Howard T, and Lindsay, James R., 1973. Cavansite and Pentagonite, new dimorphous calcium vanadium silicate minerals from Oregon. American Mineralogist, v.58, pp.405-411.

Stearns, Harold T., 1930. Geology and water resources of the middle Deschutes River basin, Oregon. U.S. Geol. Survey, Water Supply Paper 637-D, pp.125-212.

Steere, Margaret L., comp., 1953. Bibliography of the geology and mineral resources of Oregon (Second supplement), January 1, 1946 to December 31, 1950. Oregon Dept. Geol. and Mineral Indus., Bull.44, 61p.

-----1954. Geology of the John Day country, Oregon. Ore Bin, v.16, no.7, pp.41-46.

-----and Owen, Lillian F., comp., 1962. Bibliography of the geology and mineral resources of Oregon (Third supplement), January 1, 1951 to December 31, 1955. Oregon Dept. Geol. and Mineral Indus., Bull.53, 96p.

Stembridge, James E., 1975. Recent shoreline changes of the Alsea sandspit, Lincoln County. Ore Bin, v.37, no.5, pp.77-82.

Stevens, Calvin H., ed., 1983. Pre-Jurassic rocks in western North American suspect terranes. Soc. Econ. Paleont. and Mineralog., Pacific Section, 141p.

Stevens, Isaac I., 1853. Report of explorations for a route for the Pacific Railroad near the forty-seventh and forty-ninth parallels of north latitude from St. Paul to Puget Sound. In: Reports of Explorations and surveys to ascertain the most practicable and economical route from the Mississippi River to the Pacific Ocean...1853-4. Vol.l, 1855, 651p.

Stewart, John H., 1978. Basin-range structure in western North America: A review. Geol. Soc. Amer., Memoir 152, pp.1-28.

-----Stevens, Calvin H., and Fritsche, A. Eugene, eds.,1977. Paleozoic paleogeography of the western United States. Soc. Econ. Paleont. and Mineralog., Pacific Section, Paleogr. Symposium 1, 502p.

Stimson, Eric Jordan, 1980. Geology and metamorphic petrology of the Elkhorn Ridge area, northeastern Oregon. Univ. of Oregon, Ms., 123p.

Stock, Chester, 1946. Oregon's wonderland of the past - the John Day. Science Monthly, v.63, pp.57-65.

Stokes, William Lee, 1979. Paleostratigraphy of the Great Basin region. In: Newman, Gary W., et al., eds., Basin and Range Symposium and Great Basin field conference, pp.196-217.

Stovall, James C., 1929. Pleistocene geology and physiography of the Wallowa Mountains with special reference to Wallowa and Hurricane Canyons. Univ. of Oregon, Ms, 115p.

Strickler, Michael D., 1986. Geologic setting of the Turner-Albright massive sulfide deposit, Josephine County, Oregon. Oregon Geology, v.48, no.10, pp.115-122.

Strong, Emory, 1967. The Bridge of the gods. Geol. Soc. Oregon Country, Newsl., v.33, no.6, pp.49-53.

Swanson, Donald A., and Robinson, Paul T., 1968. Base of the John Day Formation in and near the Horse Heaven Mining District, north-central Oregon. U.S. Geol. Survey, Prof. Paper 660-D, pp.D154-161.

-----et al., 1979. Revisions in stratigraphic nomenclature of the Columbia River Basalt group. U.S. Geol. Survey, Bull. 1457-G, 59p.

Swanson, Frederick J., and James, Michael E., 1975. Geomorphic history of the lower Blue River-Lookout Creek area, Western Cascades, Oregon. Northwest Sci., v.49, no.l, pp.1-11.

Taubeneck, William H., 1955. Age of the Bald Mountain batholith, northeastern Oregon. Northwest Sci., v.29, pp.93-96.

-----1966. An evaluation of tectonic rotation in the Pacific Northwest. Jour. Geophys. Res., v.71, no.8, pp.2113-2120.

-----1987. The Wallowa Mountains, northeast Oregon. In: Hill, Mason L., Cordilleran Sect of the Geological Society of America, Centennial Field Guide, pp.327-332.

Taylor, David G., 1982. Jurassic shallow marine invertebrate depth zones, with exemplification from the Snowshoe Formation, Oregon. Oregon Geol., v.44, no.5, pp.51-56.

-----1988. Middle Jurassic (late Aalenian and early Bajocian) ammonite biochronology of the Snowshoe Formation, Oregon. Oregon Geol., v.50, no.11/12, pp.123-138.

Taylor, Edward M., 1965. Recent volcanism between Three Fingered Jack and North Sister Oregon Cascade Range. Ore Bin, v.27, no.7, pp.121-147.

-----1968. Roadside geology Santiam and McKenzie Pass highways, Oregon. Oregon, Dept. Geol. and Mineral Indus., Bull. 62, pp.3-33.

-----1978. Field geology of S.W. Broken Top Quadrangle, Oregon. Oregon, Dept. Geol. and Mineral Indus., Special Paper 2, 50p.

-----1981. Central High Cascade roadside geology, Bend, Sisters, McKenzie Pass, and Santiam Pass, Oregon. U.S. Geol. Survey, Circular 838, pp.55-58.

----- 1981. A mafic dike system in the vicinity of Mitchell, Oregon, and its bearing on the timing of Clarno-John Day volcanism and early Oligocene deformation in central Oregon. Oregon Geol., v.43, no.8, pp.107-112.

-----1987. Late High Cascade volcanism from summit of McKenzie Pass, Oregon: Pleistocene composite cones on platform of shield volcanoes: Holocene eruptive centers and lava fields. In: Hill, Mason L., Cordilleran Section of the Geological Society of America, Centennial Field Guide, v.1, pp.313-315.

-----1990. Volcanic history and tectonic development of the central High Cascade Range, Oregon. Jour. Geophys. Res., v.95, no.B12, pp.19,611-19,622.

Thayer, Thomas P., 1939. Geology of the Salem Hills and the North Santiam River basin, Oregon. Oregon Dept. Geol. and Mineral Indus., Bull.15, 40p.

-----1977. The Canyon Mountain complex, Oregon, and some problems of ophiolites. Oregon Dept. Geol. and Mineral Indus., Bull.95, pp.93-106.

Thompson, G.G., Yett, J.R., and Green, K.E., 1984. Subsurface stratigraphy of the Ochoco Basin, Oregon. Oregon Dept. Geol. and Mineral Indus. Oil and Gas Inv.8, 22p.

Thompson, George A., and Burke, Dennis B., 1974. Regional geophysics of the Basin and Range province. Ann. Rev. of Earth and Planet. Sci., v.2, pp.213-238.

Throop, Allen H., 1986. Fishing and placer mining: are they compatible? Oregon Geology, v.48, no.3, pp.27-28,34.

-----1989. Cyanide in mining. Oregon Geology, v.51, no.1, pp.9-11,20.

-----1991. Should the pits be filled? Geotimes, v.36, no.11, pp.20-22. Also: Oregon Geology, v.52, no.4, pp.82-84.

Thunderegg; Oregon's state rock, 1965. Ore Bin, v.27, no.10, pp.189-194.

Thwaites, Reuben G., ed, 1955. Original journals of the Lewis and Clark Expedition, 1804-1806. 8 vols. New York, Antiquarian Press.

Tolan, Terry L., and Beeson, Marvin H., 1984. Intracanyon flows of the Columbia River Basalt Group in the lower Columbia River Gorge and their relationship to the Troutdale Formation. Geol. Soc. Amer., Bull., v.95, pp.463-477.

-----Beeson, Marvin H., and Vogt, Beverly F., 1984. Exploring the Neogene history of the Columbia River: discussion and geologic field trip guide to the Columbia River gorge. Part I: discussion. Oregon Geology, v.46, no.8, pp.87-97.

-----Beeson, Marvin H., and Vogt, Beverly F., 1984. Exploring the Neogene history of the Columbia River: discussion amd geologic field trip guide to the Columbia River gorge. Part II: Road log and comments. Oregon Geology, v.46, no.8, pp.103-112.

-----et al., 1989. Revisions to the estimates of the areal extent and volume of the Columbia River Basalt group. In: Reidel, S.P., and Hooper, P.R., eds. Volcanism and tectonism in the Columbia River flood-basalt province. Geol. Soc. Amer., Special Paper 239, pp.1-20.

Treasher, Ray C., 1942. Geologic history of the Portland area. Oregon Dept. of Geol. and Mineral Indus., GMI Short Paper 7, 17p.

Trimble, Donald E., 1963. Geology of Portland, Oregon and adjacent areas. U.S. Geol. Survey, Bull. 1119, 119p.

Tyler, David B., 1968. The Wilkes Expedition; the first United States Exploring Expedition (1838-1842). Philadelphia, The American Philos. Soc., 435p.

U.S. Bureau of Mines, 1990. Mineral industry surveys; the mineral industry of Oregon in 1990. 3p.

Vallier, Tracy L., 1977. The Permian and Triassic Seven Devils Group, western Idaho and northeastern Oregon. U.S. Geol. Survey, Bull.1437, 58p.

-----and Brooks, Howard C., 1970. Geology and copper deposits of the Homestead area, Oregon and Idaho. Ore Bin, v.32, no.3, pp.37-57.

-----and Brooks, Howard C., eds., 1986. Geology of the Blue Mountains region of Oregon, Idaho, and Washington. U.S. Geol. Survey, Prof. Paper 1435, 93p.

-----and Brooks, Howard C. eds., 1987. Geology of the Blue Mountains regions of Oregon, Idaho, and Washington: The Idaho batholith and its border zone. U.S. Geol. Survey, Prof. Paper 1436, 196p.

-----Brooks, Howard C., and Thayer, T.P., 1977. Paleozoic rocks of eastern Oregon and western Idaho. In: Stewart, J.H., et al., eds. Paleozoic paleogeography of the western United States. Soc. Econ. Paleont. and Mineralogists, Pacific Coast Paleogeography Symposium 1, pp.455-466.

Vander Meulen, Dean B., et al., 1990. Mineral resources of the Blue Canyon and Owyhee breaks wilderness study areas, Malheur County, Oregon. U.S. Geol. Survey, Prof. Paper 1741-G, 28p.

Verplanck, Emily P., and Duncan, Robert A., 1987. Temporal variations in plate convergence and eruption rates in the Western Cascades, Oregon. Tectonics, v.6, no.2, pp.197-209.

Vokes, H.E., Snavely, P.D., and Myers, D.A., 1951. Geology of the southern and southwestern border areas, Willamette Valley, Oregon. U.S. Geol. Survey Oil and Gas Inv. Map

Wagner, N.S., 1952. Catlow Valley crevice. Ore Bin, v.14, no.6, pp.37-41.

-----1956. Historical notes on the Standard Mine, Grant County, Oregon. Ore Bin, v.18, no.9, pp.75-77.

-----1958. Limestone occurrences in eastern Oregon. Ore Bin, v.20, no.5, pp.43-47.

-----1959. Mining in Baker County, 1861-1959. Ore Bin, v.21, no.3, pp.21-27.

-----1966. The Armstrong nugget. Ore Bin, v.28, pp.211-219.

Walker, George W., 1974. Some implications of late Cenozoic volcanism to geothermal potential in the High Lava Plains of south-central Oregon. Ore Bin, v.36, no.7, pp.109-119.

-----1977. Geologic map of Oregon east of the 121st Meridian. U.S. Geol. Survey, Misc. Inv. Series Map I-902.

-----1979. Revisions to the Cenozoic stratigraphy of Harney Basin, southeastern Oregon. U.S. Geol. Survey, Bull. 1475, 34p.

-----1981. High Lava Plains, Brothers fault zone to Harney Basin, Oregon. In: Johnston, David, and Donnelly-Nolan, Julie, eds. Guides to some volcanic terranes in Washington, Idaho, and Northern California, U.S. Geol. Survey, Circular 838, pp.105-111.

-----1990. Miocene and younger rocks of the Blue Mountains region, exclusive of the Columbia River basalt group and associated mafic lava flows. U.S. Geol. Survey, Prof. Paper 1437, pp.101-118.

-----ed., 1990. Geology of the Blue Mountains region of Oregon, Idaho, and Washington: Cenozoic geology of the Blue Mountains. U.S. Geol. Survey, Prof. Paper 1437, 135p.

-----and Duncan, Robert A., 1989. Geologic map of the Salem 1 by 2 Quadrangle, western Oregon. U.S. Geol. Survey, Misc. Inv. Ser. Map I-1893.

-----and McLeod, Norman S., (in press). Geologic map of Oregon. U.S. Geol. Survey.

-----and Robinson, Paul T., 1990. Cenozoic tectonism and volcanism of the Blue Mountains region. U.S. Geol. Survey, Prof. Paper 1437, pp.119-134.

-----and Robinson, Paul T., 1990. Paleocene(?), Eocene, and Oligocene(?) rocks of the Blue Mountains region. U.S. Geol. Survey, Prof. Paper 1437, pp.13-27.

-----and Swanson, D.A., 1968. Summary report on the geology and mineral resources of the Harney Lake and Malheur Lake areas...Oregon. U.S. Geol. Survey, Bull. 1260-L,M, 17p.

-----Greene, Robert C., and Pattee, Eldon C., 1966. Mineral resources of the Mount Jefferson Primitive area, Oregon. U.S. Geol. Survey, Bull. 1230-D, pp.D1-D32.

Walker, Jeffrey R., and Naslund, H. Richard, 1986. Tectonic significance of mildly alkaline Pliocene lavas in Klamath River gorge, Cascade Range, Oregon. Geol. Soc. Amer., Bull., v.97, pp.206-212.

Wallace, Andy B., and Roper, Michael W., 1981. Geology and uranium deposits along the northeastern margin, McDermitt Caldera complex, Oregon. In: Goodell, P.C., and Waters, Aaron C., Uranium in volcanic and volcaniclastic rocks. Amer. Assoc. Petroleum Geol., Studies in Geology No.13, pp.73-79.

Wallace, Robert E., 1984. Patterns and timing of late Quaternary faulting in the Great Basin province and relation to some regional tectonic features. Jour. Geophys. Res., v.89, no.B7, pp.5763-5769.

Walsh, Frank K., and Halliday, William R., 1976. Oregon caves; discovery and exploration. Oregon Caves National Monument. Te-Cum-Tom Publ., Grants Pass, Ore., 27p.

Wardlaw, Bruce R., Nestell, Merlynd K., and Dutro, J. Thomas, 1982. Biostratigraphy and structural setting of the Permian Coyote Butte Formation of central Oregon. Geology, v.10, pp.13-16.

Waring, Gerald A., 1908. Geology and water resources of a portion of south-central Oregon. U.S. Geol. Survey, Water Supply Paper 220, 83p.

-----1965. Thermal springs of the United States and other countries of the world - a summary. U.S. Geol. Survey, Prof. Paper 492, 383p.

Warren, W.C., and Norbisrath, Hans, 1946. Stratigraphy of upper Nehalem River basin, northwestern Oregon. Amer. Assoc. of Petroleum Geol., v.30, no.2, pp.213-237.

Waters, Aaron C., 1961. Stratigraphic and lithologic variations in the Columbia River Basalt. Amer. Jour. Sci., v.259, pp.583-611.

-----1966. Stein's Pillar area, central Oregon. Ore Bin, v.28, no.8, pp.137-144.

-----1973. The Columbia River gorge: basalt stratigraphy, ancient lava dams, and landslide dams. Oregon, Dept. Geol. and Mineral Indus., Bull.77, pp.133-162.

Watkins, N.D., and Baksi, A.K., 1974. Magnetostratigraphy and oroclinal folding of the Columbia River, Steens, and the Owyhee basalts in Oregon, Washington, and Idaho. Amer. Jour. Sci., v.274, pp.148-189.

Weaver, C.E., 1937. Tertiary stratigraphy of western Washington and northwestern Oregon. Univ. Wash. Publ. in Geol., v.4, 266p.

-----1942. Paleontology of the marine Tertiary formations of Oregon and Washington. Univ. Wash. Pub. in Geol., v.5, pts.I, II, and III, 790p.

Weaver, C.S., Green, S.M., Iyer, H.M., 1982. Seismicity of Mount Hood and structure as determined from teleseismic P wave delay studies. Jour. Geophys. Res., v.87, no.B4, pp.2782-2792.

Webb, Thompson, Bartlein, Patrick J., and Kutzbach, John E., 1987. Climatic change in eastern North America during the past 18,000 years; comparisons of pollen data with model results. In: Ruddiman, W.F., and Wright, H.E., eds., North America and adjacent oceans during the last deglaciation. Geol. Soc. Amer., The Geology of North America, v.K-3, pp.447-462p.

Weidenheim, Jan Peter, 1981. The petrography, structure, and stratigraphy of Powell Buttes, Crook County, central Oregon. Oregon State Univ., Ms., 95p.

Weis, Paul L., et al., 1976. Mineral resources of the Eagle Cap Wilderness and adjacent areas, Oregon. U.S. Geological Survey, Bulletin 1385-E, 84p.

Weissenborn, A.C., ed., 1969. Mineral and water resources of Oregon. Oregon Dept. Geol. and Mineral Indus., Bull.64, 462p.

Wells, F.G., and Peck, D.L., 1961. Geologic map of Oregon west of the 121st Meridian. U.S. Geol. Survey, Misc. Geol. Inv. Map I-325.

Wells, Ray E., 1984. Paleomagnetic constraints on the interpretation of early Cenozoic Pacific Northwest paleogeography. In: Nilsen, Tor, ed., Geology of the upper Cretaceous Hornbrook Formation, Oregon and California. Soc. Econ. Paleont. and Mineralogists, Pacific Section, v.42, pp.231-237.

-----and Heller, Paul L., 1988. The relative contribution of accretion, shear, and extension to Cenozoic tectonic rotation in the Pacific Northwest. Geol. Soc. Amer., Bull. v.100, pp.325-338.

-----et al., 1984. Cenozoic plate motions of the volcano-tectonic evolution of western Oregon and Washington. Tectonics, v.3, no.2, pp.275-294.

-----et al., 1989. Correlation of Miocene flows of the Columbia River Basalt group from the central Columbia River plateau to the coast of Oregon and Washington. In: Reidel, S.P., and Hooper, P.R., eds., 1989. Volcanism and tectonism in the Columbia River flood-basalt province. Geol. Soc. Amer., Special Paper 239, pp.113-128.

Werner, Kenneth Stefan, 1990. I. Direction of maximum horizontal compression in western Oregon determined by borehole break-outs. II. Structure and tectonics of the northern Willamette Valley, Oregon. Ms., Oregon State Univ., 159p.

West, Donald O., and McCrumb, Dennis R., 1988. Coastline uplift in Oregon and Washington and the nature of Cascadia subduction-zone tectonics. Geology, v.16, pp.169-172.

Westhusing, J.K., 1973. Reconnaissance surveys of near-event seismic activity in the volcanoes of the Cascade Range, Oregon. Bull. Volcanology, ser.2, pp.258-285.

Whalen, Michael T., 1988. Depositional history of an upper Triassic drowned carbonate platform sequence: Wallowa terrane, Oregon and Idaho. Geol. Soc. Amer., Bull. v.100, pp.1097-1110.

Wheeler, Greg, 1981. A major Cretaceous discontinuity in north-central Oregon. Oregon Geol., v.43, no.2, pp.15-17.

-----1982. Problems in the regional stratigraphy of the Strawberry Volcanics. Oregon Geol., v.44, no.1, pp.3-7

Wheeler, Harry E., and Cook, Earl F., 1954. Structural and

stratigraphic significance of the Snake River capture, Idaho-Oregon. Jour. Geol., v.62, pp.525-536.

White, Craig McKibben, 1980. Geology and geochemistry of Mt. Hood volcano. Oregon, Dept. Geol. and Mineral Indus., Special Paper 8, 26p.

------1980. Geology and geochemistry of volcanic rocks in the Detroit area, Western Cascade Range, Oregon. Ph.D., Univ. Oregon, 178p.

-----1980. Geology of the Breitenbush hot springs quadrangle, Oregon. Oregon, Dept. Geol. and Mineral Indus., Special Paper 9, 26p.

White, D.J., and Wolfe, H.D., 1950. Report of reconnaissance of the area from Panther Butte to Tellurium Peak, Douglas County, Oregon. Ore Bin, v.12, no.12, pp.71-76.

Wiley, Thomas J., 1991. Mining and exploration in Oregon during 1990. Oregon Geology, v.53, no.3, pp.63-70.

Wilkening, R. Matthew and Cummings, Michael L., 1987. Mercury and uranium mineralization in the Clarno and John Day Formations, Bear Creek Butte area, Crook County, Oregon. Oregon Geology, v.49, no.9, pp.103-110p.

Wilkerson, William L., 1958. The geology of a portion of the southern Steens Mountains, Oregon. Univ. Oregon, Ms., 89p.

Wilkinson, W.D., and Oles, Keith F., 1968. Stratigraphy and paleo-environments of Cretaceous rocks, Mitchell Quadrangle, Oregon. Amer. Assoc. Petrol. Geol., Bull., v.52, no.l, pp.129-161.

-----et al., 1959. Field guidebook, geologic trips along Oregon Highways. Oregon Dept. Geol. and Mineral Indus., Bull.50, 148p.

Williams, David L., and Von Herzen, Richard P., 1983. On the terrestrial heat flow and physical limnology of Crater Lake, Oregon. Jour. Geophys. Res., v.88, no.B2, pp.1094-1104.

-----et al., 1982. The Mt. Hood region; volcanic history, structure, and geothermal energy potential. Jour. Geophys. Res., v.87, no.B4, pp.2767-2781.

Williams, Howel, 1976. The ancient volcanoes of Oregon. Oregon State System of Higher Ed., Condon Lectures, 70p.

-----and Compton, Robert R., 1953. Quicksilver deposits of Steens Mountain and Pueblo Mountains southeast Oregon. U.S. Geol. Survey, Bull.995-B., pp.B1-B76.

-----and Goles, Gordon G., 1968. Volume of the Mazama ash-fall and the origin of Crater Lake caldera. In: Dole, Hollis M., ed. Andesite conference guidebook - international mantle project; Science report 16-S. Oregon Dept. Geol. and Mineral Indus., Bull.62, pp.37-41.

Wilson, Douglas, and Cox, Allan, 1980. Paleomagnetic evidence for tectonic rotation of Jurassic plutons in Blue Mountains, eastern Oregon. Jour. Geophys. Res., v.85, no.B7, pp.3681-3689.

Winterer, E.L., Hussong, Donald M., and Decker, Robert W., 1989. The Eastern Pacific Ocean and Hawaii. Geol. Soc. Amer., The geology of North America, v.N, 563p.

Wise, William S., 1968. Geology of the Mount Hood volcano. In: Dole, Hollis M., ed., Andesite conference guidebook -international Mantle project; science report 16-2. Oregon, Dept. Geol. and Mineral Indus., Bull.62, 81-98.

-----1969. Geology and petrology of the Mt. Hood area: a study of High Cascade volcanism. Geol. Soc. Amer., Bull.80, pp.969-1006.

Wolfe, Jack A., 1969. Neogene, floristic and vegetational history of the Pacific Northwest. Madrona, v.20, pp.83-110.

Wong, Ivan G., and Silva, Walter J., 1990. Preliminary assessment of potential strong earthquake ground shaking in the Portland, Oregon, metropolitan area. Oregon Geology, v.52, no.6, pp.131-134.

Woodward, John, White, James, and Cummings, Ronald, 1990. Paleoseismicity and the archaeological record: areas of investigation on the northern Oregon coast. Oregon Geology, v.52, no.3, pp.57-65.

Wright, H.E., and Frey, D.G., eds., 1965. The Quaternary history of the United States. Princeton University Press, Princeton, N.J., 723p.

Wright, James E., and Fahan, Mark R., 1988. An expanded view of Jurassic orogenesis in the western United States Cordillera: Middle Jurassic (pre-Nevadan) regional metamorphism and thrust faulting within an active arc environment, Klamath Mountains, Oregon. Geol. Soc. Amer., Bull. 100, pp.859-876.

Wright, Thomas L., Grolier, Maurice J., and Swanson, Donald A., 1973. Chemical variation related to the stratigraphy of the Columbia River Basalt. Geol. Soc. Amer., Bull.84, pp.371-386.

Wynn, Jeffrey C., and Hasbrouck, Wilfred P., 1984. Geophysical studies of chromite deposits in the Josephine peridotite of northwestern California and southwestern Oregon. U.S. Geol. Survey Bull. 1546-D., pp.D63-D86.

Yeats, Robert S., 1987. Summary of symposium on Oregon's earthquake potential held February 28, 1987, at Western Oregon State College in Monmouth. Oregon Geology, v.49, no.8, pp.97-98.

-----1989. Current assessment of earthquake hazard in Oregon. Oregon Geology, v.51, no.4, pp.90-92.

-----et al., 1991. Tectonics of the Willamette Valley, Oregon. U.S. Geological Survey, Open-file Report 91-441-P, 47p.

Zierenberg, Robert A., et al., 1988. Mineralization, alteration, and hydrothermal metamorphism of the ophiolite-hosted Turner-Albright sulfide deposit, southwestern Oregon. Jour. Geophys. Res., v.93, no.B5, pp.4657-4674.

Zoback, Mary Lou, Anderson, R.E., and Thompson, G.A., 1981. Cainozoic evolution of the state of stress and style of tectonism of the Basin and Range province of the western United States. Philos. Trans. Royal Soc. London, no.300, pp.407-434.

Glossary

Aa-A Hawaiian term for basaltic lava flows with a rough, blocky surface.

Accretion-A tectonic process by which exotic rock masses (terranes) are physically annexed to another landmass after the two collided.

Accretionary wedge-A prism of sediment that builds up on the leading edge of a moving tectonic plate.

Agate-A cryptocrystalline quartz (chalcedony) which is commonly banded and colored.

Agglomerate-A pyroclastic rock made up of coarse, usually angular fragments, much like volcanic breccia.

Alluvium-A general term for all detrital deposits resulting from the operations of modern rivers, thus including the sediments laid down in river beds, flood plains, lakes, fans at the foot of mountain slopes, and estuaries.

Antecedent stream-A stream that was established before local uplift began and incised its valley at the same rate the land was rising.

Andesite-A fine-grained volcanic rock containing plagioclase feldspar (andesine) and one or more mafic constituents such as pyroxene, amphibole or biotite.

Anticline-Arched beds with limbs dipping away from the axis. Simple anticlines have older beds exposed in the center of the arch.

Aphanitic rock-This term pertains to a texture of rocks in which the crystals are too small to distinguish with the unaided eye.

Archipelago-A curved volcanic chain on land or an island arc in an ocean setting.

Arkose-A sandstone containing 25 percent or more of feldspars usually derived from coarse-grained silicic igneous rocks.

Bacarc basin-A sedimentary trough situated between a volcanic island arc and the mainland.

Basalt-A dark-colored, fine-grained igneous rock containing mostly calcic fledspar and pyroxene with or without olivine.

Basin and range structure-Regional structure typified by parallel fault-block mountain ranges, separated by sediment-filled basins. These occur in southeastern Oregon.

Batholith-A large body of coarse-grained intrusive igneous rock with a surface area greater than 40 square miles. Smaller intrusive bodies are called stocks.

Bed-A layer or stratum of sedimentary rock.

Brackish-Water with very low salinity that is usually derived from fluvial runoff into near-shore marine waters.

Breccia-Fragmental rock whose pieces are angular unlike water worn material. There are fault breccias, talus breccias and eruptive volcanic breccias.

Calcite-A mineral composed of calcium carbonate (CaCO3) and the chief constituent of limestone.

Caldera-A large basin-shaped, generally circular, volcanic depression, the diameter of which is many times greater than that of the included volcanic vent or vents. It is formed by collapse, such as at Crater Lake and the Newberry caldera.

Chert-A compact siliceous rock of varying color composed of microorganisms or precipitated silica grains. It occurs as nodules, lenses, or layers in limestone and shale.

Cinder cone-A cone-shaped peak built up almost wholly of volcanic fragments erupted from a vent.

Cirque-The rounded saucer-shaped glacially carved basin at the head of an alpine glacial valley which commonly contains a cirque lake or tarn.

Clastic rock-A consolidated sedimentary rock composed of the cemented fragments broken or eroded from pre-existing rocks of any origin by chemical or mechanical weathering. Examples are conglomerate, sandstone, and siltstone.

Columnar jointing-Cracks that separate rock into columns commonly six-sided but generally ranging from 4-8 sides. Such jointing usually forms in volcanic rocks due to shrinkage when cooling.

Concretion-A nodular or irregular mass developed by localized deposition of material from solution generally around a central nucleus. It is harder than the enclosing rock such as limey concretions in shale.

Conformable-Accumulation of generally parallel strata, without appreciable break in deposition.

Conglomerate-A sedimentary rock made up of rounded pebbles and cobbles coarser than sand.

Conodont-Phosphatic microfossils of fish-teeth used to age date many of the older Oregon marine rocks.

Contact-The surface between two different types of rocks or formations which is shown as a line on geologic maps.

Correlation-A means of determining equivalent ages of formations or geologic events by studying physical similarities or differences, by comparing fossils or by comparing absolute dates from radioactive minerals.

Crust-The outer layer of the earth above the Mohorovicic discontinuity usually basaltic in composition beneath ocean basins and granitic in composition under continents.

Crystal-The regular polyhedral form, bounded by plane surfaces which is the outward expression of a periodic or regularly repeating internal arrangement of atoms.

Crystalline rock-An inexact term for igneous and metamorphic rocks as opposed to sedimentary rocks.

Delta-A fan-shaped alluvial deposit where a stream enters a standing body of water, usually the ocean or a large lake.

Density-Mass per unit volume.

Diatom-Microscopic, single-celled plants (algae) growing in marine or fresh water. Diatoms secrete siliceous skeletons that may accumulate in sediments in enormous numbers and which are referred to as diatomite or diatomaceous earth.

Dike-A tabular body of igneous rocks that cuts across the structure of adjacent rocks or cuts massive rocks.

Dip-The inclination of rock layers or surfaces measured from the horizontal. Dip is at right angles to the strike.

Disconformity-A type of unconformity wherein the older and younger beds are parallel but separated by an erosion surface.

Eolian-Applied to deposits arranged by the wind such as the sand in dunes. Wind-blown silt is called loess.

Epicenter-A point on the ground over the center of maximum earthquake intensity.

Epoch-A unit of geologic time which is a subdivision of a period, such as the Eocene Epoch of the Tertiary Period.

Era-A major division of geologic time which contains periods, such as the Paleozoic Era.

Erosion-Several processes by which rock is loosened or dissolved and removed. It includes the processes of weathering, solution, abrasion and transportation of the resulting particles or solution.

Erratic-Those random rocks that are located far from their source. Glacial erratics are ice-borne or ice-rafted boulders or blocks scattered in regions glaciated or flooded by breaking of glacial dams.

Eustatic-Pertains to worldwide changes in sea level, commonly caused by glaciation and deglaciation.

Extrusive igneous rocks-Molten rocks that poured out as flows or are ejected as fragmental particles on the earth's surface.

Facies-A stratigraphic body as distinguished from other contemporaneous bodies of different appearance or composition.

Fan-A large-scale triangular-shaped sedimentary body that has been deposited at the mouth of a stream in a subaerial or submarine environment.

Fault-A break in the earth's crust along which movement has taken place.

Fault, lateral-A fault in which displacement is predominantly horizontal. When the block across the fault from the observer has moved to the right it is called a right lateral fault. It may also be called a strike-slip fault.

Fault, normal-A tensional fault, with the plane of slippage dipping toward the down-thrown block.

Fault, reverse-A compressional fault, with the plane of slippage dipping toward the up-thrown block.

Fauna-The animals of any place or time that live in mutual association.

Flora-The plants collectively of a given formation, age or region.

Foraminifera-A single-celled protozoa with cytoplasm and pseudopodia. Used to age date sedimentary rocks.

Forearc basin-A sedimentary trough situated on the oceanic side of a volcanic island arc between the archipelago and the major ocean basin.

Formation-Rock masses forming a unit convenient for description and mapping. Most formations possess certain distinctive or combinations of distinctive lithic features. Formations may be combined into groups or divided into members. They are usually named for some nearby geographic feature.

Fossil-The remains or traces of animals or plants which have been preserved by natural causes in the earth's crust.

Gastropod-A class of mollusks commonly called snails.

Geochronology-The study of time in relation to the history of the earth.

Geology-The science which deals with the earth, the rocks of which it is composed, and the changes which it has undergone or is undergoing. Geology includes physical geology and geophysics, stratigraphy and paleontology, mineralogy, petrology, petroleum geology and phases of engineering and mining.

Geode-Hollow globular bodies with a crystal-lined interior. A "thunder egg" is a type of geode that is filled with banded chalcedony.

Geothermal-The term pertains to the heat within the earth. Heatflow refers to the transferal of the earth's heat through rock.

Gneiss-A coarse-grained, banded metamorphic rock.

Geyser-A hot spring that erupts water or steam into the air. Some erupt on a fairly regular schedule whereas others erupt sporadically. Some drilled wells have geyser-like eruptions.

Glacier-A large mass of ice, accumulated as snow that flows because of gravity. Glaciers confined to the mountains are alpine glaciers, but those that spread out over large parts of the northern hemisphere are continental glaciers, or ice sheets. Ice caps, unlike glaciers, are usually stationary due to their horizontal plane of deposition.

Graben-A structural depression formed as a down-dropped fault block sinks between normal (gravity) faults.

Graded bedding-A type of stratification in which each bed displays gradation in grain size from coarse below to fine above.

Graywacke-A poorly-sorted sandstone containing angular quartz and feldspar fragments and some rock fragments with clay-sized matrix between sand grains. It is generally a hard dark-colored rock.

Greenstone-A field term applied to metamorphically altered basic igneous rocks which owe their greenish color to the presence of chlorite and hornblende.

Horst-Fault block uplifted between normal faults and the opposite of a graben.

Ice Age-Refers to the Pleistocene epoch when ice covered large parts of the northern hemisphere. Other periods of glaciation are recorded earlier in geologic time.

Igneous rock-Rock formed by crystallization of once molten material called lava or magma.

Ignimbrite-A welded tuff which has been fused or partly fused by the combined action of the heat retained in the particles and the enveloping hot gases.

Intrusive rock-Molten rock penetrating into older rocks and solidifying. Relatively thin tabular masses cutting across structures are dikes, whereas those paralleling stratification are called sills. Huge masses without definite shape are called batholiths.

Island arc-Curved chain of islands, often volcanic in origin, generally convex to the open ocean and bordered by a deep submarine trench.

Isostasy-Theoretical balance of all large portions of the earth's crust as though they were floating on a denser underlying layer; thus areas of less dense crustal material rise topographically above areas of more dense material.

Joint-A fracture without appreciable movement parallel to the plane.

Lahars-A mudflow of volcanically derived debris.

Laterite-A soil profile formed under tropical conditions where decomposition has been extreme leaving very little of the parent rock.

Lava-Molten rock that is extruded upon the earth's surface. Even after cooling it may be referred to as a lava flow.

Limb-One side or flank of a fold.

Limestone-A bedded sedimentary deposit consisting largely of calcium carbonate (CaCO3). It may include fragments of shells composed of calcium carbonate.

Lineament-A regular large-scale continuation of linear features visible at the earth's surface.

Lithification-Induration or hardening of soft sediments to compact rock by cementation and pressure.

Loess-A homogeneous, nonstratified, unindurated deposit consisting largely of silt. Loess has a rude vertical parting in many places. Most loess is thought to be eolian or wind-blown in origin.

Maar-A crater formed by violent explosion not accompanied by igneous extrusion. It is commonly occupied by a small circular lake. A relatively shallow, flat-floored explosion crater, the walls of which consist largely or entirely of loose fragments of the country rock and only partly of essential, magmatic ejecta. Maars are apparently the result of a single violent volcanic explosion, probably from steam generated when groundwater comes in contact with hot rock.

Mafic-Dark-colored igneous rocks high in magnesium and iron and low in silica. Peridotites and serpentinite are referred to as ultramafic rocks and are even lower in silica content.

Magma-Molten silicate material in the crust or mantle from which igneous rocks are derived.

Mantle-Thick layer between the overlying crust and the underlying core composed of ultrabasic (ultramafic) rock of a composition much like peridotite or serpentinite.

Marine deposits-Any rock laid down in the ocean of which sedimentary rocks are the most common, but it also includes submarine volcanic extrusive rocks.

Matterhorn-A sharply pointed alpine peak formed by glacial erosion when 3 or more glaciers meet by headward erosion.

Meander-One of a series of somewhat regular and looplike bends in the course of a stream, developed when the steam is flowing at grade

through lateral shifting of its course toward the convex sides of the original curves.

Melange-Mixture of irregular blocks of rock of diverse origin and size commonly formed by shearing during tectonic movements.

Metamorphic rocks-Rocks altered by heat and pressure causing recrystallization and loss of original characteristics. Includes those rocks which have formed in the solid state in response to pronounced changes in pressure, temperature, and chemical environment, below the zone of weathering and cementation.

Mineral-A naturally occurring solid having a definite atomic structure and chemical composition. It may have characteristic color, cleavage, luster or hardness.

Moraine-Debris left by a glacier and composed of unsorted material called till. Descriptive terms such as end or terminal moraine, medial (middle) moraine, lateral (side) moraine denote relative position of the glacial debris to the glacier.

Mudflow-Downslope movement of saturated weathered rock and soil. Sometimes applied to wet volcanic ash that moved down slope.

Obsidian-Volcanic glass formed by quick cooling of high-silica lava.

Olistolith-Olistostromes-Large chaotic rocks or rock masses that have broken from a marine and coastal headland and slid whole downslope into deep water.

Ophiolites-An ocean crust sequence of rocks characterized by basal ultramafics overlain by gabbro, dikes, pillow lavas, and deep-sea clay and chert sediments.

Ore-Minerals or aggregrates of minerals (usually metalliferous) that may become ores with changing economic conditions.

Orogeny-Mountain building associated with uplift, folding and faulting.

Outcrop-An exposure of bedrock often in roadcuts, along streams or in cliffs.

Outwash-Glacial-fluvial material carried by streams from the glaciers.

Overthrust-Rock shoved over another rock mass along a low angle (thrust) fault. Some thrusts have been traced for miles.

Pahoehoe-A Hawaiian term for lava flows with a ropy, somewhat smooth surface, which can be contrasted with the rough blocky surface of aa flows.

Peneplain-A land surface worn down by erosion to a nearly flat or broadly undulating plain. The surface may be uplifted to form a plateau and become subjected to renewed erosion.

Peridotite-A coarse-grained ultramafic rock consisting of olivine and pyroxene with accessory minerals It is thought to make up much of the mantle, and when altered is called serpentinite.

Period-A major, worldwide, geologic time unit corresponding to a system such as the Cambrian Period.

Phyllite-A medium-grade metamorphic rock with planar structure (foliation).

Pillow basalt-Volcanic extrusion of lava underwater

that produced discrete plump blobs of cooled basalt.

Placer-An ore deposit where the host rock is stream or beach gravel or sand.

Plunge-The dip of a fold axis. Anticlines and syncline may plunge.

Plutonic rocks-Igneous rocks formed at depth. The term includes rocks formed in a batholith.

Porosity-The ratio of the aggregate volume of void space or interstices in a rock or soil to its total volume. It is usually stated as a percentage.

Porphyry-Igneous rocks containing conspicuous phenocrysts (large crystals) in a fine-grained or aphanitic groundmass.

Pressure ridge-Large wave-like forms on the surface of solidified lava flows caused by movement and wrinkling of the cooling surface lava above the molten rock below.

Pumice-A sort of volcanic froth made up of excessively cellular, glassy lava, generally of the composition of rhyolite.

Pyroclastic rock-A general term applied to fragmental volcanic material that has been explosively or aerially ejected from a volcanic vent.

Radiolaria-A marine planktonic protozoan. A single-celled microfossil with a skeleton of opaline silica that is used to age date oceanic rocks.

Reverse fault-A fault whose hanging wall (that part above one standing on the fault plane) moved up relative to the footwall.

Reverse or inverted topography-A common occurrence in volcanic terrains where streambeds become filled with lava then stand out in relief after erosion has removed tha adjacent softer soil and rock.

Rhyolite-An extrusive rock with a high-silica content which is the fine-grained equivalent of a granite. When the lava cools quickly it may form obsidian and when gassy it may form a froth called pumice.

Rock-Any naturally formed aggregate of mineral matter constituting an essential and appreciable part of the earth's crust.

Scarp (also escarpment)-A steep slope or cliff. It is called a fault scarp if caused by faulting.

Schist-A medium to coarse-grained metamorphic rock that is composed largely of tabular minerals such as mica and hornblende.

Scoria-Vesicular basaltic lava, dark gray to dark red in color. The vesicles make up a significant proportion of the rock.

Sediment-Any particles settling from suspension in water or air. Sedimentary rocks are those made up of sediment, loosely applied to all stratified rocks.

Series-A time-stratigraphic unit ranked next below a system, such as the Eocene Series.

Serpentinite-A rock consisting almost entirely of the mineral serpentine derived from the alteration of previously existing olivine and pyroxene.

Shale-A compacted mud or clay, generally possessing a thinly bedded or laminated structure.

Silicate-Minerals or rocks containing crystalline compounds of silicon and oxygen.

Sill-An intrusive body of igneous rocks of approxi-

mately uniform thickness. A sill is relatively thin compared with its lateral extent, which has been emplaced parallel to the bedding or schistosity of the intruded rocks.

Siltstone-A consolidated sedimentary rock consisting of fragments between the size of sand and clay grains.

Slate-A fine-grained metamorphic rock which tends to split in slabs parallel to the alignment of mica rather than along bedding planes.

Slickenside-Polished and grooved rock surface where rock masses have rubbed together under pressure along a fault plane.

Spit-A long peninsula of unconsolidated sand in a bar projecting from a headland.

Stack-A steep-sided off-shore pinnacle, isolated by wave action from a sea cliff of which it was once a part.

Stratification-Parallel structure produced by deposition of sediments in beds or layers.

Strike-The course or direction of the intersection of an inclined bed or fault with the horizontal plane surface.

Strike-slip-Term applies to faults that have a prominent horizontal movement parallel to the plane of the fault.

Subduction zone-A curved planar surface where one major plate, usually the denser oceanic plate, is under-thrusting the less dense continental plate.

Superposition-The laying down of beds in order one on top of the other with the younger bed being on top.

Superimposed stream-A stream that was established on a new surface, usually a sedimentary cover, which maintained its course despite different preexisting lithologies and structures encountered as it eroded downward into the underlying rocks.

Syncline-A trough-shaped downfold with limbs rising from the axis, in which the youngest beds make up the axial part of the fold.

System-A world-wide time-rock division containing rocks formed during a period. Rock laid down during the Permian Period would be the Permian System.

Talus-Rock debris that accumulated in a slope at the foot of a cliff.

Tectonics-Study of the broader structural features of the earth and their causes. Plate tectonics relates to movements of large segments (plates) of the earth's crust and their interrelationship.

Tephra-A collective term for all clastic volcanic debris which during an eruption is ejected from a crater or vent and transported through the air, including bombs, blocks, dust, ash, cinders, lapilli, scoria, and pumice.

Tephrachronology-Relative ages of strata based on layers of volcanic ash.

Terrane-A suite of rocks bounded by fault surfaces that has been displaced from its point of origin.

Tethyan-A term for rocks and environments of the Mesozoic Tethys Sea which was a tropical ocean extending from the Mediterranean through the Himalayas to the Pacific.

Thunderegg-A term applied by rock collectors to geode-like masses that are generally filled with banded agate. They form in silicic volcanic rocks.

Till-Unconsolidated and unsorted glacial deposits which contain particles ranging from clay to boulder size.

Tuff-A rock formed of compacted volcanic ash whose particles are generally finer than 4mm in diameter.

Turbidity current-A muddy, relatively dense current laden with sediments that moves in response to gravity along the bottom slope of a body of standing water.

Ultramafic (see mafic)

Unconformity-The break or gap in depositional sequence in sedimentary rocks. It usually represents an erosion surface cut upon older rocks upon which later sedimentary beds are deposited. An angular unconformity shows a measurable discordance in dip between the older and younger beds. A disconformity is an erosional break between parallel beds.

Vein-A tabular occurrence of ore, usually disseminated through a gangue and having relatively regular development in length, width and depth. The term may also apply to non-ores such as quartz and calcite veins in rocks.

Vent-The opening from which volcanic material is erupted.

Vesicle-A small cavity in a fine-grained or glassy igneous rock, formed by the expansion of gas or steam during the solidification of the rock.

Volcanic rocks-Igneous rocks that have poured out or were ejected at or near the earth's surface. The term is synonymous with extrusive rocks.

Volcano-A mountain which has been built by material ejected from within the earth through a vent that includes lava, pyroclastic materials and volcanic gases. Volcanoes composed of flows with gentle slopes are shield volcanoes; those made up almost entirely of volcanic ash with or near the maximum angle of repose are cinder cones; and volcanoes containing both flows and fragmental rock may be called composite cones.

Water table-The top of the saturated zone.

Weathering-The processes, such as chemical action of air and water, plants and bacteria, and the mechanical action of changes of temperature, whereby rocks on exposure to weather change in character, decay, and finally crumble to form soil.

Welded tuff see ignimbrite

Index